Studies in Fuzziness and Soft Computing

Volume 327

Series editor

Janusz Kacprzyk, Polish Academy of Sciences, Warsaw, Poland
e-mail: kacprzyk@ibspan.waw.pl

About this Series

The series "Studies in Fuzziness and Soft Computing" contains publications on various topics in the area of soft computing, which include fuzzy sets, rough sets, neural networks, evolutionary computation, probabilistic and evidential reasoning, multi-valued logic, and related fields. The publications within "Studies in Fuzziness and Soft Computing" are primarily monographs and edited volumes. They cover significant recent developments in the field, both of a foundational and applicable character. An important feature of the series is its short publication time and world-wide distribution. This permits a rapid and broad dissemination of research results.

More information about this series at http://www.springer.com/series/2941

Luis Argüelles Méndez

A Practical Introduction
to Fuzzy Logic using LISP

 Springer

Luis Argüelles Méndez
European Centre for Soft Computing
  (ECSC)
Mieres, Asturias
Spain

ISSN 1434-9922   ISSN 1860-0808   (electronic)
Studies in Fuzziness and Soft Computing
ISBN 978-3-319-23185-3   ISBN 978-3-319-23186-0   (eBook)
DOI 10.1007/978-3-319-23186-0

Library of Congress Control Number: 2015947397

Springer Cham Heidelberg New York Dordrecht London

Printed on acid-free paper

Springer International Publishing AG Switzerland is part of Springer Science+Business Media
(www.springer.com)

To my wife Ana
To my mother Esperanza

Preface

The origins of this book are rooted back to the 1990s of the last century. In those times I was asked to prepare an introductory course on Fuzzy Logic for engineering students who were eager to take the fundamentals of the theory, especially from a practical approach. The students (already in their last year at the university) were certainly bored of theoretical issues and heavy mathematical treatises since they had suffered a hard curriculum at college. Additionally, some students had varying programming experience in C language, some others in Pascal and then some others simply had no programming experience at all. Simultaneously, the most available publications and books on the subject at that time were either at a popular science level or excessively theoretical, usually far from the practical approach course they were expecting.

The solution was to prepare readings from selected material on the subject while using a teaching vehicle specifically suited to the task. This vehicle was the Lisp programming language. In that seminal course students would learn both fuzzy logic theory and the minimum Lisp required to use FuzzyLisp, a set of Lisp functions that I designed from scratch in such a way that students could easily understand the theory and at the same time build simple fuzzy models.

More than 15 years later, the overall situation has not changed significantly. The quantity of information on the Internet is nowadays overwhelming, but requires time and dedication in order to filter and put into order the essential concepts of the theory. Excellent books are nowadays on the market on Fuzzy Logic, but they continue being located either at the academic side or at the popular science level. The mission of this book is to fill the gap between these two shores of complexity, always under a practical approach and using Lisp as a sort of computing gateway that will allow the reader to reach two destinations: To perfectly understand the basics of Fuzzy Logic and to design and develop from small to medium complexity models based on this powerful artificial intelligence paradigm. This book is neither the best book on Fuzzy Logic nor the best available book on Lisp. However, it tries to be the only book that offers the reader (I sincerely hope it) the perfect balance for getting the aforementioned mission accomplished. In the first four chapters the reader will learn the required level of Lisp and then even more, using a dialect of this language, NewLisp, that can be freely downloaded from the Internet and runs in the mainstream computer operating systems as of 2015. From Chap. 5, the theory is gradually introduced, seizing the opportunity to elaborate every FuzzyLisp language function (the complete set of functions that make up the FuzzyLisp programming toolbox can be downloaded from the companion book's website: http://www.fuzzylisp.com). At a higher level of resolution, this book is based on the following structure:

Chapter 1 puts Lisp in context, explaining its advantages as a modern computer language and offering a historical perspective of its development from its inception back in the last fifties of the twentieth century until nowadays. Then the NewLisp implementation is introduced and immediately the reader finds the first dialogues with the language by means of using it as a powerful electronic calculator.

Chapter 2 deals with lists. Lisp derives in fact from the words LISt Processing, so lists are the building blocks of the language. The fundamental list management functions are introduced there and soon the reader finds as an example a simple practical model of queue for a highway toll station.

Chapter 3 is all about user-defined functions in Lisp and explains how to structure and organize functions, extending first the available set of list management functions and then showing how to incorporate conditional structures. Later some loop structures introduce the concept of iteration, a common paradigm to many other programming languages. Finally, recursion is shown in detail, at a level not usually shown in Lisp introductory books.

Chapter 4 can be seen almost as a Lisp celebration where all the material exposed in the previous chapters is assembled in order to show how to build real applications in Lisp. The first one is a simulation of a French roulette that aside from dealing with random numbers allows the user to bet and then lose or win. The second application is a simple but powerful collection of functions for CSV (comma separated values) database management. This approach of Lisp programming preludes the architecture of FuzzyLisp.

With Chap. 5 the book enters into the realm of fuzzy sets theory. It starts with a quick review of classic set theory and soon transits toward the foundations of fuzzy sets. The central pages of the chapter express the nuclear concepts and ideas of the theory. If the reader understands well this section then he or she will probably not have difficulties to digest the rest of the material in the book. For assisting the reader in its travel through the chapter, plenty of Lisp functions are introduced in order to test every new concept at the keyboard. Finally, and as a practical application, Life Illness Curves are presented, a new approach for interpreting human health evolution in medicine.

Chapter 6 introduces a number of additional material about the theory of fuzzy sets and includes also a big share of the Lisp functions that make up FuzzyLisp. The code of every FuzzyLisp function is discussed and exhaustively commented in those rare occasions where some tricky code is used. An important section is presented about fuzzy numbers, including the notion of intervals and interval arithmetic for introducing fuzzy numbers arithmetic and then fuzzy averaging, together with a first view on defuzzification. After this, linguistic variables are introduced. The chapter ends with a practical application on fuzzy databases.

Chapter 7 deals with fuzzy logic. Since fuzzy logic is a special and powerful type of logic it deals also with propositions and logical inferences. The theory exposed in the chapter is dense but the abundance of Lisp examples helps the reader to understand every hidden corner in the way. Fuzzy hedges are introduced and Fuzzy Rule-Based Systems (FRBS) are presented in detail. As practical applications, an air-conditioner control model is developed in FuzzyLisp as well as an intelligent model for evaluating performance in regularity rallies, a modality of car racing.

The last chapter is entirely dedicated to practical applications of Fuzzy Logic using FuzzyLisp. These applications are exposed as "projects" that try to stimulate the creativity of the reader. Three applications are developed at an increased complexity level with respect to models from previous chapters. The first one merges simulation and fuzzy control, creating a simplified, yet intelligent model for landing the Apollo XI Lunar Module on the Moon. The second project deals with speech synthesis where double stars in astronomy are the excuse for elaborating on the required architecture for the project, but the theoretical aspects described in the chapter are of direct application in practically every field in science. The last section introduces Floating Singletons, an advanced modeling technique that shows its potentiality in an example model for getting an index of pulmonary chronic obstruction. The book is tail-complemented by two appendices: The first appendix shows the main differences between NewLisp and ANSI Common Lisp, while the second is a complete reference of the FuzzyLisp functions developed along this book.

The audience for this book covers readers not only from the field of computer science, but also those from the world of engineering and science. The focus is on undergraduate students and practicing professionals of technical or scientific branches of knowledge, including engineering, medicine, biology, geology, etc. It also can serve as a textbook for and introductory course on Fuzzy Logic. The book assumes basic tertiary mathematical knowledge from the intended reader. Any first-year student from college should be able to read it without special efforts. Needless to say, professionals in engineering and scientific fields will find it easy to follow. No previous programming knowledge is needed in any computer language since, as already said, it includes a gentle introduction to LISP programming.

Incidentally, the book can also be of huge interest to software developers. Certain FuzzyLisp functions can be understood as a bridge between FuzzyLisp and any other programming language. The text files produced by these functions can be loaded into any software project and then all the expert knowledge from the previously developed fuzzy models can be incorporated in those software projects. As an example, this opens the possibility for using FuzzyLisp as a software tool for developing intelligent apps for smartphones and other mobile devices.

Finally, from these pages I would like to seize the opportunity to thank the many persons who in some way have helped me to write this book, being well aware that it is impossible to mention them all, so I shall try to at least thank the people closest to this work. First, I would like to show my most intimate gratitude to my wife Dr. M.D. Ana Fernández-Andres not only for always being there with her patience and support while writing the manuscript but also for her always useful advice in the medical examples used in this book. I wish also to express my deepest appreciation to Dr. Rudolf Seising for giving me the ultimate boost for starting this project when I shared with him my preliminary ideas of an introductory book on Fuzzy Logic using Lisp. On the other hand, I must express my deepest gratitude to Dr. Gracián Triviño for so many conversations while writing the chapters of this book. It is also impossible to forget the inspiration from the scientific sessions at the European Centre for Soft Computing from Prof. Michio Sugeno, and last, but not least, I would like to thank the inspiration from Prof. Enric Trillas. This book owes him a very special kind of push. Much more than he would ever have imagined. Also, I would like to thank Dr. Lutz Müller (the creator of NewLisp) for his kindness when confronted with technical questions concerning his Lisp implementation. An especial mention must be made also to Dr. Janusz Kacprzyk for accepting this book in his series *Studies in Fuzziness and Soft Computing*, and the Springer Verlag (Heidelberg), in particular to Dr. Leontina di Cecco and Mr. Holger Schaepe for their support and valuable comments while developing this work.

Oviedo, Spain   Luis Argüelles Méndez
June 2015

Contents

1 **Discovering Lisp** 1
   1.1   Introduction 1
   1.2   Why Lisp? 1
   1.3   A Short History on Lisp 4
   1.4   The NewLisp Implementation 7
   1.5   A Quick Start Using NewLisp 8
   1.6   Using Variables 13
   1.7   As a Summary 17
   References 17

2 **Lists Everywhere** 19
   2.1   Introduction 19
   2.2   Atoms and Lists 20
   2.3   First and Rest 24
   2.4   Building Lists 26
   2.5   Some Geometry and then Some Art, Too 30
   2.6   A World Full of Queues 38
   2.7   Rotate, Extend and Flat 42
   2.8   As a Summary 45
   References 48

3 **Functions in Lisp** 49
   3.1   Introduction 49
   3.2   Starting Functions with Three Simple Examples and then
         Came Map 50
   3.3   Managing Lists with User Defined Functions 55
   3.4   The Discovery of Conditional Structures 58
         3.4.1   From Predicates to If-Then-Else Structures 58
         3.4.2   A Note About Functional Programming 62

3.4.3   Robust Functions from the Use of Conditional
            Programming   64
    3.4.4   Solving Multiple Conditions Without Using If   67
3.5   The Discovery of Loop Structures.   71
    3.5.1   While Computing   71
    3.5.2   Other Looping Structures   77
3.6   Recursion Is Based on Recursion   81
3.7   A Note on Lambda Expressions   87
3.8   As a Summary   89
References   91

4   Lisp Programming.   93
4.1   Introduction   93
4.2   From Montecarlo with Love   94
    4.2.1   Declaring Global Variables.   95
    4.2.2   Throwing the Ball and Checking Results   96
    4.2.3   Betting and Playing.   99
    4.2.4   Building a Simple User Interface   102
    4.2.5   Putting It All Together.   105
4.3   Messier Was a French Astronomer   109
    4.3.1   Opening and Loading Databases in CSV Format.   111
    4.3.2   Querying the Database.   113
    4.3.3   Updating the Database.   115
    4.3.4   Modifying the Database   116
    4.3.5   Filtering the Database   120
    4.3.6   Performing Simple Statistics.   122
    4.3.7   Saving the Database   123
4.4   A Word on Function Design for This Chapter   125
4.5   As a Summary   125
References   127

5   From Crisp Sets to Fuzzy Sets   129
5.1   Introduction   129
5.2   A Review of Crisp (Classical) Sets   129
    5.2.1   Definition of Sets and the Concept of Belonging.   130
    5.2.2   Subsets   131
    5.2.3   Union, Intersection, Complement and Difference.   134
    5.2.4   Set Properties   139
    5.2.5   Cartesian Product and Relations   140
5.3   Moving Towards Fuzzy Sets   143
    5.3.1   The "Fuzzy Sets" Paper.   144
    5.3.2   Union, Intersection and Complement of Fuzzy Sets.   149
    5.3.3   Fuzzy Sets Properties.   155
    5.3.4   Fuzzy Relations   158

5.4   Membership Degrees: An Example Application in Medicine.   161
5.5   As a Summary   166
References   167

6   **From Fuzzy Sets to Linguistic Variables**.   169
6.1   Introduction   169
6.2   Towards Geometrical Characteristic Functions   170
6.3   From Geometry to FuzzyLisp.   174
6.4   Support, Nucleus and Alpha-Cuts   180
6.5   Fuzzy Sets with Discrete Characteristic Functions   186
6.6   Revisiting Complement, Union and Intersection of Fuzzy Sets . . .   193
6.7   Fuzzy Numbers.   197
6.7.1   Fuzzy Numbers Arithmetic.   198
6.7.2   More Numerical Operations on Fuzzy Sets.   206
6.7.3   Fuzzy Averaging.   209
6.8   Linguistic Variables.   212
6.9   Fuzzy Databases   220
6.10   As a Summary   225
References   227

7   **Fuzzy Logic**.   229
7.1   Introduction   229
7.2   The Beginning of Logic.   230
7.3   Modern Bivalent Logic   231
7.4   Fuzzy Logic.   234
7.5   Logical Connectives in Fuzzy Propositions.   236
7.6   Fuzzy Hedges.   242
7.7   Fuzzy Systems from Fuzzy Propositions   248
7.7.1   Fuzzy Rule-Based Systems.   251
7.7.2   Defuzzification   252
7.8   Modeling FRBS with FuzzyLisp.   258
7.9   Fuzzy Logic in Motor Racing: Scoring in Regularity Rallies   270
7.10   FRBS Using Fuzzy Sets with Discrete Membership Functions . . .   279
7.11   As a Summary   283
References   287

8   **Practical Projects Using FuzzyLisp**.   289
8.1   Introduction   289
8.2   Landing the Eagle: Simulation and Fuzzy Control
in Engineering   290
8.2.1   Fuzzy Control.   294
8.2.2   Controlling the Eagle.   296
8.2.3   Interpreting Results   303
8.3   Double Stars in Astronomy: Speech Synthesis   306
8.3.1   Generating Suitable Linguistic Variables   310
8.3.2   Managing Linguistic Expressions   323

8.4   Spirometry Analysis in Medicine: Floating Singletons 330
       8.4.1   Introducing Floating Singletons. 332
       8.4.2   Modeling Pulmonary Obstruction with FuzzyLISP 334
8.5   As a Summary 344
References 346

Appendix A: NewLisp Versus Common Lisp 347

Appendix B: Glossary of FuzzyLisp Functions 353

About the Author

Luis Argüelles Méndez holds an M.Sc. degree in Astronomy (High Distinction) from the Swinburne University of Technology, Melbourne, Australia. He received a Master of Business Administration degree (High Distinction) from the ITEAP Institute, Malaga, Spain, in March 2010. He has been working as the R+D department's director in a private coal mining company in Spain for 20 years, directing and coordinating more than 30 research projects, within regional, national, and European frames of research. As part of these projects, he has developed discrete event simulation-based models, as well as virtual reality systems and soft-computing-based models for geology and mining engineering. In the field of observational astronomy, Luis Argüelles is not only considered a pioneer in developing intelligent models for double stars, but has also explored the applications of fuzzy-sets theory in galactic morphology. Moreover, he is interested in the applications of fuzzy-logic-based techniques in both industrial processes and organizational management, especially in the field of total quality management (TQM). Currently, he is also involved in studies related to the theory of artificial consciousness.

Chapter 1
Discovering Lisp

1.1 Introduction

As the old Chinese proverb says, every travel begins with a simple step. This chapter is our first step in a long voyage that will take us ultimately towards a good understanding of fuzzy-logic theories. This first step is neither complex nor dense, as every first step should be when beginning ambitious undertakings, being in fact a gently introduction to Lisp. We shall start exposing some key features of the language in order to appreciate what makes it unique among all the existing programming languages, exposing at the same time its unique philosophy. Later on we shall dedicate a well-deserved space to enjoy the history of Lisp because I think this subject is not sufficiently covered in traditional Lisp books and it helps to give the reader a good computer science perspective before moving forwards in our voyage. Finally, the reader will have the opportunity to get their feet wet into the language by means of a quick practical first session with NewLisp.

1.2 Why Lisp?

Java, Python or PHP are hot these days as programming languages. Other languages such as C or even actual improved versions of it such as C# or C++ seem to have lost some momentum when in comparison to the former ones, not to mention Pascal or Basic, well buried in the eighties last century, although Pascal is still offered in some schools as a programming language for teaching "correct programming" as they say. Java is highly portable. This means that a program written in Java under Windows can be executed under Mac OS, or even in your smartphone with some slight additions of XML and some tricky organization. Java, Python and PHP work extremely well in today's world of Internet. The money is on the Internet

© Springer International Publishing Switzerland 2016
L. Argüelles Méndez, *A Practical Introduction to Fuzzy Logic using LISP*,
Studies in Fuzziness and Soft Computing 327,
DOI 10.1007/978-3-319-23186-0_1

while I write these lines in the second decade of century XXI. So, the question seems obvious: why Lisp?

As we shall see in the next section, Lisp is the second older high-level language in the history of computer languages, so it can seem a strange decision at first to use Lisp as a vehicle for teaching Fuzzy Sets theory and Fuzzy Logic. In some way it could be seen almost as a contradiction: Using a very old language for teaching and spreading forefront computing concepts that are applied today in robotics, machine learning, adaptive systems and other fields of Artificial Intelligence (AI).

It can be argued that Lisp was the preeminent computer language for research in Artificial Intelligence since its inception. Expectations were extremely high in the sixties and very high in the seventies, before falling in what is known as the AI Winter. By 1991, the Fifth Generation Computer Project in Japan resulted into a complete failure. The same happened to similar advanced undertakings in Europe and US. Expectations in AI had been so high and results were so poor in comparison that the term "Artificial Intelligence" has been since used with prudence and moderation. Even today. However, from my personal point of view, it was more a case of bad algorithms than bad programming tools. We shall have again the opportunity to discuss this subject in other sections of this book, but we can say from now on that Lisp is a computer language extremely well suited to manage fuzzy sets and fuzzy logic techniques, and, since fuzzy logic is a branch of AI, it follows that Lisp continues being a great tool for developing Artificial Intelligence applications.

But there are some strong technical reasons for selecting Lisp as the ideal programming tool for learning and understanding Fuzzy Logic in a practical way, too. First of all, Lisp offers the user or programmer (hereafter the terms "programmer" and "Lisp user" will be interchangeable in this book) an automatic system of memory management. Other computer languages demand from programmers a special and constant attention for seizing and liberating memory while writing programs. Traditionally, this is made with a programming concept named "pointers". The C family of languages is full of pointers. If only one pointer is badly managed by the programmer, the whole program will crash, sooner or later. Java does not require the use of pointers from the programmer, but still demands from him the careful allocation of memory for any object in the program. Lisp manages pointers and memory allocations automatically for you. It does it without notice for the Lisp user. Smoothly. Securely. This means more confidence in the code and more speed in writing programs. It means superior productivity.

Lisp is relatively simple. I could show you the basic concepts of the language in two afternoons of class (well, maybe three or four) and you could start to write from simple to medium complexity useful programs in Lisp very quickly. As with chess, the mechanics of the play are simple and powerful, yet the possibilities of play are practically infinite. Needless to say, as it happens with chess, it takes time to be a sophisticated Lisp programmer, but the essential movements are easy to grasp. In this book you will learn all the "movements", "apertures" and basic strategies needed to reach a level where you could have written FuzzyLisp and then a bit beyond.

Lisp is elegant. Paraphrasing and playing with the famous words from Winston Churchill, never in the history of computer languages so few powerful concepts have derived into so many wonderful applications. Take recursion as an example: Recursion is a programming paradigm (speaking properly it is a mathematical paradigm, but someone told once that computing is no other thing that living mathematics) where a function written by the user in Lisp is called inside the same function. The funny thing is that recursion does happen in nature. It happens even in human behavior. While recursion is not needed to write Lisp programs, it is an art in itself and we shall dedicate space for it in this book. Yes, it is true that other computer languages support recursion, but no one of them does so efficiently.

Lisp builds Lisp-thinking on you. In some way, it happens the same after learning a human language: Learning the language allows you to understand the culture of the country that speaks the language. Then, if you are an open-minded individual, you end up adopting part of that culture in your way of thinking, in your very own culture. Surprisingly it happens with Lisp, too. It helps you to think on creating programs with Lisp style, with Lisp organization, with Lisp freedom. This philosophical aspect of Lisp is not present in any other computer language. After a while learning the language you will realize, probably with surprise, that Lisp code is even aesthetically attractive and fine looking, both in the screen and after being printed, on paper.

Lisp is highly interactive. While you write code you can take apart one function or a set of them and try how they do behave. You can test functions or fragments of your program without effort. Even better, you can interactively modify functions in a Lisp session, make some improvements and then incorporate the transformed functions into the main program, if you like. This is described as the "read-eval-print loop" paradigm. You type Lisp expressions at the keyboard, Lisp interprets and evaluates what you have entered and after evaluation, it prints the results. You cannot only run complete programs, but parts of it. You feel that you are talking with the system, and that talk creates a special relationship with the system and the Lisp language itself. Later in this chapter we will have our first Lisp session in order to allow you to have a first experience with the language.

Finally, and as previously suggested, Lisp is the best general purpose computer language for representing fuzzy-sets. You still don't know, but for describing the fuzzy set "young" relative to the age of a human being, we could simply write the following expression in Lisp:

$$(\text{setq young } '(12\ 21\ 28\ 35))$$

It doesn't matter in this moment that you don't understand this line of Lisp code, but it assigns a membership function representing the fuzzy set "young" to the Lisp symbol "young", just created in the expression. We shall dedicate enough time to describe symbols, membership functions and lots of new concepts, but suffice is to say for now that no other computer language has at the same time this level of abstraction and naturalness.

1.3   A Short History on Lisp

LISP is one of the oldest high-level languages in the history of computer science and only FORTRAN, a language whose main goal is to write technical programs where numbers are the main data type, is older. In fact, it's highly probable that you, dear reader, are younger than the LISP language. LISP was born from the works of John McCarthy and his colleagues at the Massachusetts Institute of Technology, MIT, back in the last fifties of century XX in an attempt to develop a language for symbolic data processing, deriving in an extraordinary tool for **LIS**t **P**rocessing. As we shall see in future sections of this book, the main data type of the LISP programming language are lists. Needless to say, the concept of list will be introduced in the very first next pages of this work.

The term "high-level" mentioned in the above paragraph is important. It means that the programmer uses, aside a logical structure and a well-defined grammar, words from natural language in order to develop programs. It can seem natural nowadays to think that computers should be able to understand some language at least not far from the languages we, humans, use to communicate ourselves, but before the apparition of high-level languages, the usual way to program computers was to "speak" to them using machine specific code, either writing directly numeric code or using an assembler, a low-level language that produced a set of instructions specific to a given computer's CPU (Central Processing unit). Using machine code means that every instruction used for programming a computer in the forties and first half of the fifties last century was expressed in binary code, that is, sets of zeroes and ones representing either a number, an instruction for a specific CPU, or a position in memory. As an example, the number 201 is expressed in binary as 11001001. For subtracting 55 (00110111) from 201 some binary expressions were usually required:

$$11001001 \quad 00110111 \quad 01111001$$

where the last sequence of zeroes and ones, 01111001, would be the instruction, let's say, number 121 (subtraction) for a hypothetical CPU. The result of the subtraction is 10010010 (146). Additionally, some other binary sequences were still required to access memory positions, and reading and storing the used numerical values.

Sound complex, right? Let's do an analogy using human language. Just imagine a father and his six years old son are having a walk along a park. Suddenly, the young boy exclaims: "Dad, why trees are so big?" As usually, children like to make not easy at all questions, so after a while, the father finally replies: "oh, that's because they live long lives, so they have plenty of time to grow". We are not going to decide in this moment if the answer is a good one. The important thing here is that the father has transmitted to his son a concept, a chunk of knowledge that will be stored on the child's brain. We could use some symbolic language to express this situation with the question at left of the colon and the answer and short explanation at right:

Why trees big: long-live → plenty-of-time-to-grow

This expression is not pure natural language, but it is still a very high level language. Very interestingly, and without being even aware of it, the brains of both human beings, father and son, have triggered a complex set of mechanisms in order to first elaborate a question, supply a reply and then store the new knowledge in memory. Let's go a bit more in depth about this reflection: Both brains have used brain cells, that is, neurons, for getting the job done. Using biochemical energy some neurons (in the order of thousands or tens of thousands, the exact number doesn't matter here) have established a set of connections in their synapses for manipulating the transfer of information. In computing jargon this is what we call a low level process. We can go at an even lower level, the lowest of all in our brains, in order to see what is happening. Observing the chemical processes in the synapses between neurons, we would see, probably with wonderful awe, how all the processes are ruled by the extremely quick exchange of electrons between atoms of calcium and molecules of neurotransmitters in a huge number of individual chemical reactions. This chemical reaction, in this analogy, is machine code in our brains. Natural language, the highest level language we know, is the most important tool that nature has developed through evolution in order to manage learning and cognitive abilities. We shall revisit this analogy later in this book when discussing some artificial intelligence topics. For now, we can appreciate how easy is to use natural language for us, while the low-level equivalent of managing chemical processes in the synapses between neurons in a one-by-one basis is far from our technological abilities. Now, and probably forever.

Albeit using a strong exaggeration, we have used this analogy in order to appreciate the huge leap in computer science that happened with Fortran and Lisp languages. From working with sequences of zeroes and ones to using expressions related to human language, programmers could start to focus their energies on solving more ambitious problems. The first programs in Lisp were in fact used to solve symbolic calculations in differential and integral calculus, electrical circuit theory, mathematical logic, game playing and other projects that required manipulations of symbols.

The seminal paper of LISP, "Recursive Functions of Symbolic Expressions and Their Computation by Machine" was written by McCarthy himself as a programming system for the IBM704 (the first mass-produced computer with floating point hardware), in order to facilitate experiments with a proposed system called "Advice Taker", that could exhibit some class of "common sense" in carrying out its instructions (McCarthy 1960). Soon after this theoretical paper, the first real implementation of Lisp was created by Steve Russell on the 704 using its own machine language, and not much later, in 1962, Tim Hart and Mike Levin wrote the first Lisp compiler, written in Lisp, for the IBM7090, a more advanced (and really expensive) computer at that time.

Two important documents were published by M.I.T Press afterwards: "LISP 1.5 Programmer's Manual" (McCarthy et al. 1962) and "The Programming Language

LISP: Its Operations and Applications" (Abrahams et al. 1964). These documents, together with the seminal paper by McCarthy are easily found today on the Internet and constitute the real cradle of Lisp.

No doubt those were exciting times. Russia had launched the Sputnik satellite in October of 1957 and soon took the lead on the spatial race. In the midst of Cold War, the reaction of the United States was manifold and we should mention at least two strategic measures that the North American government immediately announced. First, it sparked a massive federal education funding program, significantly called the "National Defense Education Act", to stimulate better teaching of math and science as well as foreign languages to more students throughout the country, and second, the "National Space Act" of 1958 that would give birth to the National Agency of Space Administration, NASA, on the same year. Funding was immense and computers were high on the buying list of NASA, especially IBM704s and the IBM709X family of systems.

NASA hired technicians and computer scientists with experience in these computers. This is equivalent to say that at least some people from the MIT started to work in the spatial agency. Aside specific machine code, Fortran and Lisp were already available programming languages, so the Mercury, Gemini and Apollo missions were developed with the help of these languages. We are not going to say that Lisp ultimately won the space race, but it was indeed a reliable and creative tool for accomplishing the mission of landing on July 20th, 1969, on the Moon.

In the late sixties, people at MIT developed an enhanced version of Lisp named MACLISP for the PDP-6/10, a family of 32 bits computers manufactured by Digital Equipment Corporation. MACLISP included new data types such as arrays and strings of characters. It helped to enhance the development of Artificial Intelligence (AI) until the early 80s. The Reference manual for MACLISP (Moon 1974) has been preserved and can be consulted freely in the Internet.

While MIT was the cradle of LISP and even today is considered an in-house development for completing a curriculum in computer science, it soon spread to other regions, both academic and commercial. Interlisp (Teitelman and Masinter 1981) was, seen with the perspective of an historian, a transfer of computer language from MIT to the Palo Alto Research Center (PARC) in California, a division of Xerox Corporation at mid seventies last century. Also from coast to coast in the United States, another dialect of Lisp based on MACLISP and named Franz Lisp, appeared in Berkeley in the seventies and eighties, becoming one of the most commonly available Lisp dialects on machines running under the Unix operating system (Gabriel 1985).

The language didn't take too much time to reach Europe. One of the first dialects flourished in France under the name of VLISP in 1971. It was developed at the University of Paris VIII at Vincennes, giving name to the dialect, V(incennes) LISP. Interestingly, the first book I read about Lisp used VLISP for describing and teaching the language (Wertz 1985) so in some way, if you are reading this book you are getting a connection in time with the dialect of Lisp developed at that nice French village and if you read some French it is still a nice download from the Internet. After VLISP soon came Le-LISP a close variation from VLISP, also in

France. Under an historic point of view Le-LISP is a milestone that deserves some attention because it was one of the first implementations of the Lisp language on the IBM PC, thus, inaugurating an epoch of, let's say, personal programming on microcomputers (Chailloux et al. 1984).

Since Lisp can be written using Lisp itself, it is easy to understand the easy and quick blooming of versions and dialects of the language, so soon it became evident that some type of standardization was needed in the community of Lisp programmers and users. Due to this, the American National Standard Institute published in 1994 a language specification document that gave birth to Common Lisp, also known as ANSI Common Lisp. As of this writing, several companies sell their Common Lisp products for several operating systems. Some of them even offer "personal" or "evaluation" editions that can be freely downloaded from the Internet.

1.4   The NewLisp Implementation

Lutz Müller, the designer of NewLisp, is a computer scientist always avid of new knowledge. Born in Germany, he earned first a Master degree in Psychology and then a Ph.D. (Magna Cum Laude) in statistical patter recognition, with postgraduate studies in Computer Science at the University of Hamburg. Psychology, cognitive processes and computer science tools soon brought him to the Lisp language. However, Müller quickly realized that Lisp had evolved into large implementations that usually required a long learning curve for mastering them. At the same time, the mainstream programming paradigm had deeply changed in the last decades due to the common accessibility of the microcomputer in the 80s and specially because the popularization of the Internet since the 90s.

Lisp pioneered concepts such as interpretation, that is, it does not need to be compiled for having a working application. Another important feature of Lisp is that is a language with dynamic typing, that is, a Lisp programmer does not need to declare the type of data he or she is using. The most important thing today in software development is productivity, and it is not clear that strong typing contributes to this goal (Ousterhout 1998). Müller has created with NewLisp a dialect of the Lisp language that introduces scripting features, that is, a dialect that give up some execution of speed (not so much) and strength of typing when in comparison to system programming languages such a C++, but it provides significantly higher programmer productivity and software reuse. It is small and has built in most of the technologies needed in today's networked and distributed applications. Tomorrow's software systems will not be constituted by big monolithic systems but small agile agents running and cooperating on distributed architectures. For this new horizon, NewLisp is ideal, and the small loss of execution speed is easily compensated by the huge evolution in hardware. Last, by not least, NewLisp has been the ideal programming platform for developing FuzzyLisp.

1.5   A Quick Start Using NewLisp

As already stated, NewLisp can be downloaded free of charge from Lutz Müller's site on the Internet: http://www.newlisp.org/ for different computing platforms, including Mac OS X, GNU Linux, Unix and Windows. After downloading and installing it on your computer has the appearance shown in Fig. 1.1, at least on a Mac OS computer. If you are using other operating system it can show a slight different appearance.

Aside the usual menus and icons on the upper area of the application window, we can observe two rectangular areas, the bigger one in white and other, smaller and light gray colored at the bottom of the window. Between the two areas there is a bar that we can drag with the mouse in order to resize the white and gray areas until approximately getting the appearance shown in Fig. 1.2.

Now the gray area is bigger. As in the previous image, the following appears on the gray area, depending a bit on the NewLisp version and operating system:

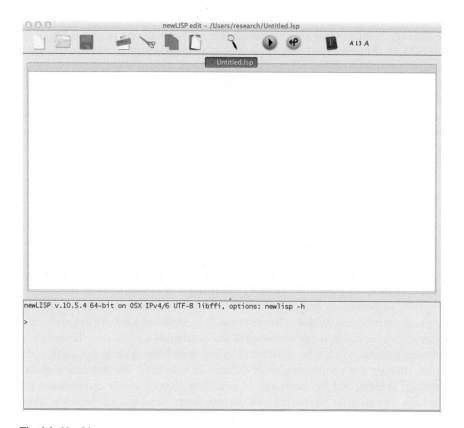

Fig. 1.1 NewLisp appearance

Fig. 1.2 NewLisp, ready for a Lisp session

newLISP v.10.5.4 64-bit on OSX IPv4/6 UTF-8 libffi, options: newlisp –h

This indicates that NewLisp is ready for a Lisp session, that is, some computing time where you "speak" to the system and it answers you. In a Lisp session you enter Lisp expressions, NewLisp evaluates them and then it provides an answer. Here "to speak" means "to type". This is a key concept that deserves some attention.

Until now, we have exposed material about the history of Lisp and some theoretical concepts about the language. Now it is time to use NewLisp. As already stated, this is a practical book and readers usually feel eager to put his or her fingers at the keyboard and start typing code. This is the best approach to learn a computer language. I would like to urge you to type the code in the next sections. You can read the expressions of the book and learn from the explanations, of course, but this is not the best approach. In some way it is the same as learning a foreign language, let us say, Spanish. You can read books about Spanish grammar (boring, Spanish grammar is huge), you can hear recordings from CDs or mp3 files stored on your favorite device, you can hear Spanish songs and so on. Even you can travel to Spain in order to get a correct pronunciation (take care, only in the central areas of the country the language is spoken with a neutral, pure accent). All of this will help, of course, but if you don't try to speak you will never be able to learn Spanish well, nor German, nor any language. Here "typing" is the same and works the same as "speaking". Without any trace of doubt you are going to make mistakes, and this is great because making mistakes is the best way to learn more quickly. If you do not type and only read you will not make mistakes, and the learning curve will be a lot longer. If you do not try to speak Spanish you will never learn it well. Even more: the joy of learning a (computer) language is in fact in the mistakes.

Let us begin. From now on, we shall call "session area" to the gray area on the NewLisp environment. You can see a cursor blinking at the right of a "bigger than" (>) sign called "prompt". In a Lisp session you type expression at the prompt and the system answer you in the next line. In this very first session I am going to omit the replies from NewLisp because I want you to feel like an explorer. You are going to type and you are going to observe the answers. Later we shall revise and comment the replies. Ready? Let us use NewLisp as an electronic calculator!

```
> (+ 1 1)
:

> (+ 1 1 1)
:

> (− 3 2)
:

> (− 5 1 1 1)
:

> (* 6 5)
:

> (* 2 2 2 2)
:

> (/ 8 2)
:

> (/ 8 2 2 2)
:

> (/ 3 2)
:

> (/ 2 3)
:
```

> (/ 2.0 3)
:
> (+ 5.5 7.3)
:
> (add 5.5 7.3)
:
> (sub 7.3 4)
:
> (mul 2.5 2.5)
:
> (div 7 2)
:
> (div 17 2)
:
> (div 17 2 2 2 2 2 2 2)
:

Congratulations! You have already made your first Lisp session. There have been some surprises, of course, but we have broken the usual scenic fear and learnt some very important things on Lisp. Before commenting the expressions one by one, we are going to seize the opportunity to express two fundamental aspects, or better said, rules of Lisp. The first one is the following:

Every Lisp expression contains a balanced number of parentheses

Every expression in this first session has two parentheses: One left parenthesis and one right parenthesis. It does not matter if we are dealing with the simplest Lisp expression or with a Lisp program composed by thousands of lines of code: The number of left parenthesis will always be the same as the number of right parenthesis. This is of a so capital importance that when we write Lisp programs the program editor in the system helps us to remember what parenthesis matches which.

Functions are introduced first in a Lisp expression and then the arguments are located at right from the name of the function

Every function in Lisp, either implemented on the language or defined by the programmer, must be located immediately after a left parenthesis. Some functions need a fixed number of arguments, other ones admit a variable number of them, but the positions always adhere to the same grammatical syntax: First the name of the function and then the arguments. In this session we have used the basic arithmetical functions of addition, subtraction, multiplication and division and then numbers have been used as arguments, with some caveats, as we are going to see in the next lines:

```
> (+ 1 1)
: 2

> (+ 1 1 1)
: 3
```

It cannot be simpler. The first expression is the simplest arithmetic calculation. The second expression is more interesting. It shows that the operator '+' can be used with several numbers (arguments), resulting into the sum of all of them. The same happens with subtraction, as can be seen in the following two expressions:

```
> (− 3 2)
: 1

> (− 5 1 1 1)
: 2
```

For multiplying and dividing it works the same:

```
> (* 6 5)
: 30

> (* 2 2 2 2)
: 16

> (/ 8 2)
: 4

> (/ 8 2 2 2)
:1

> (/ 3 2)
: 1

> (/ 2 3)
: 0
```

Here is our first surprise. Dividing 3 by 2 should yield 1.5, not 1, and dividing 2 by 3 should yield 0.666667, not 0. What has happened? Well, it is simple: The operators '+', '−', '*' and '/' in NewLisp are defined for integer addition, subtraction, multiplication and division, respectively. It does not matter if we try to force them using real numbers by means of writing numbers with decimal points:

```
> (/ 2.0 3)
: 0

> (+ 5.5 7.3)
: 12
```

NewLisp uses other functions for addition, subtraction, multiplication and division of real numbers, as we can see:

> **(add 5.5 7.3)**
: *12.8*

> **(sub 7.3 4)**
: *3.3*

> **(mul 2.5 2.5)**
: *6.25*

> **(div 7 2)**
: *3.5*

> **(div 17 2)**
: *8.5*

> **(div 17 2 2 2 2 2 2 2)**
: *0.1328125*

These functions are easy to remember, because their names are formed by the first three letters of their respective functions. Since Fuzzy-Logic, as we shall see, is ultimately a question of real numbers, only the add, sub, mul, and div functions will be used in Fuzzy-Lisp and the applications developed with it.

1.6 Using Variables

Any mid-level handheld calculator is able to use variables to store numerical values, and since NewLisp is more sophisticated that any electronic calculator ever built it can manage variables, too. In fact, the internal management of variables in Lisp in general and NewLisp in particular is rather sophisticated, being able to store any type of data in memory, including lists, symbols, arrays and so on, but for now we shall concentrate in variable management using real numbers. For this we shall introduce a very important function named "setq" that has the following syntax:

(setq name-of-variable value-to-store)

This function serves to store almost any thing you can think of while using NewLisp, but in this first approach to setq, we shall use it to assign numerical values to numerical variables. Let us start a new Lisp session at the prompt, typing the following expressions:

> **(setq pi 3.141592654)**
: *3.141592654*

Pi is not predefined in NewLisp, so this expression will be present in almost any technical program you write in this language. Since setq is used to store values in memory, we can recall any stored value at any time simply using its variable name:

> pi
: 3.141592654

The following expression adds some mathematical sophistication and shows the first blink of how NewLisp evaluates expressions:

> (setq pi-squared (mul pi pi))
: 9.869604404

NewLisp scans first for the innermost expression, in this case it is *(mul pi pi)*, then it looks for the first element in the list and finds the function "mul". Since it is a function for multiplying real numbers, the system looks for the arguments to the function, finding pi and pi. In this moment, NewLisp finds in memory a numerical value stored in memory for the symbol pi, so it gets it and makes the multiplication in such a way that the evaluation of (mul pi pi) produces the real number 9.869604404. Freezing the action at this time, the system now sees this expression: *(setq pi-squared 9.869604404)*. Finally, the function setq assigns the value 9.869604404 to the just created variable "pi-squared", returning the result of the last evaluation, that is, 9.869604404.

> pi-squared
: 9.869604404

As it is well known, the Pythagoras Theorem for a right triangle is given by the following formula:

$$h^2 = a^2 + b^2 \tag{1-1}$$

where a and b are the sides of the triangle and h is the hypotenuse. For a triangle where a = 4.0, and b = 3.0, we can type the following:

> (setq h-squared (add (mul 4.0 4.0) (mul 3.0 3.0)))
: 25

alternatively:

> (setq a2 (mul 4.0 4.0))
: 16

> (setq b2 (mul 3.0 3.0))
: 9

> (setq h-squared (add a2 b2))
: 25

Or, using only a line for getting the length of the hypotenuse, h:

```
> (setq h (sqrt (add (mul 4.0 4.0) (mul 3.0 3.0))))
: 5

> h
: 5
```

By the way, "sqrt" is a mathematical function supplied in NewLisp that returns the squared root of a real number. NewLisp contains an impressive number of floating point functions: trigonometric, hyperbolic, logarithmic, etc. All of them are well documented in the NewLisp User manual, always available from the NewLisp menu: Help → NewLisp Manual and Reference.

A word of caution must be said about trigonometric functions in NewLisp. All of them use radians as parameters. If we, expecting grads, try to calculate tan(45) typing the following:

```
> (tan 45)
```

NewLisp returns:

```
: 1.619775191
```

Very different from the expected value of 1, so we must do some calculations first for passing a correct value expressed in radians to any trigonometric function from the language. Since a complete circumference covers an angle of 2π radians, the following formula converts from an angle α to r radians, as it is well known:

$$r = \frac{\alpha\pi}{180} \tag{1-2}$$

Some paragraphs above we used an "ad hoc" value for pi. Now we'll seize the opportunity to extract pi using the math libraries embedded in NewLisp. As previously stated, no value comes predefined for pi, but it is sure hidden in the code for managing trigonometric functions. Remembering that:

$$\mathrm{acos}(1) = 90 \text{ degrees} = \pi/2$$

then we can type:

```
> (setq pi (mul 2.0 (asin 1.0)))
3.141592654
```

Now, for getting the value of tan(45), we can type:

```
> (tan (div (mul 45.0 pi) 180.0))
: 1
```

Using this strategy, we can use the trigonometric functions without effort. For calculating sin(45) we should type:

> (sin (div (mul 45.0 pi) 180.0))
: 0.7071067812

And for cos(60):

> (cos (div (mul 60.0 pi) 180.0))
: 0.5

Needless to say, if we type now:

> (acos 0.5)
: 1.047197551

The inverse trigonometric functions returns values expressed, yes, you have guessed, in radians. From (1-2), we can write:

$$\alpha = \frac{180r}{\pi} \tag{1-3}$$

And now, armed with our solid foundations learnt in this chapter, we can express:

> (setq r (acos 0.5))
: 1.047197551

and as expected:

> (div (mul 180.0 r) pi)
: 60

Putting it all into a one line of code:

> (div (mul 180.0 (acos 0.5)) pi)
: 60

You can say, and I am almost sure you are thinking it in this moment, that you can obtain the same results with an electronic calculator without so much typing. It is true, but first I have tried that you get used to type code in Lisp in order to feel comfortable with the language from the very beginning, and second, we shall learn to design and write functions in later chapters that will allow you to perform any calculation at speeds not even dreamt for any electronic calculator.

1.7 As a Summary

In this chapter we have had a gently introduction to the Lisp language. We have not forgotten to put it into an historical perspective and we have quickly moved into our first practical movements with it.

Two structural concepts in Lisp are contained in the following sentences: "Every Lisp expression contains a balanced number of parentheses" and "Functions are introduced first in a Lisp expression and then the arguments are located at right from the name of the function".

About arithmetic operators: NewLisp uses the symbols "+", "−", "*" and "/" for integer addition, subtraction, multiplication and division, respectively. When handling real numbers (the usual scenario in technical and scientific uses of the language) the functions *add*, *sub*, *mul* and *div* must be used instead.

References

Abrahams, P., et al.: The Programming Language LISP: Its Operations and Applications. The M.I.T. Press, Cambridge (1964). http://www.softwarepreservation.org/projects/LISP/book/III_LispBook_Apr66.pdf

Chailloux, J., Devin, M., Hullot, J.M.: Le_Lisp, a portable and efficient lisp system. INRIA (1984). Retrieved Jan 2014

Gabriel, R.P.: Performance and Evaluation of Lisp Systems. MIT Press, Computer Systems Series. (1985). ISBN 0–262-07093-6. LCCN 85-15161

McCarthy, J.: Recursive functions of symbolic expressions and their computation by machine. Commun. ACM (1960). http://www.cs.berkeley.edu/∼christos/classics/lisp.ps

McCarthy, J., et al.: LISP 1.5 Programmer's Manual. The M.I.T. Press, Cambridge (1962). http://www.softwarepreservation.org/projects/LISP/book/LISP%201.5%20Programmers%20-Manual.pdf

Moon, D.: MACLISP Reference Manual. The M.I.T. Press, Cambridge (1974). http://www.software preservation.org/projects/LISP/MIT/Moon-MACLISP_Reference_Manual-Apr_08_1974.pdf

Ousterhout, J.: Scripting: higher level programming for the 21st century. IEEE, Mar 1998

Teitelman, W., Masinter, L.: The interlisp programming environment. IEEE Computer, Apr 1981. http://larry.masinter.net/interlisp-ieee.pdf

Wertz, H.: LISP: Une Introduction a La Programmation. Masson Editions (1985). http://www.ai.univ-paris8.fr/∼hw/lisp/letout.pdf

Chapter 2
Lists Everywhere

2.1 Introduction

As we saw in Chap. 1, Lisp is an acronym for **LIS**t **P**rocessing, that is, all the data and programming in Lisp is based on an organizational structure characteristic of this programming language called "list". This chapter is entirely dedicated to this pivotal aspect of Lisp and we shall dedicate space enough for showing how to create, manage, modify, reset and eliminate lists.

The previous chapter gave you a quick introduction to simple numerical calculations using Lisp as a way of getting a first touch with the language. Every Lisp expression exposed was genuinely Lisp, but all the exposed material was even far from scratching the surface of it. Now it is time to ask you for a bit of concentration in order to undertake the new material. The concepts you are going to find in the sections of this chapter will give you a solid understanding of the conceptual pillars of Lisp. In fact, when you feel you have understood entirely all the ideas exposed in this chapter you will start to get comfortable with the language, being able to appreciate its philosophy, elegance, and flexibility.

As usual, we shall use a practical approach, getting help in this case from the use of examples taken from meteorology, geometry, art and discrete event simulation. There is a lot of material in this chapter and sometimes you will find that the information supplied is a bit dense. Be patient and take the time you need. As Italian people usually say, *"piano piano si va lontano"*, that is, *"going slowly you'll reach distant places"*. This is not a literal translation, of course, but gives you an idea about the attitude you should take while reading this chapter.

© Springer International Publishing Switzerland 2016   19
L. Argüelles Méndez, *A Practical Introduction to Fuzzy Logic using LISP*,
Studies in Fuzziness and Soft Computing 327,
DOI 10.1007/978-3-319-23186-0_2

2.2   Atoms and Lists

Berlin has a relatively cold climate. Summer is something between mild and warm, with cold winters, while the rest of the months are more chilly than mild. In any case, the *Berliner Luft* (Berlin air) is famous, and Berliners are really proud of it. This paragraph, almost a small meteorology report, serves us for introducing the following Lisp expression:

$$(-3 \ -2 \ 1 \ 4 \ 8 \ 11 \ 13 \ 12 \ 9 \ 6 \ 2 \ -1)$$

This is a list formed by the monthly average of low temperatures in Berlin, from data obtained from Climatedata (2014). Another example of list could be:

$$(Jan \ Feb \ Mar \ Apr \ May \ Jun \ Jul \ Aug \ Sep \ Oct \ Nov \ Dec)$$

This list contains the months of the year. A list is anything enclosed between an open and a closed parenthesis. If an element from a list can not be expressed in a simpler way, that element is called an **atom**. An example of atom is "-3", as is also "Jan". Generalizing we can say:

A list is a formal collection of elements enclosed by parenthesis whose elements are either atoms or lists

However, a list can not live isolated from the Lisp environment. If we supply a list to the prompt in a Lisp session we obtain the following:

> (-3 -2 1 4 8 11 13 12 9 6 2 -1)
: *ERR: illegal parameter type : -2*

This error message means that Lisp does not know what to do with a solitary list. Lisp needs that some activity must be performed on lists, or better said, the *grammar* of Lisp uses a structure where solitary lists are not allowed. Here, what we call structure can be as simple as a function that takes a list as an argument or as complex as an entire program written in the language. For better fixing ideas we should remember again that the acronym for Lisp is LISt Processing, and a list always needs some processing. When we add some structure, things start to work perfectly well:

> (setq Berlin-lows '(-3 -2 1 4 8 11 13 12 9 6 2 -1))
: *(-3 -2 1 4 8 11 13 12 9 6 2 -1)*

As we already know, this is a well-formed Lisp expression because, among other things, it contains a matched number of parentheses. The *setq* function is also known from the previous chapter, and it serves to assign values to a variable. At least this is what we have learnt so far. In the above expression it seems clear that something is assigned to the just created variable "Berlin-lows". Typing it we can observe that NewLisp returns:

> **Berlin-lows**
: *(-3 -2 1 4 8 11 13 12 9 6 2 -1)*

Now the symbol *Berlin-lows* stores the list *(-3 -2 1 4 8 11 13 12 9 6 2 -1)*, but an important detail remains unexplained and it is the quote operator that is usually represented in Lisp by the quote sign '. This important operator prevents Lisp from evaluating lisp expressions located at its right. In order to know how many expressions are not evaluated, we must distinguish two cases, depending on the lisp expression located immediately after the quote operator:

- If an atom is found just after the quote operator, Lisp stops its evaluation processing until finding a new atom or list. That is, only the quoted atom is not evaluated.
- If an open parenthesis is found just after the quote operator, Lisp stops its evaluation processing until finding its matching right parenthesis.

We shall understand it better with several examples:

> **'(+ 1 2)**
: *(+ 1 2)*

Without the quote, Lisp evaluates the list immediately, as you already know:

> **(+ 1 2)**
: *3*

By the way, alternatively you can also type:

> **(quote (+ 1 2))**
: *(+ 1 2)*

> **(setq a-nested-list '(a b (p1 (x1 y1) p2 (x2 y2))))**
: *(a b (p1 (x1 y1) p2 (x2 y2)))*

The *quote* operator in Lisp has a powerful counterpart function named *(eval)* that is imperative to introduce in this moment. Using it forces Lisp to evaluate the expression(s) located at its right. Let us type the following expression:

> **(setq operation '(+ 7 3))**
: *(+ 7 3)*

Calling *(eval)* in this moment we shall have:

> **(eval operation)**
: *10*

This function is extremely powerful, and although purist Lisp programmers tended to consider its use as an abuse, it certainly helps to save code in certain occasions. We shall have the opportunity to observe how does it work in future chapters of this book.

Again at our list representing the monthly average of low temperatures in Berlin, the expression:

(setq Berlin-lows '(-3 -2 1 4 8 11 13 12 9 6 2 -1))

processes the list of temperatures by means of the adequate grammatical use of setq and quote. As a result the list is bound to the symbol Berlin-lows, converting it into a variable. The distinction between symbol and variable is important and we are going to discuss it immediately.

The list introduced some paragraphs above:

(Jan Feb Mar Apr May Jun Jul Aug Sep Oct Nov Dec)

is a list composed by the symbols Jan, Feb, … etc. These symbols are not variables, since they are not still bound to any value. Lisp itself helps us to identify symbols, lists and atoms by some useful functions called *Predicates*. Predicates are special functions that always return either *true* or *nil*, the equivalent of "false" in Lisp. The name of every predicate always ends with a question mark. Let us type the following Lisp expressions with attention:

> (atom? a)
: *true*

Since "a", either containing a value or not, is indivisible, is clearly an atom.

> (symbol? 'a)
: *true*

Here "a" is quoted, that is, it is not evaluated by Lisp, so it is a completely legal symbol for Lisp. It does not matter if it holds any associated value.

> a
: *nil*

The direct evaluation of "a" returns nil because it does not contains anything. We have not yet assigned any value to it by means of the function *setq* so Lisp evaluates it to nil. Now let us assign a value to "a", numerical in this case:

> (setq a 10)
: *10*

if now we evaluate it:

> a
: *10*

it is bounded to the value 10, and not only that: after being bounded to a numerical value "a" is a variable of numeric type. But not only is a variable, it continues being a symbol:

> **(symbol? 'a)**
: *true*

Note again that in this expression we have quoted "a". If we evaluate again it without quoting:

> **(symbol? a)**
: *nil*

Lisp immediately evaluates "a". Since it is now a variable, it stores the numerical value 10 so this expression is equivalent to:

> **(symbol? 10)**
: *nil*

This is absolutely correct since "10" is a numerical value, not a symbol. We can follow using the number? predicate:

> **(number? a)**
: *true*

and as you can easily imagine:

> **(number? 'a)**
: *nil*

Speaking about predicates, The language itself provides a function in order to know if something is a list:

> **(list? '(-3 -2 1 4 8 11 13 12 9 6 2 -1))**
: *true*

> **(list? Berlin-lows)**
: *true*

> **(symbol? 'Berlin-lows)**
: *true*

As you can see, this is a wonderful computing game although maybe a bit confusing at first. Do not worry if these concepts are a bit difficult to grasp in this moment. As they say, practice makes perfect and my best advice is to enjoy several Lisp sessions where you create lists, symbols and, by assigning values, variables. Incidentally, I suspect it is harder if you already know other computer languages as C++ or Java, languages that are strongly typed. As we wrote in Chap. 1, Lisp automatically takes care of memory allocation and management so you can concentrate in program design and thus, being more productive. Anyway, for C programmers we could give a hint: symbols in Lisp are in fact pointers in disguise. The good thing is that Lisp manages both pointers (memory addresses) and values in a transparent way for the Lisp user.

After these theoretical concepts, and as a small relaxation, we can seize the opportunity to introduce a practical function embedded in NewLisp for calculating

Table 2.1 Calculating basic statistics with the NewLisp built-in function *(stats)*. See text

Statistical element	Calculated values for Berlin-lows
Number of values	12
Mean of values	5
Average deviation from mean value	4.833333333
Standard deviation	5.640760748
Variance	31.81818182
Skew	0
Kurtosis	-1.661427211

simple statistics. The name of this function is *stats*, and we can apply it to the list of low temperatures in Berlin for observing how does it work:

> **(stats Berlin-lows)**
: *(12 5 4.833333333 5.640760748 31.81818182 0 -1.661427211)*

The function *(stats)* takes as argument a list containing a set of numerical values and after some almost instantaneous calculations returns statistical values ordered in the following sequence: Number of values, mean of values, average deviation from mean value, standard deviation, variance and skew and kurtosis of distribution, as shown in Table 2.1. The actual version of NewLisp (v.10.5.4 as of this writing) contains several built-in statistical and probabilistic functions that are useful in science and engineering. Since the mission of this book is to allow the reader to understand the theories of fuzzy-sets and fuzzy-logic we shall not dedicate more space to calculate statistics, but the reader should be aware that these NewLisp built-in functions deserve to be explored. The NewLisp Manual and Reference provide all the needed information to start doing statistics with this Lisp implementation.

For calculating the mean of values with the knowledge learnt from Chap. 1 we would have used, for example the following expression:

> **(setq average-min (div (add -3 -2 1 4 8 11 13 12 9 6 2 -1) 12))**
: *5*

As promised, Lisp offers us a way of making things in a very powerful way. And we are only starting with it.

2.3   First and Rest

The first element in a list has an enormous importance in Lisp since its combination of symbol and value determines if it is a list of data or a list with a call to a function. In fact, when Lisp finds a list, it immediately evaluates its first element in order to know what to do with it. For example, the first element in the list of low monthly temperatures for Berlin (-3 -2 1 4 8 11 13 12 9 6 2 -1) is "-3", so Lisp identifies it as a list where no function is called. On the other hand, the first element of the list (add -3 -2 1 4 8 11 13 12 9 6 2 -1) is "*add*" and since it is included in the list of symbols of Lisp that represent functions then treats the rest of the list as arguments that must be provided to the function call. Having into account this grammatical rule of the language, it is not a surprise that Lisp incorporates two functions for accessing the first element of a list and the rest of elements. In the NewLisp implementation, these functions are named *first* and *rest*, respectively (in Common Lisp and many other Lisp implementations these functions are named *car* and *cdr* as an homage to their first use on the IBM704 computer). Its use is more than intuitive:

> **(first '(Jan Feb Mar Apr May Jun Jul Aug Sep Oct Nov Dec))**
: Jan

Please note that the list of months is quoted. Let us observe how the function *rest* does it work:

> **(rest '(Jan Feb Mar Apr May Jun Jul Aug Sep Oct Nov Dec))**
: (Feb Mar Apr May Jun Jul Aug Sep Oct Nov Dec)

The natural question now is: Ok, it is easy to understand the use of the functions *first* and *rest*, in fact it is one of the more easy concepts exposed so far in the book, but what kind of things can we do with them aside obtaining the first element and the rest of elements in a list? Important answer: Using *(first)* and *(rest)* we can access any element on any list. In order to type less code, let us begin typing the following in a Lisp session:

> **(setq months '(Jan Feb Mar Apr May Jun Jul Aug Sep Oct Nov Dec))**
: (Jan Feb Mar Apr May Jun Jul Aug Sep Oct Nov Dec)

> **(first months)**
: Jan

> **(rest months)**
: (Feb Mar Apr May Jun Jul Aug Sep Oct Nov Dec)

You should already know why the variable *months* is not quoted in the above two expressions, but the important thing is that the first element from the list returned after the evaluation of *(rest months)* corresponds to the second month, February. Let us rewrite the previous expression in order to store the list it returns on the variable *rest-1*:

> **(setq rest-1 (rest months))**
: (Feb Mar Apr May Jun Jul Aug Sep Oct Nov Dec)

Now, let us continue typing:

> **(setq second-month (first rest-1))**
: Feb

> (setq rest-2 (rest rest-1))
: *(Mar Apr May Jun Jul Aug Sep Oct Nov Dec)*

> (setq third-month (first rest-2))
: *Mar*

Reiterating this procedure by using several calls to *(first rest-i)* for *i* going from 1 to 11 we can access any month from the list from February to December. For January, in order to generalize the algorithm just exposed, we could have written:

> (setq rest-0 months)
: *(Jan Feb Mar Apr May Jun Jul Aug Sep Oct Nov Dec)*

Now rest-0 contains the full list of months. Obtaining the first element in the sequence of lists *rest-0, rest-1, ..., rest-11*, we can access any month in the year. This type of procedure for traversing lists is typical of Lisp and we shall use them frequently. Yes, at this stage it is a lot of typing, but our agreement for learning Lisp and ultimately fuzzy logic at the beginning of the book requires typing Lisp code. In the next chapter we shall learn to write functions and that will allow us to go from interactive Lisp sessions to Lisp programming. For now, the typing will continue. However, and almost without noticing it, you are learning Lisp at a good pace.

2.4   Building Lists

Without doubt, *(first)* and *(rest)* are powerful functions, but accessing elements from a list is so frequent in Lisp programming that Lisp designers soon realized that a special function should be created for accessing any element from a list without using *first* and *rest*. The name of this wanted function is *n*th. Having into account the previous expression:

(setq months '(Jan Feb Mar Apr May Jun Jul Aug Sep Oct Nov Dec))

Now we can write:

> (nth 0 months)
: *Jan*

> (nth 5 months)
: *Jun*

> (nth 11 months)
: *Dec*

Please note the special indexing for the function *n*th: the first element in a list is acceded by the number 0, not by the number 1, so the last element in a list of n elements can be acceded by using n − 1 as the index for *n*th. By the way, in the same manner there is a function for accessing the first member of a list, there is also another one for directly accessing its last element:

> (last months)
: Dec

When using *first*, *last* and especially *n*th, it is usual, if not mandatory before making a call to *n*th, to use the Lisp function *length*. As its name suggests, it returns the number of elements in a list:

> (length months)
: 12

then, another way to access the last element of a list could be:

> (nth (- (length months) 1) months)
: Dec

Nobody does it that way, of course, but it is a safe way to avoid errors while programming. If we write:

> (nth 12 months)
*: ERR: invalid list index in function n*th

This error would break the execution of any Lisp program, so as a general rule:

Always use the function *length* before using *n*th and remember that indexing in lists starts at zero

Figure 2.1 helps to shed more light into the relationships between *(first)*, *(rest)*, *(last)* and *(nth)*:

The trick to understand it all is as simple as observing the first block, where we do *list = rest$_0$*. Understanding *rest$_0$* as the whole list help us to get used to zero indexing for the function *n*th. This is exactly what we did with the expression *(setq rest-0 months)* some paragraphs above.

Fig. 2.1 A graphical representation of the relationships between the Lisp functions *(first)*, *(rest)* and *(nth)*

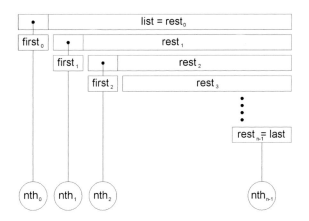

After arriving to this point, we are ready to learn how to build lists. Until now we have used the assignment function *setq* for linking a list to a symbol, and the symbols *Berlin-lows* and *months* were built this way. From these two symbols we are going to build a more complex list where every month and its correspondent low temperature will be paired in a sublist. Let us type the following:

> **(setq mt1 (cons (nth 0 months) (nth 0 Berlin-lows)))**
: (Jan -3)

Here the new function is *(cons)*. It takes two arguments, either atoms or lists, and then constructs a new list with them. Let us see other examples:

> **(cons (first months) (rest months))**
: (Jan Feb Mar Apr May Jun Jul Aug Sep Oct Nov Dec)

> **(cons (first months) (last months))**
: (Jan Dec)

But take care with the arguments. If we type *(cons Jan Feb)* Lisp returns the list *(nil nil)*. Why? Because both *Jan* and *Feb* are not yet bounded to any value and consequently Lisp evaluate these symbols to *nil*. It is a different story if we type *(cons 'Jan 'Feb)*, then Lisp returns the list *(Jan Feb)*.

We have chosen *mt1* as the name of the symbol representing January and its temperature almost like an acronym for **m**onth and **t**emperature, appending a figure one for linking it to the number of the month. It is always a good strategy to choose the name of symbols and variables with a clear meaning for us as Lisp programmers. We shall revisit this idea in the next chapter. For now let us continue for the rest of months:

> **(setq mt2 (cons (nth 1 months) (nth 1 Berlin-lows)))**
: (Feb -2)

> **(setq mt3 (cons (nth 2 months) (nth 2 Berlin-lows)))**
: (Mar 1)
…
…

and after some heavy typing (not at all, you can copy, past and modify the code), we reach December:

> **(setq mt12 (cons (nth 11 months) (nth 11 Berlin-lows)))**
: (Dec -1)

And now, another new function, named *list*, makes the magic:

> **(setq months-temps (list mt1 mt2 mt3 mt4 mt5 mt6 mt7 mt8 mt9 mt10 mt11 mt12))**
: ((Jan -3) (Feb -2) (Mar 1) (Apr 4) (May 8) (Jun 11) (Jul 13) (Aug 12) (Sep 9) (Oct 6) (Nov 2) (Dec -1))

Now, every element of the list represented by the symbol *months-temps* is in itself a list of two elements. The *list* function takes any number of arguments (12 arguments have been used in this example) and then, every argument is evaluated and then is used as an element for building a new list. After creating the list *months-temps* we can type, for example:

> **(nth 5 months-temps)**
: *(Jun 11)*

But this is not the end of the story. This is the perfect moment for introducing another function named *assoc*. Let us see how does it works with an example:

> **(assoc 'Aug months-temps)**
: *(Aug 12)*

If you suspect we are entering the realms of database programming you are completely right. The function *assoc* takes the first argument passed to the function and then uses it as a key for searching the list passed as the second argument. If a match is found, then it returns the member list. If no match is found, it returns nil. See now how easy is to access any data from the list *months-temps*:

> **(first (assoc 'May months-temps))**
: *May*

> **(last (assoc 'May months-temps))**
: *8*

This function not only serves for accessing elements in a list, it also permits to directly change information in lists using *setq*. Let us see an example for changing the minimum temperature in July from data from, for example, Canary Islands:

> **(setq (assoc 'Jul months-temps) '(Jul 21))**
: *(Jul 21)*

after *setq*, *assoc* uses *Jul* as the key symbol for finding the sublist *(Jul temperature)* and then substitutes it by the sublist given in the second argument, in this case, *(Jul 21)*. Now we can observe the changes made:

> **(nth 6 months-temps)**
: *(Jul 21)*

We could have written, for example:

> **(setq (assoc 'Jul months-temps) '(Jul 21 Canary-Islands))**
: *(Jul 21 Canary-Islands)*

and now, the contents for the *July* sublist are, obviously:

> **(nth 6 months-temps)**
: *(Jul 21 Canary-Islands)*

Using *assoc* in this way has a destructive effect on the original list, in this case, on *months-temps*. While some Lisp functions preserve the original list (*nth* or *length*, for example, come quickly to mind) other ones modify the original data. In many cases it is a safe strategy to make first a copy of the original data and then manipulate the copy. For now, we can restore the original data for minimum temperature in Berlin in July simply writing:

> **(setq (assoc 'Jul months-temps) '(Jul 13))**
: *(Jul 13)*

2.5 Some Geometry and then Some Art, Too

In general, lists of the type:

$$((atom_{11}\ atom_{12})\ (atom_{21}\ atom_{22})\ ...\ (atom_{n1}\ atom_{n2}))$$

are extremely useful in Lisp, since they can be an excellent representation of many observational data and phenomena in science, engineering or technical fields. Needless to say, it is the natural way in Lisp for representing pairs of geometric coordinates, giving access to the wonderful world of graphics. Since fuzzy-logic can be seen, as we shall see in the second part of this book, as a matter of geometry, we must seize the opportunity to play a bit with geometric forms in this moment.

As a general rule, we can represent rectangles, trapeziums and squares by means of four coordinates, triangles by means of three coordinates and circles by means of a coordinate for the centre and another number for its radius. Table 2.2 shows some list structures for representing geometrical shapes in a flat, two dimensions space:

If the shape is a rectangle or a square, its list representation is even simpler by using the coordinates of the two points of one of its diagonals. Let us type some values in order to create some rectangles:

> **(setq pt1 '(5 -4) pt2 '(8 12)) (setq R1 (list pt1 pt2))**
: *(8 12)*
: *((5 -4) (8 12))*

The rectangle identified by the symbol R1 now storages two points, defined by its coordinates. By the way, we have written several Lisp expressions in the same

Table 2.2 List representation of simple 2D geometrical forms

Geometric form	Lisp coordinate representation
Rectangle, trapezium, square	$((x_1\ y_1)\ (x_2\ y_2)\ (x_3\ y_3)\ (x_4\ y_4))$
Rectangle, square (by diagonal)	$((x_1\ y_1)\ (x_2\ y_2))$
Triangle	$((x_1\ y_1)\ (x_2\ y_2)\ (x_3\ y_3))$
Circle	$((x_1\ y_1)\ r)$

line after the prompt in order to save some space. Now let us repeat the operation for some other rectangles:

> **(setq pt3 '(0 0) pt4 '(10 10)) (setq R2 (list pt3 pt4))**
: (10 10)
: ((0 0) (10 10))

> **(setq pt5 '(2 -1) pt6 '(16 2)) (setq R3 (list pt5 pt6))**
: (16 2)
: ((2 -1) (16 2))

> **(setq pt7 '(6.5 1) pt8 '(13.5 8)) (setq R4 (list pt7 pt8))**
: (13.5 8)
: ((6.5 1) (13.5 8))

After having well defined rectangles R1, R2, R3 and R4, now we can build a geometrical composition with them named "*Mondrian-style*" as an homage to Piet Mondrian, the famous Dutch painter:

> **(setq Mondrian-style (list R1 R2 R3 R4))**
: (((5 -4) (8 12)) ((0 0) (10 10)) ((2 -1) (16 2)) ((6.5 1) (13.5 8)))

Figure 2.2 is a visual representation of the list *Mondrian-style*.

Pablo Picasso was a bit more complex than Mondrian. Aside rectangles we must include now circles and triangles following the structure depicted in Table 2.2. Let us start with a rectangle:

Fig. 2.2 Four *rectangles* representing a composition inspired in the style of Piet Mondrian

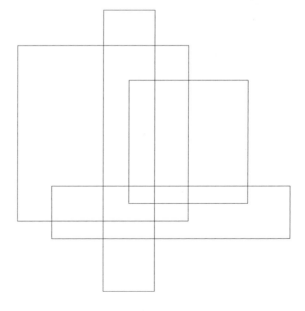

> **(setq pt9 '(5 -4) pt10 '(8 12)) (setq R5 (list pt9 pt10))**
> *: (8 12)*
> *: ((5 -4) (8 12))*

Now it is the time for circles C1 and C2. We shall define them in only one line of Lisp code:

> **(setq C1 '((8 2.5) 1.5) C2 '((6 9) 0.5))**
> *: ((6 9) 0.5)*

And now it is time for the triangles:

> **(setq T1 '((6 9) (10 6.75) (6 4.5)))**
> *: ((6 9) (10 6.75) (6 4.5))*

> **(setq T2 '((1.438 2.5) (7.5 -1) (7.5 6)))**
> *: ((1.438 2.5) (7.5 -1) (7.5 6))*

Finally let us join the shapes in order to build the composition:

> **(setq Picasso-style (list R5 C1 C2 T1 T2))**
> *: (((5 -4) (8 12)) ((8 2.5) 1.5) ((6 9) 0.5) ((6 9) (10 6.75) (6 4.5)) ((1.438 2.5)*
> *(7.5 -1)*
> *   (7.5 6)))*

In Fig. 2.3 we can see now our personal, Lisp based Picasso composition.

We have forgotten an important detail: The name of the paintings! For this we shall use a new function named "*append*" that, as its name implies, appends data to the tail of an existing list. Let us use it:

> **(setq Mondrian-style (append Mondrian-style '("Composition #968")))**
> *: (((5 -4) (8 12)) ((0 0) (10 10)) ((2 -1) (16 2)) ((6.5 1) (13.5 8)) "Composition*
> *#968")*

> **(setq Picasso-style (append Picasso-style '("Pregnant woman")))**
> *: (((5 -4) (8 12)) ((8 2.5) 1.5) ((6 9) 0.5) ((6 9) (10 6.75) (6 4.5)) ((1.438 2.5)*
> *(7.5 -1) (7.5 6)) "Pregnant woman")*

This seems a bit complex at first sight, because aside the function *(append)* we have introduced at the same time a new type of data named "*string*". Let us start with the function *append*. It appends anything to an existing list, but first the new thing(s) to append must be enclosed into a list and second, the function *append* is not destructive, meaning that the original list remains unchanged. Let us see this important detail with a simpler example:

> **(setq my-list '(a b c))**
> *: (a b c)*
> **(append my-list '(d))**
> *: (a b c d)*

Fig. 2.3 A *rectangle*, two
circles and two *triangles* for
representing a Picasso style
composition

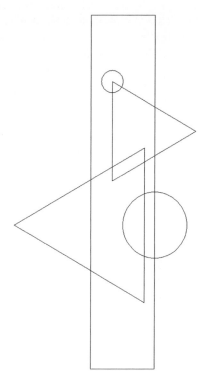

but if we evaluate what is now pointed by the symbol *my-list,* a surprises appears:

> **> my-list**
> *: (a b c)*

After applying *append* to a list it returns <u>another</u> list with an appended new element at its tail, but it does not even touch the original list. If we want the original list to be modified we need to use *setq* as follows: *(setq my-list (append my-list '(d)))* After introducing this Lisp expression at the prompt then *my-list* contains (a b c d), as desired. This is the explanation for using *setq* in the expressions:

> (setq Mondrian-style (append Mondrian-style '("Composition #968")))
> (setq Picasso-style (append Picasso-style '("Pregnant woman")))

"Composition #968" and "Pregnant woman" are strings, that is, a type of data composed by characters. Interestingly, strings are not symbols:

> **> (symbol? (last Mondrian-style))**
> *: nil*

We shall dedicate more time to strings in a future chapter. This type of date is important because when we write Lisp programs that ask an user for data, either numerical or alphanumerical, Lisp reads the input (usually from the keyboard) and returns strings.

Returning to our peculiar paintings, yes, I am hearing you about the issue of having the name of the paintings at the end of the lists. Usually, a painting is identified by its name, so it seems natural to have it located at the first position. Well, this is a very easy to solve problem by means of using the function *(reverse)*. As you have imagined, it reverses the order of elements in a list:

> (reverse Mondrian-style)
: ("Composition #968" ((6.5 1) (13.5 8)) ((2 -1) (16 2)) ((0 0) (10 10)) ((5 -4) (8 12)))

(reverse Picasso-style)
: ("Pregnant woman" ((1.438 2.5) (7.5 -1) (7.5 6)) ((6 9) (10 6.75) (6 4.5)) ((6 9) 0.5)
((8 2.5) 1.5) ((5 -4) (8 12)))

This function is destructive, modifying the original list. However the use of *reverse* over *reverse* restores the list at its original status, that is: *(reverse (reverse a-list))* → *a-list*

Now it is time to finally make a small gallery, a collection of two paintings:

> (setq my-gallery (list Mondrian-style Picasso-style))
: (("Composition #968" ((6.5 1) (13.5 8)) ((2 -1) (16 2)) ((0 0) (10 10)) ((5 -4) (8 12))) ("Pregnant woman" ((1.438 2.5) (7.5 -1) (7.5 6)) ((6 9) (10 6.75) (6 4.5)) ((6 9) 0.5) ((8 2.5) 1.5) ((5 -4) (8 12))))

The representation of my-gallery can be shown in Fig. 2.4.

Well, the representation of Fig. 2.4 is not exact, as any attentive reader has already noticed. First, there is a common rectangle in the two compositions and second, coordinates from the two compositions share a common space in order to

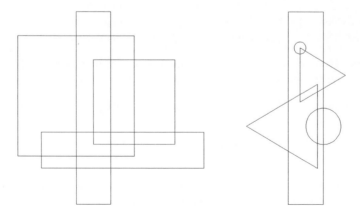

Fig. 2.4 The my-gallery collection, representing Mondrian and Picasso look-alikes

Fig. 2.5 The my-gallery
collection, true representation

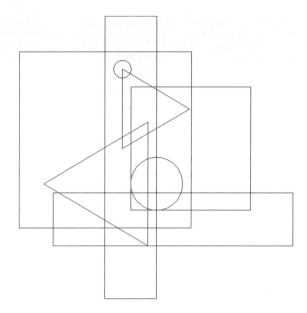

use small and simple coordinates in the Lisp expressions. The true representation of
my-gallery is shown in Fig. 2.5 that seems to suggest that Mondrian and Picasso
compositions maybe do not mix well.

A positive shift for every x value in the coordinates for the *Picasso-style* list
should have been applied for obtaining the image shown in Fig. 2.4. We shall learn
how to perform such a shift in the next chapter.

In Table 2.2 we have seen how to represent some geometric shapes in 2D. If we
assume that the geometric forms are resting over a horizontal axis, it is even easier
to represent them. As we can see in Table 2.3, in this case a rectangle only need two
values of x for representing the location of its base and a height, h. A square is even
simpler because given the base by x_1 and x_2 then its height is equal to the distance
between x_1 and x_2. The base of a triangle comes represented by its extremes at x_1 x_3
and then its third point is given by the point x_2,h. Finally, a trapezium is represented
by the extremes at its base x_1 and x_4 and the points x_2,h_1 and x_3,h_2.

Let us see a simple example of a triangle and two trapeziums resting over a
horizontal axis using some lisp expressions:

> **(setq trapezium-1 '(2 (4 4) (6 4) 8))**
: (2 (4 4) (6 4) 8)

> **(setq triangle '(6 (8 4) 10))**
: (6 (8 4) 10)

> **(setq trapezium-2 '(8 (10 4) (12 4) 14))**
: (8 (10 4) (12 4) 14)

Geometric form	Lisp coordinate representation
Table 2.3 List representation of simple 2D geometrical forms when resting over a horizontal axis	
Rectangle, square	$((x_1\ x_2)\ h)$
Square	$(x_1\ x_2)$
Triangle	$(x_1\ (x_2\ h)\ x_3)$
Trapezium	$(x_1\ (x_2\ h_1)\ (x_3\ h_2)\ x_4)$

as we know, for joining these geometric forms into a single Lisp expression, we can type:

(setq simple-composition (list trapezium-1 triangle trapezium-2))
: ((2 (4 4) (6 4) 8) (6 (8 4) 10) (8 (10 4) (12 4) 14))

Figure 2.6 shows the graphic representation of the just created symbol *simple-composition*. This type of geometry seems simple at first, but as we shall see in this book, has a big conceptual importance.

Histograms and other related graphics are also based on using both vertical and horizontal axes where values can be absolute or relative. For example, our list of minimum temperatures by month in Berlin can be shown easily on a two-dimensional chart. As previously seen in Sect. 2.2, the symbol *Berlin-lows* points to the list *(-3 -2 1 4 8 11 13 12 9 6 2 -1)*. The range of stored temperatures goes from -3 to 13, resulting into a whole range of 16°. Since there are twelve values corresponding to all the months in the year, we can choose a space between months of two units for the horizontal axis, yielding a 2D region of 24 × 16 units for representing the data. Manipulating the list as we saw in Sect. 2.4 by means of the functions *cons* and *list*, or directly by *setq*, we can easily arrive at the following list of points $(x_i\ y_i)$ where x_i are the abscissa values and y_i the temperature (ordinate) values:

((0 -3) (2 -2) (4 1) (6 4) (8 8) (10 11) (12 13) (14 12) (16 9) (18 6) (20 2) (22 -1))

Joining the points stored in this list with lines we obtain a representation of the minimum temperatures in Berlin, as shown in Fig. 2.7.

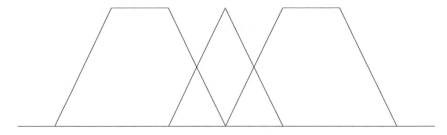

Fig. 2.6 Graphic representation of the list *simple-composition*

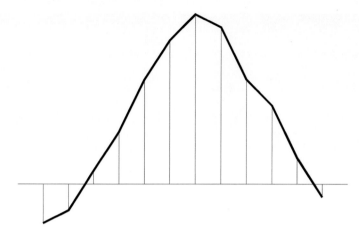

Fig. 2.7 Minimum temperatures in Berlin by month

Going from two dimensions to three dimensions (3D) graphics implies an additional degree of complexity. Some simple 3D geometric forms are shown as lists in Table 2.4.

One of the causes of the bigger complexity of 3D objects is that its orientation in space must be defined. Hence, for example, the eight points needed for representing a truncated pyramid. However, if a 3D object rests or hovers over a flat horizontal surface, some simplification results, especially for parallelepipeds. In this situation, a parallelepiped can be expressed in list form by:

$$((x_1\ y_1)\ (x_2\ y_2)\ (x_3\ y_3)\ (x_4\ y_4)\ h_1\ h_2)$$

where h_1 and h_2 are, respectively, the free height over the plane XY (floor), and the intrinsic height of the parallelepiped. This geometrical element is one of the most important ones in architecture. Figure 2.8 shows the Barcelona Pavilion (1929) by the famous architect Mies Van der Rohe. Practically all the components of the Pavilion could be expressed by means of Lisp expressions as the one shown above.

Before ending this section on geometry and Lisp lists, it deserves to be commented that one of the most popular Computer Aided Design (CAD) applications in the industry, AutoCAD, uses Lisp as an embedded programming language. Named

Table 2.4 List representation of simple 3D geometrical forms

Geometric form	Lisp coordinate representation
Parallelepiped	$((x_1\ y_1)\ (x_2\ y_2)\ (x_3\ y_3)\ (x_4\ y_4)\ (x_5\ y_5)\ (x_6\ y_6))$
Sphere	$((x_1\ y_1)\ (x_2\ y_2)\ (x_3\ y_3)\ r)$
Pyramid	$((x_1\ y_1)\ (x_2\ y_2)\ (x_3\ y_3)\ (x_4\ y_4)\ h)$
Truncated pyramid	$((x_1\ y_1)\ (x_2\ y_2)\ (x_3\ y_3)\ (x_4\ y_4)\ (x_5\ y_5)\ (x_6\ y_6)\ (x_7\ y_7)\ (x_8\ y_8))$

Fig. 2.8 Barcelona Pavilion, 1929. Architect: Ludwig Mies van der Rohe *Photograph* by the author

"AutoLISP", it is a close Lisp dialect to NewLisp, so, after reading this book you would be in a good position to use Lisp in a 3D graphic environment. And not only that, you could apply fuzzy logic in Computer Aided Design and graphics, too. Sounds exciting? I can assure you it really is.

2.6   A World Full of Queues

Everyday we experience queues in our lives. Before paying in the supermarket, just before entering inside a plane at the airport or while being trapped in a traffic jam in the highway, just to name a few examples. Queue theory is important in factories, in production management, data processing, financial strategies and so on. Lisp has two functions, named *pop* and *push* that are especially suited to manage and model queues. They were not created as a direct solution to problems derived from queue theory but for list management, but since this is a practical book we are going to learn to use them while modeling a toll station in a highway.

As it is well known, a toll station is basically a facility located on a fixed point in a highway where cars arrive, are serviced by a human being or a machine (the driver gets a ticket and/or pays a variable sum of money) and after some service time, they leave the queue. Let us say we are interested in two things: first, to record all the cars that enter the facility, their car plate and the enter and exit time from the toll station. This is very useful to the organism, either public or private, that manages the highway and it also serves for supplying information in special cases, let us say, as an example, for a police requirement. Second, we are interested in modeling the facility itself, that is, to represent how many cars are waiting in the toll station in real time. Before entering into the details, let us take a quick view at how *pop* and *push* do work. Let us start creating an empty list:

> **(setq queue '())**
: *()*

Now three elements, a, b and c, enter in the list sequentially:

> **(push 'a queue)**
: *(a)*

> **(push 'b queue)**
: *(b a)*

> **(push 'c queue)**
: *(c b a)*

As it can be seen, the *push* function adds elements into a list "pushing" them sequentially from the frontal part of the list to its back. Under this point of view, it is a list builder, but also a function that modifies the original contents of the list just after being applied. Note also that the list is built in reversal order. This is an important detail. Let us use now the function pop. It does not need any argument:

> **(pop queue)**
: *c*

Let us now observe what remains in *queue* after using *pop*:

> **queue**
: *(b a)*

In other words, while the function *push* introduces elements into a list, the use of *pop* returns the first element of a list and at the same time eliminates it from the list. As already suggested, it is useful for queue representation, but only for queues where its discipline is Last-In-First-Out, or LIFO, as it is usually named in queue theory. This is not what we were waiting for to apply to our model of toll station whose queue discipline is in fact First-In-First-Out, or FIFO, that is, the first car that enters the toll station is the first car that leaves it. Sadly, Lisp does not provide a built-in function for managing elements in a list in a FIFO way, so we shall need to do some list manipulation management in order to perform the desired behavior in the facility.

To better fix ideas let us assume that some devices are already installed in the toll station: a camera that reads the car plates from cars by means of an artificial vision system and another sophisticated device composed by a chronometer and a set of photoelectric cells for registering the arrival and departure time for every car. We also assume that the facility starts its morning shift at 8:00. Table 2.5 shows some car plate data and arrival and departure times for several cars being served at the toll station.

Table 2.5 Example data for cars entering a toll-station in a highway

Car plate	Arrival time	Departure time
CKT8623	08:02:12	08:02:54
GWG2719	08:02:18	08:04:32
MKA8772	08:02:25	08:05:55
DYN2140	13:15:21	13:16:22

For simplicity, we shall represent the arrival and departure times as simple numbers, i.e., 08:02:12 will be represented by 80212, 13:15:21 by 131521, and so on. Our desired working model is that every car enters the queue from the "left" of the list and departs at its "right". First let us create the queue and then add some cars to it.

```
> (setq queue '())
: ()
```

```
> (push 'CKT8623 queue)
: (CKT8623)
```

```
> (push 'GWG2719 queue)
: (GWG2719 CKT8623)
```

```
> (push 'MKA8772 queue)
: (MKA8772 GWG2719 CKT8623)
```

Between 08:02:12 and 08:02:54 we have three cars in the queue. At 08:02:54 the first car departs, so it must leave the list. The actual contents of *queue* are:

```
> queue
: (MKA8772 GWG2719 CKT8623)
```

Now the car that must leave the queue must be the one with the car plate CKT8623, so a call to the function *pop* is not possible. In order to perform a correct working model of the queue we must proceed with caution. The first step is to use the function *reverse*:

```
> (reverse queue)
: (CKT8623 GWG2719 MKA8772)
```

Now we can apply the function *pop*:

```
> (pop queue)
: CKT8623
```

```
> queue
: (GWG2719 MKA8772)
```

We are almost done. A second call to the function *reverse* will finish the algorithm for representing a FIFO queue in Lisp:

> **(reverse queue)**
> *: (MKA8772 GWG2719)*

Reiterating the procedure, car MKA8772 would leave the queue at 08:05:55. At this moment, the list *queue* would be equal to the empty list, *()* until the car DYN2140 arrives at 13:16:22.

Meanwhile, and at the same time, we must record all the cars that enter the facility. This is easier to do:

> **(setq toll '())**
> *: ()*

> **(push '(CKT8623 80212 80254) toll)**
> *: ((CKT8623 80212 80254))*

> **(push '(GWG2719 80218 80432) toll)**
> *: ((GWG2719 80218 80432) (CKT8623 80212 80254))*

> **(push '(MKA8772 80225 80555) toll)**
> *: ((MKA8772 80225 80555) (GWG2719 80218 80432) (CKT8623 80212 80254))*

> **(push '(DYN2140 131521 131622) toll)**
> *: ((DYN2140 131521 131622) (MKA8772 80225 80555) (GWG2719 80218 80432) (CKT8623 80212 80254))*

The list *toll* does not need any treatment with the function *reverse*, because it is used to register all the vehicles served at the facility, but at the end of the day, it can help to make the recorded data easier to read:

> **(reverse toll)**
> *: ((CKT8623 80212 80254) (GWG2719 80218 80432) (MKA8772 80225 80555) (DYN2140 131521 131622))*

Now let us assume the police tries to track a car with car plate MKA8772 and calls the highway company in order to know if that car has used the facility. At the company, they only need to type:

> **(assoc 'MKA8772 toll)**
> *: (MKA8772 80225 80555)*

For confirming the police that a car with that plate has been observed at toll station X from 08:02:25 to 08:05:55. A call to all the toll stations in the country would result in a map with the route of the car. By the way, NewLisp incorporates a function named *now* that can be very useful when we try to record time in an application. Let us observe the results of a call to *now* just as I'm writing:

> **(now)**
> *: (2014 2 26 13 34 40 592460 56 3 -60 0)*

Table 2.6 Structure of the numerical information returned by the NewLisp function *(now)*

Element	Example	Observations
Year	2014	From the Gregorian calendar
Month	2	From 1 to 12
Day	26	From 1 to 31
Hour	13	From 0 to 23, (UT)
Minute	34	From 0 to 59
Second	40	From 0 to 59
Microsecond	592,460	From 0 to 999,999
Day of current year	56	Begins at 1 for January, 1st
Day of current week	3	From 1 to 7. Starts at Monday
Time zone offset (min)	-60	West of Greenwich meridian
Daylight saving time	0	From 0 to 6

The meaning of the elements of the list returned by a call to the function *(now)* is shown in Table 2.6.

With the use of *(now), (pop), (push) and (random)* (we shall introduce this function in the next section) and some other programming resources, Lisp can be used for solving Discrete Event Simulation problems. See for example Banks et al. (2009), Sturgul (1999). Until recently, queue studies have been studied mainly by simulation, but relatively recent works seem to suggest that Fuzzy-Logic represents a good approach to this kind of problems, too (Zhang 2005).

2.7 Rotate, Extend and Flat

For finishing this chapter we shall introduce yet three NewLisp functions for managing lists. They are not, comparatively, so nuclear to Lisp as the previously ones, that is, they are not present in all the Lisp dialects, but they sure have an additional interest.

For explaining the function *(rotate)*, we should visualize a list as a circular structure where the first and last element would be contiguous elements. For a better and complete understanding of this structure there is not a better example than a French roulette wheel of unique zero located into a casino. Its list representation is as follows:

> **(setq roulette '(0 1 2 3 4 5 6 7 8 9 10 11 12 13 14 15 16 17 18 19 20 21 22 23 24 25 26 27 28 29 30 31 32 33 34 35 36))**

: (0 1 2 3 4 5 6 7 8 9 10 11 12 13 14 15 16 17 18 19 20 21 22 23 24 25 26 27 28 29 30 31 32 33 34 35 36)

The function *rotate* uses an integer n as parameter and it moves every element for the list n positions towards its right if n is positive or towards left if n is negative. For example:

> (rotate roulette 3)
: (34 35 36 0 1 2 3 4 5 6 7 8 9 10 11 12 13 14 15 16 17 18 19 20 21 22 23 24 25 26 27 28 29 30 31 32 33)

As can be easily seen every element has moved three positions towards right, and the elements 34, 35 and 36 have rotated from their original positions at the tail of the list to the first ones. Typing *(rotate roulette -3)* would restore the original position of the roulette.

While we are speaking about roulettes, casinos and chance, it seems convenient to introduce the function *random*. This function simply returns a real number between 0 and 1 at random (strictly speaking no computer is able to generate pure random numbers, but the use of pseudo-random numbers is usually enough for representing random events in the real world). Let us see a simple call to random:

> (random)
: 0.283314746

In order to obtain a random integer number from 0 to 36 we should type:

> (setq alpha (integer (mul (random) 37)))
: 24

Now, we can give a spin with a value *alpha* to our roulette and obtain the first element after the rotation of the roulette is applied:

> (first (rotate roulette alpha))
: 13

Now you can argue that a simple call to *(mul (random) 37)* could produce the same roulette simulation, and you are true, but our model not only gets a random number between 0 and 36 but it also models the position of the roulette wheel:

> roulette
: (13 14 15 16 17 18 19 20 21 22 23 24 25 26 27 28 29 30 31 32 33 34 35 36 0 1 2 3 4 5 6 7 8 9 10 11 12)

Another application for *rotate* could be to save a programming step in our example of toll station. As we have seen, after adding some cars to *queue* it pointed to the list *(MKA8772 GWG2719 CKT8623)*. Now if we apply *(rotate queue 1)*, it turns to *(CKT8623 MKA8772 GWG2719)* and therefore, the car with car-plate CKT8623 can exit from the queue graciously by typing *(pop queue)*. As we can see, by using *(rotate)* we have avoided a double use of the function *(reverse)*.

The functions *(extend)* and *(flat)* are simple in their working, but help to make the life of a Lisp user easier. The first one takes several lists as arguments and returns a single list formed by all the elements from the supplied lists. Let's type as an example:

> (setq first-semester '(Jan Feb Mar Apr May Jun))
: (Jan Feb Mar Apr May Jun)

> (setq second-semester '(Jul Aug Sep Oct Nov Dec))
: (Jul Aug Sep Oct Nov Dec)

(setq year '())
: ()

> (extend year first-semester second-semester)
: (Jan Feb Mar Apr May Jun Jul Aug Sep Oct Nov Dec)

The function *(flat)* analyzes the list supplied as its unique argument and then eliminates any sub-list included in it, returning a list where all the elements are atoms. As an example, remembering Sect. 2.4, we saw that the symbol *months-temps* pointed to the list:

((Jan -3) (Feb -2) (Mar 1) (Apr 4) (May 8) (Jun 11) (Jul 13) (Aug 12) (Sep 9)
(Oct 6) (Nov 2) (Dec -1)))

Now, typing:

> (flat month-temps)
: (Jan -3 Feb -2 Mar 1 Apr 4 May 8 Jun 11 Jul 13 Aug 12 Sep 9 Oct 6 Nov 2 Dec -1)

The function *flat* is not destructive, so it lefts intact the function supplied as argument. If we wish that *month-temps* adopts the new, flattened structure, we should type:

> (setq months-temps (flat months-temps))
: (Jan -3 Feb -2 Mar 1 Apr 4 May 8 Jun 11 Jul 13 Aug 12 Sep 9 Oct 6 Nov 2 Dec -1)

This function is especially useful when we are creating a list by a repetitive use of the function *cons*. As we saw in Sect. 2.4, *cons* takes two arguments, but successive calls to *cons* causes a deepening of the formed list. As an example let us use *cons* by building a queue of cars at the toll facility:

> (setq queue (cons 'CKT8623 'GWG2719))
: (CKT8623 GWG2719)

> (setq queue (cons queue 'MKA8772))
: ((CKT8623 GWG2719) MKA8772)

> (setq queue (cons queue 'DYN2140))
: (((CKT8623 GWG2719) MKA8772) DYN2140)

Do you appreciate the problem with the function *(cons)* when used this way? A call to flat solves it all:

> (setq queue (flat queue))
: (CKT8623 GWG2719 MKA8772 DYN2140)

By the way, have you noticed that the combination of the functions *(cons)* and *(flat)* are still another way to model a queue with FIFO discipline? Multiple solutions to a problem are usual when using programming languages, but it is especially true with Lisp due to its high flexibility, as you are starting to appreciate.

In the Sect. 2.1 of this chapter we have learnt how to create lists. Now, at the end of it, we should learn how to reset a list and how to eliminate it from memory. Let us imagine the list *toll* contains now thousands of elements, that is, cars that have entered the facility along the day. In order to reset it, we only need to type:

> **> (setq toll '())**
> *: ()*

Thus becoming an empty list. If we wish to completely eliminate the list *toll* from memory, it is even easier:

> **> (setq toll nil)**
> : nil

If what you want is to eliminate a list from memory, it is not necessary, of course, to reset it previously. In any case Lisp takes care for handling these memory issues automatically.

2.8 As a Summary

All the nuclear functions of Lisp, and yet some more, needed for developing FuzzyLisp have been exposed in this chapter, and what is even more important: you are now into a position where you are able to understand the basics of the language. We have made an extensive use of examples, using monthly temperatures in Berlin, abstract art compositions and some examples of queues from the real world in order to progressively introduce the most important functions of Lisp. If you have followed our advice we know it has been a lot of typing at the keyboard, and if you have experimented with examples created by your own imagination, the typing has been certainly impressive. Many typing or conceptual errors have happened you while cruised along this chapter. If this has been so, congratulations because you are now more than ready to explore the next step in being fluent in Lisp: Writing you own functions. We shall dedicate the entire next chapter to functions defined by the Lisp user. Now, a last reminder to all the functions visited in this section of the book will help you to have a good view of your position in the travel of learning Lisp:

- The function *(quote)* or its more often used *quote sign*, prevents Lisp from evaluating Lisp expressions located at its right. If an atom is found just after the quote operator, Lisp stops its evaluation processing until finding a new atom or list. If an open parenthesis is found just after the quote operator, Lisp stops its evaluation processing until finding its matching right parenthesis. Example: *(quote a)* or *'a* → *a.*

- The *quote* operator in Lisp has a powerful counterpart function named *(eval)*. It forces Lisp to evaluate the expression(s) located at its right. As an example, *(setq expression '(first (this is great)))* → *(first (this is great))*, and then *(eval expression)* → *this*.
- The function *(atom?)* is a predicate that returns *true* if its argument is an atom. Otherwise it returns *nil*, the Lisp equivalent to false. Predicates always end its name with a question mark and only return either *true* or *nil*. As an example *(atom? bird)* → *true*, but *(atom? '(bird fish))* The *quote* operator in Lisp has a powerful counterpart function named *(eval)* that is imperative to introduce in this moment. Using it forces Lisp to evaluate the expression(s) located at its right. *il*.
- The function *(symbol?)* is another predicate that returns *true* if the supplied argument is a symbol. Otherwise it returns *nil*. Example: *(symbol? 'bird)* → *true*, but note that *(symbol? bird)* → *nil*. The use of the quote operator is essential in Lisp.
- The function *(number?)* is just another predicate that returns *true* if its argument is a number, either integer or real. Otherwise, the predicate *number?* returns *nil*. Example: *(number? 3.14159265359)* → *true*.
- The function *(list?)* is the last of the predicates seen in this chapter. It returns *true* when the supplied argument is a list and returns *nil* if its argument is not a list. Example: *(list? '(a b c))* → true, but *(list? 'a)* → *nil*.
- The function *(stats)* is a function contained in the mathematical library of NewLisp. It takes a list of numbers as its argument and returns another list whose elements are statistical values corresponding to the statistical operations performed on the numbers of the list used as argument such us mean value, standard deviation, variance, etc. Example: *(stats '(-3 -2 1 4 8 11 13 12 9 6 2 -1))* → *(12 5 4.833333333 5.640760748 31.81818182 0 -1.661427211)*. This function is an extended feature of NewLisp and may not be available in other Lisp dialects.
- The function *(first)* takes a list as its argument and returns its first element. Example: *(first '((a b c) x y z))* → *(a b c)*. This function is named *(car)* in the Common Lisp dialect.
- The function *(rest)* takes a list as its argument and returns a list composed by all the elements contained in the original list with the exception of its first one. Example: *(rest '(a b c d))* → *(b c d)*. This function is named *(cdr)* in the Common Lisp dialect.
- The function *(nth)* takes an index *i* and a list of *n* elements as arguments. It returns the element at position *i* in the list. The first element in the list is indexed as zero by *(nth)*. In other words, *(nth 0 list)* returns the first element of *list*, while *(nth (- n 1) list)* returns the last element of *list*. Example: *(nth 1 '(this (seems tricky) at first))* → *(seems tricky)*.
- The function *(last)* takes a list as its argument and returns the last element of the list supplied as argument. Example: *(last '(seems tricky))* → *tricky*.

- The function *(length)* takes a list as its argument and returns the number of elements in it. Example: *(length '(this (seems tricky) at first))* → *4*, but note that *(length '(this seems tricky at first))* → *5*.
- The function *(cons)* is used for building lists. It takes two arguments, either atoms or lists, and then constructs a new list with them. Example: *(cons 'not 'hard)* → *(not hard)* but take care with the use of reiterated *conses* because it quickly builds nested lists: *(cons '(this is) '(not hard))* → *((this is) not hard)*.
- The function *(list)* is another function for building lists. It takes any number of atoms or lists and then joins it all, returning another list. As an example, *(setq a-list (list 'a 'b 'c 'd))* and then *(list '1 '2 '3 a-list '(x y z))* → *(1 2 3 (a b c d) (x y z))*.
- The function *(assoc)* takes the first argument passed to the function and then uses it as a key for searching the list passed as the second argument. If a match is found, then it returns the member list. If no match is found, it returns nil. Example: *(setq mountains '((Everest 8848) (Kilimanjaro 5895) (Mont-Blanc 4695) (Aconcagua 6962))* and then: *(assoc 'Kilimanjaro mountains))* → *(Kilimanjaro 5895)*.
- The function *(append)* takes a list as its first argument and then appends the second argument, which becomes the last element of the list given by the first argument. As an example: *(setq a-list '())* then *(setq a-list (append a-list '(first-element)))* and then *(setq a-list (append a-list '(second-element third-element)))* → *(first-element second-element third-element)*. Note that *append* is not destructive, hence the continuous use of *setq* for *a-list*
- The function *(reverse)* simply reverses all the elements inside a list. Example *(reverse '(a b c)* → *(c b a)*. Remember that this function is destructive.
- The function *(push)* adds elements into a list "pushing" them sequentially from the frontal part of the list to its back. It is a destructive function. Example: *(setq names '())* then *(push 'John names)* and then *(push 'Anna names)* → *(Anna John)*.
- The function *(pop)* extracts and then returns the first element in the list supplied as its argument. In the previous example, *names* → *(Anna John)* and then (pop names) → *Anna*. Needless to say, it is also a destructive function.
- The function (now) returns data from the computer's internal clock when it is called. Example: *(now)* → *(2014 2 26 13 34 40 592460 56 3 -60 0)*. This function is an extended feature of NewLisp and may not be available in other Lisp dialects.
- The function *(rotate)* takes a list as its first parameter and then moves every element contained in it *n* positions towards its right if the parameter *n* is positive or towards left if *n* is negative. Example: *(setq a-list '(a b c d))* then *(rotate a-list 3)* → *(b c d a)* and *(rotate a-list -3)* → *(a b c d)*. This function is destructive.
- The function *(random)* returns a real number between 0 and 1. Example: *(random)* → *0.7830992238*. This function is an extended feature of NewLisp and may not be available in other Lisp dialects.
- The function *(extend)* takes several lists as arguments and returns a single list formed by all the elements from the supplied lists. Example: *(setq whole-list '())* then *(setq list-1 '(a b c))* and *(setq list-2 '(d e f))*, then typing *(extend whole-list list-1 list-2)* → *(a b c d e f)*. Since this function modifies the original list, it is a destructive function.

- Finally, the function *(flat)* analyzes the list supplied as its unique argument and then eliminates any sub-list included in it, returning a list where all the elements are atoms. Example: *(flat '(a (b (c d) (e f) g) h))* → *(a b c d e f g h)*.

A more comprehensive description of these functions and their respective parameters can be obtained consulting NewLisp's on-line help from the Menu: Help → Newlisp Manual and Reference. In this section of the chapter we have only described the simplest use of some of the functions.

References

Banks, J., et al.: Discrete-Event Simulation, Prentice Hall, Upper Saddle River (2009)
Climatedata.eu.: Climate Berlin, Germany (2014). http://climatedata.eu/climate.php?loc=gmxx0007 &lang=en. Accessed Feb 2014
Sturgul, J.: Mine Design: Examples Using Simulation. Society for Mining Metallurgy and Exploration (1999)
Zhang, R., et al.: Fuzzy Control of Queuing Systems. Springer, Berlin (2005)

Chapter 3
Functions in Lisp

3.1 Introduction

I have been in doubts when thinking about the title for this chapter. A serious alternative was "Introduction to functional programming" but that title seems to be better suited for an entire book and not for a single chapter. However, functions and functional programming have a close interrelationship in Lisp. Both concepts, together with list structures for representing data, already introduced in the previous chapter, are the solid core of the language.

This chapter is entirely dedicated to user-defined functions. First we shall introduce them with simple examples in growing complexity. Then, logical tests and conditional structures will be added, without forgetting the use of logical connectives. Later, looping structures are introduced, with multiple examples that will allow the reader to understand how powerful these programming constructions really are.

Recursion is introduced in another section of the chapter. This is a relatively sophisticated concept that can be categorized as "advanced" for a text at an introductory level on Lisp. However, recursion is of such importance, both historically and computationally in Lisp that I have felt absolutely obligated to include it. This section is probably the most complex of the chapter under a conceptual point of view, especially if the reader has some previous experience with other traditional programming languages. Due to this, special care has been put in offering a gentle start on recursion without renouncing to show some interesting examples. Finally, some general ideas are exposed about lambda expressions, that is, anonymous functions.

© Springer International Publishing Switzerland 2016
L. Argüelles Méndez, *A Practical Introduction to Fuzzy Logic using LISP*,
Studies in Fuzziness and Soft Computing 327,
DOI 10.1007/978-3-319-23186-0_3

3.2 Starting Functions with Three Simple Examples and then Came Map

We have already discovered functions in the previous chapters. Functions like *(div)*, *(first)*, *(reverse)*, or *(append)*, to name only a few, perform interactions with data, either atoms or lists, and then return an atom, a list or modify the supplied data. In this way, Lisp supplied functions can be seen as black boxes where some data is accepted as an input and a result is obtained as an output. We do not need to know what happens inside these black boxes and in a first approach we can say that we are not very much interested in the inner workings of the functions supplied in the language because all we need to know is what data is needed to be supplied to the functions (arguments) and what type of information results from a function call to them. The problem is, neither Lisp nor any other programming language, supplies all the functions needed for all sorts of imaginable computer processing require-ments. There is not a supplied function for calculating the deformation of a steel beam under the weight of some loads, nor another function to calculate the Body Mass Index (BMI) of a person, for example, not to mention a function for imple-menting a Genetic Algorithm, or a Neural Network. Here is where user defined functions go into play.

User defined functions mean that the Lisp user must design new black boxes. The usual starting point is to define the goal of the function (its output) and what information is available (its input). Hereafter the Lisps user must design the inner workings of the black box. Through this design, the contents of the black box become clear for the Lisp programmer: Every black box is full of Lisp expressions. With these ideas in mind we can now express what a function defined by user is:

A Lisp function defined by user is a set of Lisp expressions that, working together, performs a programming task

The anatomy of a user defined Lisp function always uses the following grammar:

(define (function-name arguments)
     (lisp-expression$_1$)
     (lisp-expression$_2$)
     .....
     (lisp-expression$_n$)
)

As can be observed, the first step for creating a new function is to use the keyword "*define*". When Lisp finds it in the code of a program it understands that all the Lisp expressions contained until finding a matching closed parenthesis form part of the function. Just after the keyword *define*, and starting with an open parenthesis, comes the name of the function that the user wish to create, and then, separated by an space, the name of the argument(s) that will be used as the supplied information (input) to the function. The set of Lisp expressions *(lisp-expression$_1$)* *(lisp-expression$_2$)... (lisp-expression$_n$)* constitute what is known as the *body of the function*. The function will always return the evaluation of the last Lisp expression contained in the body function, that is, *(lisp-expression$_n$)* in our representation of function anatomy. Needless to say, *(lisp-expression$_n$)* is usually the result of all the data processing performed by the previous Lisp expressions contained in the body of the user-defined function. Additionally, the set of Lisp expressions contained in the body can optionally modify external data in the program. This last feature will be covered in the next chapter. In this one, all the functions we are going to create will always return a single output, either an atom or a list.

It is also important to note that in this chapter our Lisp sessions will include typing at the Lisp prompt in the "session area" (gray background) of the NewLisp environment, as in the previous chapters, and also typing in the "programming area" (white background), as can be seen in Fig. 3.1. In other words, we shall create our functions in the programming area and will test them in the session area, at the usual Lisp prompt. Additionally we shall use from time to time two icons from the

Fig. 3.1 Programming and session areas in NewLisp, revisited

icons bar located at the bottom of the NewLisp environment: The blue icon representing a computer disk for saving our functions on the computer's hard disk and the green "play" icon. Clicking on this green icon have the effect of triggering an event that tells NewLisp to read and "understand" all the Lisp expressions contained in the programming area and store them in memory in such a way that all the written expressions and functions will be ready to be used at the session area. As it usually happens, describing this procedure in words is many times more complex than doing it and you will soon feel at home with the NewLisp environment.

For writing our first function we are going to remember than in Chap. 1 we saw how tedious was to use trigonometric functions because the need of supplying values in radians to them. Specifically, for translating an angle *alpha* in grads to radian we used the expression *(div (mul alpha pi) 180.0))*. Now we shall write a function named *(grad-to-rad)* that will automatically do all the required calculations for us.

The previous analysis for designing the intended function is easy: (a) the name of the function is *grad-to-rad*, (b) the supplied information to the function, that is, its argument, is a numerical value named, for example, *alpha*, and (c) the last Lisp-expression in the body of the function must return the desired output, that is, the value of *alpha* expressed in radians. The function is shown in code 3-1.

```
;code 3-1
(define (grad-to-rad alpha)
    (div (mul alpha 3.1415926535) 180.0)
)
```

It is easy to appreciate that the required grammatical structure for defining a function, previously exposed as the anatomy of a Lisp function, is present in Code 3-1. We have the mandatory keyword *define*, the function-name is *grad-to-rad*, the argument, unique in this case, is *alpha*, and the body of the function contains only one expression: *(div (mul alpha 3.1415926535) 180.0)*. Now, after clicking on the green icon of the NewLisp environment let's try the function at the session area just to the right of the Lisp prompt in order to check if it works as expected:

> **(grad-to-rad 45)**
: 0.7853981634

and now we can test it with any NewLisp built-in trigonometric function:

> **(tan (grad-to-rad 45))**
: 1

Two important observations deserve to be made at this point. First, the just created *grad-to-rad* function is, after clicking the green icon on the NewLisp environment, just another Lisp function, that is, it converts automatically itself in an extension of the language and thus, can be combined with any other Lisp function. The second observation is the use of the semicolon signs in code 3-1: In Lisp, semicolons are used for writing comments in the programs. When Lisp sees a semicolon in a program it treats all the material located at its right in a line as a comment and thus it's not included in the code of the program.

Comments are extremely important when using a programming language, and Lisp is not an exception. They help the Lisp user or other programmers to explain the code just written or to remark some special data manipulation, procedure or algorithm in order to make it easy to understand it in the future. I use lots of comments in my programs, maybe in excess. Although it takes some additional time, the payoff is that I am able to read and understand code I wrote twenty years ago. Software maintenance is extremely important, so take the following sentence as a serious advice:

Always use comments in your Lisp programming

Let us see now another function. This time for converting Celsius degrees to Fahrenheit. The formula is as follows:

$$F = 1.8C + 32 \tag{3-1}$$

and the corresponding function in Lisp is shown in code 3-2:

```
;code 3-2
(define (celsius-to-fahr celsius)
     (add (mul 1.8 celsius) 32.0)
)
```

Now, in order to check it, let us type in the session area of the NewLisp environment:

> **(celsius-to-fahr 40)**
: *104*

As expected, the function *(celsius-to-fahr)* has converted 40 °C into 104 °F. The interesting thing now is the existence of a built-in Lisp function named *(map)*. This functions uses a function as its first argument and then applies its working to all the elements contained in a list supplied as a second argument. If the function supplied as argument has itself more than one argument, then *(map)* applies the function to several elements, taken one-by-one from each list, where the number of lists as arguments match the number of arguments of the function to be mapped. In the following example we shall use *(map)* with the just created function *(celsius-to-fahr)*. In Sect. 2.2 of this book, for describing the minimum monthly tempera-tures in Berlin we wrote:

> **(setq Berlin-lows '(-3 -2 1 4 8 11 13 12 9 6 2 -1))**
: *(-3 -2 1 4 8 11 13 12 9 6 2 -1)*

Now is when the magic of Lisp start to shine:

> **(map celsius-to-fahr Berlin-lows)**
: (26.6 28.4 33.8 39.2 46.4 51.8 55.4 53.6 48.2 42.8 35.6 30.2)

In this example, the function *(map)* takes the first element in the list Berlin-lows (-3) and passes it to the function *(celsius-to-fahr)*, resulting in the first element of a new list *(26.6)*. After reiterating the process for every element in the list pointed by *Berlin-lows*, the function *(map)*, applied on the function *(celsius-to-fahr)*, creates a new list where all the monthly temperatures in Berlin are given in Fahrenheit degrees. This ability of Lisp to use functions as arguments of other functions is one of the strongest points in the language.

The functions *(grad-to-rad)* and *(celsius-to-fahr)* take one argument as input. Now, let us see how to design a function that takes two arguments. Some paragraphs above we spoke about the Body Mass Index, or BMI. It is an index used by physicians for knowing if the weight of a person is normal or if the patient suffers from overweight or obesity. It is calculated by means of using the following formula:

$$BMI = \frac{m}{h^2} \qquad\qquad (3\text{-}2)$$

where *m* is the patient's mass in kilograms and *h* his or her height in meters. The analysis for designing this function is as follows: (a) we shall give the name *BMI* as the name of the function, (b) the input for the function, that is, the arguments needed for making it working are the mass and height, *m* and *h*, and (c) the last expression in the body of the function (output) must return the calculated numerical value for the corresponding Body Mass Index. The functions is shown in code 3-3:

```
;code 3-3
(define (BMI mass height)
    (div mass (mul height height))
)
```

after the keyword *define* we use *BMI* for naming the function and then we group together *mass* and *height* in the same list *(BMI mass height)*. The body of the function is represented by only one line of code that is the Lisp representation of expression (3-2): *(div mass (mul height height))*. The definition of the function ends with a closing matching parenthesis for the keyword *define*. Let us observe now how the function *BMI* works for a man weighting 80 kg and with 1.80 m height after clicking on the green "run" icon in the NewLisp environment:

> *(BMI 80 1.8)*
: 24.69135802

Since overweight begins with a BMI equal or bigger than 25, we can say that an 80 kg man with a height of 1.80 m has a normal weight (in the next part of the book dedicated to fuzzy-logic we shall make additional and sound comments to this example).

3.3 Managing Lists with User Defined Functions

In the previous section we have seen how easy is to build user functions for making calculations. Any branch of engineering and applied science has a rich toolbox of formulae. A civil engineer could create his or her own library of functions in Lisp for calculating, let us say, structures. Such a collection of functions would be the basis for a Lisp application for structural analysis. Needless to say, this idea extends to chemistry, physics, astronomy, ... you name it.

After these initial concepts on functions we are now ready to extend the Lisp language itself. As the reader already well knows, list processing is the primordial basis of Lisp, so defining new functions for list processing is the equivalent to extend the language. As a first example, and remembering that the function *(first)* returns the first element of a list, let us build a function for returning the fourth element from a list. Such a function is shown in code 3-4:

```
;code 3-4
(define (fourth a-list)
            (first (rest (rest (rest a-list))))
)
```

As an example of use, after clicking on the green "run" icon in the NewLisp environment we can type at the Lisp prompt:

> (fourth '(a b c d e))
: d

Do you remember what we said about *(first)* and *(rest)* in Chap. 2? You can see it now in the body of the function, where after successive calls to *(rest)* we finally extract the first element of the processed list, thus obtaining the fourth element of the list passed as argument to the new function. Revisiting Fig. 2.1 can help you to better understand the function *(fourth)*.

It is now the time to write some user functions for removing elements from a list. The first one, named *(remove2-head)* removes the two first elements from a list. As can be seen in code 3-5, it simply calls the function *(pop)* two times:

```
;code 3-5
(define (remove2-head a-list)
    (pop a-list)
    (pop a-list)
  a-list
)
```

Just note the last expression in the body of the function. It is not enclosed into parenthesis! This is so because we want that Lisp return a list as the result for a function call to *(remove2-head)*. While running the function, NewLisp will remove the first element from a-list. Then, since the function *(pop)* is destructive, a second call to it will effectively remove the second element from a-list. Since the last expression is not preceded by a quote sign, we force Lisp to evaluate the symbol *a-list*, thus obtaining the desired result. Trying it at the Lisp prompt we can see how does it work:

> **(remove2-head '(a b c d e))**
: *(c d e)*

By the way, the function *(remove2-head)* is not destructive. Try, for example: *(setq test-list '(a b c d))*, then *(remove2-head test-list)* → *(c d)*, but after using the function, *test-list* → *(a b c d)*. Let us write now a symmetrical function to *(remove2-head)*, which is a function that removes the two last elements in a list as expressed in code 3-6:

```
; code 3-6
(define (remove2-tail a-list)
    (reverse a-list)
    (pop a-list)
    (pop a-list)
    (reverse a-list)
  a-list
)
```

As can be easily seen, *(remove2-tail)* works in a similar way to *(remove2-head)*. The important detail is a first call to *(reverse)* in the body function before popping two elements and then another call to *(reverse)* for restoring the original order of elements in the list. Trying it at the Lisp prompt:

> **(remove2-tail '(a b c d e))**
: *(a b c)*

By the way, NewLisp has a built-in function named *(chop)*, not destructive, which removes *n* elements from the tail of a list supplied to it. As an example, if we type: *(setq some-states '(NY AL CA NJ NE))* then *(chop some-states 2)* → *(NY AL CA)*, and *(chop some-states 4)* → *(NY)*. This is fine, since the function *(chop)* is more flexible than the user-defined function *(remove2-tail)*. And this is inspiring, too, because after observing how *(chop)* does it work we could design a user function named, for example, *(flex-chop)* for removing elements both from the head and tail of the list at the same time. The syntax of the function will be *(flex-chop a-list n1 n2)*, where *n1* is the number of elements to remove from the head and *n2* the number of elements to remove from the tail of the list. Code 3-7 shows the function *(flex-chop)*:

```
; code 3-7
(define (flex-chop a-list n1 n2, list1 list2)
    (setq list1 (chop a-list n2)) ;chop it at the tail
    (setq list2 (reverse list1))
    (setq list2 (chop list2 n1))
    (reverse list2)
)
```

Something new appears in this user-defined function. We already know that *a-list*, *n1* and *n2* are the arguments to the function, that is, its input, but there are other symbols, named *list1* and *list2*. These are **internal variables** destined to do internal calculations. Internal variables in NewLisp are declared just after writing the arguments to the function, being separated from these by a comma. As we can see, all this material comes into one only line: *(define (flex-chop a-list n1 n2, list1 list2)*. Observe again the comma, serving as a delimiter for separating the function arguments from the definition of the internal variables. Without the comma, NewLisp would understand all the symbols as arguments. At this moment some-thing remarkable must be said:

Internal variables live only inside the functions where they are defined

This must be understood perfectly well. A Lisp program, as we shall see, is just a collection of functions, and local variables can only be used inside the functions where they are defined. In other words, a function defined by user is absolutely hermetic to the rest of a Lisp program with respect to its local variables. This is a very convenient feature in a programming language because it avoids the existence of errors that could appear if a programmer mistakenly could use the same name of variable in more than a section of a program. On the other hand, Lisp programs can also use global variables, that is, variables whose content can be accessed from the inside of any function. Global variables can also be used as arguments for any function in a Lisp program. We shall have the opportunity to speak about global variables in a future chapter. Meanwhile, let us test the *(flex-chop)* function:

> **(flex-chop '(a b c d e f g h) 1 3)**
: *(b c d e)*

Since *(flex-chop)* is really, well, flexible, and in order to save space, some results from different calls are shown in Table 3.1. The list *(a b c d e f g h)* will be used by typing *(setq a-list '(a b c d e f g h))* first at the Lisp prompt:

It must be noted that some optimization can be done in the function shown in code 3-6. In fact we can re-write it eliminating the use of local variables. The new resulting function, named now *(flex-chop2)* is shown in code 3-8:

```
;code 3-8
;flex-chop2 in an optimization of the function flex-chop
(define (flex-chop2 a-list n1 n2)
    (reverse (chop (reverse (chop a-list n2)) n1))
)
```

Table 3.1 Several calls to function *(flex-chop)*

Function call	List returned by *(flex-chop)*
(flex-chop a-list 0 2)	*(a b c d e f)*
(flex-chop a-list 2 0)	*(c d e f g h)*
(flex-chop a-list 2 2)	*(c d e f)*
(flex-chop a-list 4 4)	*()*

as expected, the function *(flex-chop2)* produces exactly the same results as *(flex-chop)* and Table 3.1 is valid for the two functions. The second version need less code than the first one, but is, without doubt, more cryptic. As a safe bet we can say:

Never try to optimize a function without extensive testing of the non-optimized version

This sentence is complemented by this one

(almost) Never exchange optimization for legibility

Using optimized functions takes less space and usually they are more efficient, but many times the loss of legibility is so strong that makes the code hard to maintain along the time for a Lisp user, not to mention for other Lisp programmers. Fuzzy-Lisp is not strongly optimized because from its inception it was clear to me that it should be easy to read. The same concept will be applied in this book. My goal in these chapters is to put in your hands a tool for understanding fuzzy- logic theories. The easier to learn to use the tool, the quicker the goal will be accomplished.

3.4   The Discovery of Conditional Structures

As we saw in Chap. 1, the Lisp programming language brought revolutionary features to the realm of computer science in the last fifties last century. One of them was the implementation of conditional structures that, simply put, did not exist previously. The only other high-level programming language at that time, Fortran, only had a conditional "goto" command, a really poor feature in comparison with the new possibilities that Lisp brought to the computing community (Graham 2014).

3.4.1   From Predicates to If-Then-Else Structures

We have already experienced an introduction to some of these structures when talking about those special functions called *predicates*. A predicate returns only two possible outcomes: *true* or *nil* (the equivalent to *false* in Lisp). Table 3.2 summarizes the predicates learnt in the previous chapter.

Table 3.2 Some important predicates in Lisp

Predicate	Return value
(atom? arg)	*true* **if** *arg* is an atom, either *nil*
(symbol? arg)	*true* **if** *arg* is a symbol, either *nil*
(number?arg)	*true* **if** *arg* is a number, either *nil*
(list? arg)	*true* **if** *arg* is a list, either *nil*

Aside predicates, it is possible to use relational expressions in Lisp that, as it happens with predicates, also return either *true* or *nil*. Let us type some of them at the prompt:

> **(< 1 2)**
: *true*

> **(< 2 1)**
: *nil*

As it is well known from mathematics, the operator "<" means "less than" so *(< 1 2)* → *true*, and *(< 2 1)* → *nil*. In Table 3.3 we can see all the relational operators included in NewLisp:

All types of expressions (numbers, atoms, symbols, etc.) can be compared in NewLisp by using the relational operators from Table 3.3. For example if we type *(setq a 1)*, *(setq b 2)*, *(setq c 3)*, then we can obtain *(= (first (list a b)) (sub c b))* → *true*. *(= (list a b) (list a c))* → *nil*, and so on.

In Sect. 3.2 we developed a function for calculating the Body Mass Index of a person, named *(BMI)*, shown in code 3-3. Now we are going to create a predicate named *(obesity?)* that will return *true* if the BMI of a person is bigger or equal to 29.9 or *nil* otherwise. In order to accomplish this, we shall use an **"if-then-else"** structure, which is in fact the basis of conditionals in Lisp and in many other computer languages. This structure has the following grammar:

(if (logical-test) (lisp expressions if test is true) (lisp expressions if test is false))

The keyword of a conditional expression in Lisp, and in many other programming languages born after Lisp, is the word "**if**". It must be used immediately after an opening parenthesis. After that, a **logical-test** must be included as a list, containing a relational expression and/or a predicate. Then a **first set of Lisp expressions** must be specified. This set of Lisp expressions will be executed by Lisp if the logical-test is evaluated to *true*. If the logical test is evaluated to *nil*, then the **second set of Lisp expressions** will be used. As usually, an example will demonstrate that the practical use of a programming concept is usually easier to understand than pure theory. The code for programming the user-defined predicate *(obesity?)* is shown in code 3-9:

Table 3.3 Relational operators in NewLisp

Operator	Meaning	Example
<	less than	*(< 3 7)* → *true*
>	bigger than	*(> 7 2)* → *true*
=	equal to	*(= 3 (+ 1 2))* → *true*
<=	less or equal to	*(<= 5 (sub 7 2))* → *true*
>=	bigger or equal to	*(> = 30 (mul 5 5))* → *true*
!=	different to	*(! = 5 7)* → *true*

```
;code 3-9
(define (obesity? mass height, result)
    (setq result (div mass (mul height height))))
    (if (>= result 29.9)
            'true
            nil
    )
)
```

Let's try the new predicate before explaining its design for two different persons, one weighting 80 kg and another one weighting 120 kg. Both are 1.80 meters in height:

> **(obesity? 80 1.8)**
: nil

> **(obesity? 120 1.8)**
: true

Yes, it seems to work perfectly well. Let us go into the details. Since *(obesity?)*, aside being a predicate is also a function, the code starts with the keyword *define*, and then we have the arguments needed as input to the function: *mass* in kilograms and *height* in meters. Finally, in the first line of code we have the variable *result*, which will be used for calculating the BMI as we saw in code 3-3. This calculation is made in the second line. Let us see what we have in the third line:

```
(If (>= result 29.9)
```

As it can be observed, here is where the conditional begins by means of the use of the *if* keyword. Just after this keyword it comes the logical-test, in this case, *(> = result 29.9)*. This logical test will be evaluated to *true* if the calculated BMI is bigger or equal to 29.9, otherwise it will be evaluated to *nil*. Hereafter, the Lisp expressions are very simple: if the logical-test evaluates to *true* then Lisp finds the expression *'true* and if the logical-test evaluates to *nil*, then it finds *nil* in the following line. Since either *'true* or *nil* are the last expression found in the function after the logical-test is performed, only one of these values can be returned by the predicate *(obesity?)*

We can notice in this moment that the function *(BMI)* and the predicate *(obesity?)* inform us, respectively, about the numerical value of the Body Mass Index and if a person suffers obesity or not, but nothing is said about our weight condition. Now we are going to improve these functions by taking advantage of conditional programming in such a way that it will return an interpretation of the obtained index in natural language. The improved version is shown in code 3-10 and follows the National Heart, Lung, and Blood Institute classification for the BMI calculation (NHLB Institute 2014).

```
;code 3-10
(define (bmi2 mass height, result advice)
    (setq result (div mass (mul height height))))

    (if (< result 18.5)
            (setq advice "You are excessively thin")
    )
    (if (and (>= result 18.5) (< result 24.9))
            (setq advice "Congrats, your weight is ok")
    )
    (if (and (>= result 24.9) (< result 29.9))
            (setq advice "You should try some diet and exer-
            cise because you have some overweight")
    )
    (if (>= result 29.9)
        (setq advice "Warning, you suffer obesity. Speak
        with your doctor")
    )
    advice
)
```

After running this code in the NewLisp environment, let us try it at the Lisp prompt with two examples:

> (bmi2 80 1.8)
: "Congrats, your weight is ok"

> (bmi2 120 1.8)
: "Warning, you suffer obesity. Speak with your doctor"

Following the information available from the NHLB Institute on the Internet, we have partitioned the range of possible BMI values in four subsets that we could name this way: Underweight, Normal Weight, Overweight and Obesity, as can be seen in Fig. 3.2.

For each of them we have designed an if-then structure. For example, when the range of BMI values corresponds to a normal weight, we have the following expression:

```
(if (and (>= result 18.5) (< result 24.9))
            (setq advice "Congrats, your weight is ok")
)
```

Underweight	Normal Weight	Overweight	Obesity
	18.5	24.9	29.9

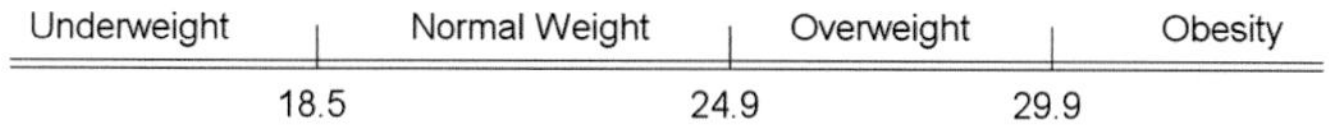

Fig. 3.2 Body mass index continuum as defined in NHLB (2014)

Table 3.4 Truth table for the logical operators and/or

Logical value$_1$	Logical value$_2$	Logical operator	Result
True	True	and	True
True	Nil	and	Nil
Nil	True	and	Nil
Nil	Nil	and	Nil
True	True	or	True
True	Nil	or	True
Nil	True	or	True
Nil	Nil	or	Nil

That is: if the result of the calculated BMI is bigger or equal to 18.5 and it is less than 24.9, then the weight/height relationship represents a normal weight. It is extremely important to note that in this case there is no "else" structure and the if statement only has an expression that is chosen by Lisp only if the logical-test evaluates to *true*. This can be seen observing the matching open and closed parentheses for the whole if expression: Only a unique Lisp expression appears if the logical-test evaluates to true: *(setq advice "Congrats, your weight is ok")*. By means of this strategy the variable *advice* always gets its adequate value, which, as can be seen, is always a string of characters, a type of data especially appropriate for generating reports in natural language.

Of no less importance is the construction we have used for the logical test. In this case we have combined two relational operators by the use of the Lisp keyword *and*. In Lisp we can combine any number of relational operators by the use of the keywords *and* and *or*. The more internal relational expressions are evaluated first and then, the obtained values of *true* or *nil* are supplied to the logical keywords *and/or*, obtaining the results given by Table 3.4.

We can express this concept by using these two simple sentences: For obtaining a result value of *true* while using the *and* logical operator, it must be observed that both logical values must be *true*. For obtaining a result value of *true* while using the *or* logical operator it is only required that at least one of the logical values is *true*.

3.4.2   A Note About Functional Programming

Functional programming in Lisp is at the same time a programming paradigm and a computing philosophy where every function, either included in the language or defined by the user, returns a single Lisp expression. Every function generates a result that depends exclusively on the input to the function. This produces clean and well behaved programs, but it generates the following question: If every function in Lisp returns only one element of information, what computing strategy must be used when we need that a function returns more than one value? For example, let us take the function *(bmi2)* described in code 3-9. The only thing it returns is a string of characters representing a human-like advice. What if we, for example, need a function that returns both human-like advice and the numerical value for the BMI? Do we need two separate functions? Code 3-11 shows an improved version to the function *(bmi2)* that answers these questions:

```
;code 3-11
(define (bmi3 mass height, result advice)
    (setq result (div mass (mul height height)))

    (if (< result 18.5)
           (setq advice "You are excessively thin")
    )
    (if (and (>= result 18.5) (< result 24.9))
           (setq advice "Congrats, your weight is ok")
    )
    (if (and (>= result 24.9) (< result 29.9))
               (setq advice "You should try some diet and
           exercice because you have some overweight")
    )
    (if (>= result 29.9)
               (setq advice "Warning, you suffer obesity.
           Speak with your doctor")
    )
    (cons result advice) ;here we construct the return-list
)
```

The improvement from *(bmi2)* is subtle but important. The internal processing is almost the same: first a numerical value is computed and assigned to the variable *result* and then, after a conditional *if* structure, a string value is assigned to the variable *advice*. The trick is on the last line of code:

(cons result advice)

The already known function *(cons)* takes two arguments and constructs a list formed by them. Since it is the last expression evaluated inside the function, *(bmi3)* will return it. Let us try an example:

> (bmi3 85 1.75)
: (27.75510204 "You should try some diet and exercise because you have some overweight")

That is great! It is true that every function in Lisp returns only one element of information, but that element can be, aside an atom or a number, a list. And any type and quantity of data can be contained into a list. Hereafter, *(first (bmi3 85 1.75))* → *27.75510204*, and *(last (bmi3 85 1.75))* → *"You should try some diet and*

exercise because you have some overweight". In Chap. 2 we saw how the built-in
NewLisp function *(stats)* returned a list containing seven numerical values. Now
Code 3-10 show us how to use Lisp for obtaining several pieces of information
integrated in only one list. This is a powerful programming strategy from the Lisp
language.

3.4.3   Robust Functions from the Use of Conditional Programming

Testing functions is part of the process of function design. Until now in this chapter
we have made simple testing of every exposed function, but we have not stressed
any function in the tests. We have not tried them at their limits. We should not only
create functions that meet the design goals, that is, to produce the intended output,
but also making them robust against failure. As an example, and again at Table 3.1,
after typing *(setq a-list '(a b c d e f g h))*, we can see that:

> **(flex-chop a-list 4 4)**
: *()*

After this expression we obtain an empty list. This is correct, since *a-list* contains
eight elements. However if we type: *(flex-chop a-list 5 5)* the function still returns
an empty list. This should not happen, since the function should inform us in some
way, for example, returning *nil*, that it is simply not possible to chop five elements
from the head and tail of a list of eight elements simultaneously. Similar mistakes
happen when we call the functions *(remove2-head)* and *(remove2-tail)* using very
short lists as arguments. Again as an example:

> **(remove2-head '(a))**
: *()*

Here the argument is a list with only one element inside, so the function should
inform us about this fact. Code 3-12 solves this problem for the function *(remove-head)*:

```
;code 3-12
(define (remove2-head a-list)
    (if (< (length a-list) 2)
            nil
            (begin
                    (pop a-list)
                    (pop a-list)
                    a-list
            )
    )
)
```

Let us try this new version of the function with two simple function calls:

> (remove2-head '(a b c d))
: (c d)

> (remove2-head '(a))
: nil

This more robust function seems at first more complex than the one shown in code 3-4, and it is indeed! The key idea in the new design is to check for the number of elements in the list supplied as argument to the function in such a way that it returns *nil* if the list contains less than two elements, or the expected result (previous behavior) if the list contains two or more elements. It must be noted in code 3-8 that we have used a new keyword named *begin*. In NewLisp, *begin* is used for telling the language that several Lisp expressions are working together under the same set of conditions. Let us analyze what happens in code 3-11 in order to have a more clear perspective.

Just after starting the conditional by means of *(if (< (length a-list) 2)*, Lisp evaluates it to *true* or *nil*, as we already know. If the result of the evaluation is *true*, then Lisp evaluates the following Lisp line in the code, in this case finding the *nil* value. But if the evaluation of the *if* sentence is *nil* then Lisp springs to the *begin* keyword, and then evaluates all the complete block of expressions enclosed by the *begin* matching parentheses:

```
(begin
      (pop a-list)
      (pop a-list)
      a-list
)
```

In fact, the complete anatomy for an if-then-else conditional structure is as follows:

(if (logical-test)
(begin
(lisp-expression₁ if logical-test is true)
(lisp-expression₂ if logical-test is true)
.....
(lisp-expressionₙ if logical-test is true)
); first begin block ends
(begin
(lisp-expression₁ if logical-test is false)
(lisp-expression₂ if logical-test is false)
.....
(lisp-expressionₙ if logical-test is false)
); second begin ends
); if –then-else structure ends

In code 3-13a we can observe a more robust version of the function *(flex-chop2)*:

```
;code 3-13a
(define (flex-chop3 a-list n1 n2)
    (if (> (+ n1 n2) (length a-list))
            nil
            (reverse (chop (reverse (chop a-list n2)) n1))
    )
)
```

The following code, code 3-13b, includes *begin* blocks, being identical in computing behavior:

```
;code 3-13b
(define (flex-chop4 a-list n1 n2)
    (if (> (+ n1 n2) (length a-list))
    (begin
            nil
    );end begin block
    (begin
            (reverse (chop (reverse (chop a-list n2)) n1))
    );end begin block
    );end if
);end function
```

Both functions *(flex-chop3)* and *(flex-chop4)* work exactly the same. Since only a single Lisp expression is needed for the corresponding *true* and *nil* evaluations of the logical-test, then there is no need to use *begin* blocks. Anyway, if you feel more comfortable, you can always use *begin* blocks for enclosing a single Lisp expression and not only when they are strictly needed. It is a question of programming style. In any case:

Always test for matching parenthesis in begin blocks

Now, for testing function *(flex-chop4)*, after making *(setq a-list '(a b c d e f g h))*, we have, for example:

> (flex-chop4 a-list 3 2)
: (d e f)

> (flex-chop4 lista 5 4)
: nil

We should comment another detail that represents a good programming practice: Indentation. This placement of text farther to the right in some lines of code in a Lisp program is not intended for getting a more aesthetic appearance, but for making things more clear and avoid mistakes, specially with the number of matching parenthesis. Code 3-14 is absolutely equivalent as code 3-11:

```
;code 3-14
(define (bmi3 mass height, result advice)
(setq result (div mass (mul height height)))
(if (< result 18.5)
(setq advice "You are excessively thin")
)
(if (and (>= result 18.5) (< result 24.9))
(setq advice "Congrats, your weight is ok")
)
(if (and (>= result 24.9) (< result 29.9))
(setq advice "You should try some diet and exercise because
you have some overweight")
)
(if (>= result 29.9)
(setq advice "Warning, you suffer obesity. Speak with your
doctor")
)
(cons result advice) ;here we construct the return-list
)
```

Code 3-11 is clearly easier to read and understand. Indentation, while not mandatory, should be always used, especially with if-then-else constructions and when using *begin* blocks inside Lisp structures of expressions. In fact, it is the first step for constructing robust functions, because indentation avoids many programming mistakes and help to fix mistakes when they appear. On the other hand, designing robust functions takes more space, time, and effort, that is, more resources, but it always pays. In this book, and for space requirements, we concede priority to show new Lisp concepts instead of heavy use of robust function design, but my advice to you is to always try to design robust, safe functions. The philosophy of the Lisp language and the paradigm of functional programming are of a huge help in this direction but is not enough. Be always alert against possible failures in your user-defined functions.

3.4.4 Solving Multiple Conditions Without Using If

When a range of values is classified or divided in sub-ranges it is usual to utilize several if-then blocks sequentially, as we did in Code 3-10/3-11 for expressing expert advice from the obtained Body Mass Index. In this type of situation, Lisp offers us an alternative conditional structure whose keyword is *cond*. In Code 3-15 we have completely eliminated the use of the conditional *if*, yet the function *(bmi4)* produces exactly the same results as the function exposed in Code 3-11:

```
; code 3-15
;bmi4 uses cond as a conditional structure
(define (bmi4 mass height, result advice)
    (setq result (div mass (mul height height)))

    (cond
            ((< result 18.5)
                    (setq advice "You are excessively thin")
            )
            ((and (>= result 18.5) (< result 24.9))
                    (setq advice "Congrats, your weight is
                    ok")
            )
            ((and (>= result 24.9) (< result 29.9))
                    (setq advice "You should try some diet
                    and exercise because you have some
                    overweight")
            )
            ((>= result 29.9)
                    (setq advice "Warning, you suffer
                    obesity. Speak with your doctor")
            )
    );end cond
    (cons result advice)
)
```

As you can observe, the *cond* structure is not very different from the *if-then* structure. After the keyword *cond*, several logical tests are used until covering all the possible options. The complete anatomy for a *cond* conditional structure is as follows:

(cond
    ((logical-test$_1$)
        (lisp-expression$_1$ if logical-test$_1$ is true)
        (lisp-expression$_2$ if logical-test$_1$ is true)
        .....
        (lisp-expression$_n$ if logical-test$_1$ is true)
    ); logical-test$_1$
    .....
    ((logical-test$_n$)
        (lisp-expression$_1$ if logical-test$_n$ is true)
        (lisp-expression$_2$ if logical-test$_n$ is true)

.....
>   *(lisp-expression$_n$ if logical-test$_n$ is true)*
>   *);end logical-test$_n$*
> *);end cond*

A comfortable feature of a *cond* structure is that there is no need to use the *begin* keyword for grouping several Lisp expressions into the same logical block. Cod 3-16 still represents another function that uses the BMI for showing these concepts. Just note that every test option has two Lisp expressions and no *begin* keyword is needed:

```
;Code 3-16
(define (bmi5 mass height, result advice subset)
    (setq result (div mass (mul height height)))
    (cond
            ((< result 18.5)
                    (setq advice "You are excessively thin")
                    (setq subset "The first subset of cond
                    has been used")
            )
            ((and (>= result 18.5) (< result 24.9))
                    (setq advice "Congrats, your weight is
                    ok")
                    (setq subset "The second subset of cond
                    has been used")
            )
            ((and (>= result 24.9) (< result 29.9))
                    (setq advice "You should try some diet
                    and exercise because you have some
                    overweight")
                    (setq subset "The third subset of cond
                    has been used")
            )
            ((>= result 29.9)
                    (setq advice "Warning, you suffer
                    obesity.
                    Speak with your doctor")
                    (setq subset "The fourth subset of cond
                    has been used")
            )
    );end cond
    (list result advice subset)
)
```

Testing the function at the Lisp prompt, we obtain, for example:

> **(bmi5 80 1.80)**
: (24.69135802 "Congrats, your weight is ok" "The second subset of cond has been used")

Another important use of the *cond* structure is quite often used in any application nowadays: Menu selection. Every menu contains several options to choose from, and a *cond* structure is the most appropriate one to use in such cases. Code 3-17 shows how to organize a three-options menu. The function expects a number from one to three in order to show what option has been chosen:

```
;Code 3-17
(define (menu option)
    (cond
              ((= option 1)
                      (setq str1 "You have chosen menu option
                      #1")
              )
              ((= option 2)
                      (setq str1 "You have chosen menu option
                      #2")
              )
              ((= option 3)
                      (setq str1 "You have chosen menu option
                      #3")
              )
              (setq str1 "You have made a bad selection")
    );end cond
)
```

Trying *(menu)* at the Lisp prompt:

> **(menu 3)**
: "You have chosen menu option #3"

It is important to note that no *else* formation happens in a *cond* structure. The mere existence of every condition/logical-test eliminates the need of an *else* alternative, or at least this is how it looks as first sight. In fact a *cond* structure always has a hidden, automatic, and optional else construction available. Let us just look at the following function calls:

> **(menu 4)**
: "You have made a bad selection"

> **(menu any-option)**
: "You have made a bad selection"

Just observe the last line of code still located inside the *cond* structure in Code 3-17:

```
(setq str1 "You have made a bad selection")
```

If all the logical tests in a *cond* structure are evaluated to nil, an optional last expression can be included as a lifeboat that works as an *else* statement. This is a convenient feature included in this type of conditional structure that represents another step in the right way of robust programming.

3.5   The Discovery of Loop Structures

Computers are not smart (yet) machines, but since its inception they have been excellent, efficient tools for performing repetitive tasks. It doesn't matter if a computer must repeat a task three times, one million times or forever in its life. While energy is available and no hardware failures do appear, a computer can perform the same process time after time. Programming languages have control structures that allow repetition as an easy to implement computing strategy. This strategy, or computing paradigm, is called looping. The Lisp programming language has several looping structures available, although we are going to focus our time on the most used one in actual programming languages: The *while* structure.

3.5.1   While Computing

The basic idea behind a *while* looping structure in Lisp is that a set of Lisp expressions must be repeated time after time while a logical-test, located at the beginning of the loop structure, evaluates to *true*. It is easy to imagine that for avoiding infinite loops, that is, loops that repeat a task forever, at least one of the internal Lisp expression in the loop must be able to modify the value of the logical-test from *true* to *nil*. Just after the logical-test evaluates to *nil*, the program flow exits the loop.

The anatomy of a *while* looping structure in Lisp is as follows:

(while ((logical-test)
      (begin
             (lisp-expression$_1$)
             (lisp-expression$_2$)
             .....
             (lisp-expression$_n$)
      ); end begin
    ); end while

As said previously, at least one *(lisp-expression_i)* inside the loop must be able to change the value of the logical-test, either a predicate-function, an if-then structure, or an assignation in a Lisp variable. In order to show practically how a *while* structure works we are going to design a function, named *(my-find)* that identifies the position of an atom in a list, returning its position and the atom itself in a small list of two elements. Code 3-18 shows this function:

```
;Code 3-18
(define (my-find atm lst, n i aux result)
     (setq n (length lst))
     (setq i 0)
     (setq result nil)

     (while (< i n)
           (begin
                     (setq aux (nth i lst))
                     (if (= aux atm)
                                    (setq  result  (cons  i
                                    atm))
                     );end if
                     (setq i (+ 1 i))
           );begin end
     );while end
     result
)
```

After defining the name of the function and its required arguments by means of the keyword *define*, the first line of the function, *(setq n (length lst))*, measures the length of the list supplied to the function and then assigns the numerical result to the internal variable *n*. The two following lines are of strategic importance for the adequate working of the loop and the function, eventually. The expression *(setq i 0)* initializes *i* to zero. This is easy to see, but it is this variable, absolutely innocent at first sight, what will allow the function to exit the loop. The second assignment expression, *(setq result nil)*, assumes that no matching element in the supplied list will be found, and only if something happens inside the loop it will change its value. In other words: the function *(my-find)*, as a default, is designed to return *nil*. This programming strategy is very common in Lisp and in other languages with looping structures.

The *while* loops starts with the line *(while (< i n)*. The logical-test is composed by the relational operator "<", that compares the values of variables *i* and *n*, so translated in plain-English we can read it as "while the value of *i* is less than the value of *n*, then...". And now the internal content of the loop starts with a *begin* keyword. If the loop had only one Lisp expression inside, then the *begin* keyword would not be necessary, but the usual case is to have several Lisp expressions inside a loop, so here the use of *begin* is mandatory.

Now the line of code *(setq aux (nth i lst))* gets an element from the list *lst* supplied as an argument to the function and assigns it to the internal variable *aux*. Here we must remember that the function *(nth)* takes zero as the required argument for reaching the first element in a list. This explains also why we have chosen *(< i n)* as the logical test for traversing all the elements in the list and not *(<= i n)*.

The following expression, *(if (= aux atm) (setq result (cons i atm)))* compares first if the *i*-element of the list is equal to *atm*, that is, the element to find in the list. If the logical-condition belonging to the *if* test evaluates to *true*, then a *cons* of the index *i* and the element *atm* is assigned to the variable *result*. If the element *atm* that we are trying to find does not exist in the list, the evaluation of *(= aux atm)* will always be *nil* and the variable *result* will maintain its initial assigned value at the beginning of the function, that is, *nil*.

The following line, *(setq i (+ i 1))* is also critical for the adequate working of the loop. Here the actual value of *i* is incremented by one unit and then assigned again to the same variable. If you are new to programming, this expression can be a bit difficult to understand, but this concept of incrementing the value of a numerical variable is strongly used in practically any programming language. Maybe a more graphical representation of this programming concept can help you to better grasp it: $i \leftarrow i + 1$.

After a few closed parenthesis, the *while* loop is entirely described in the function. When the value of the variable *i* reaches the number of elements in the scanned list minus one (remember the zero indexing of function *(nth)*) then the logical-test *(< i n)* evaluates to *nil* and the program leaves the loop. Now it finds the expression *result*, and since it is the last expression in the function it is also what the function returns. Let us try it at the Lisp prompt:

> **(my-find 'RED '(GREEN BLUE RED YELLOW MAGENTA))**
: *(2 RED)*

As we can observe, the symbol RED is found in the third position of the list, and the function *(my-find)* returns *(2 RED)*, as expected, since inside the while loop we are using the function *(nth)*, which starts its indexing at zero. We could easily transform the line of code inside the *if* structure as *(setq result (cons (+ 1 i) atm))*, and then *(my-find 'RED '(GREEN BLUE RED YELLOW MAGENTA))* would return *(3 RED)*, but this is not advisable since in that case we would be mixing different indexing concepts, and this can lead to difficulties when trying to find errors in a program.

If we try to find an element that does not exists in the list:

> **(my-find 'BLACK '(GREEN BLUE RED YELLOW MAGENTA))**
: *nil*

no iteration of the while loop gets a match in the expression *(if (= aux atm) (setq result (cons i atm))*, so the initial assignment *(setq result nil)* remains unchanged and thus the function ultimately returns *nil*. Incidentally, the function also works for finding sublists:

> (first (my-find '(a b) '(1 2 3 (a b) 4 5)))
: 3

The important concept to learn, or, better said, one of the more important ideas to understand when using a *while* structure, consists in its ability for traversing lists. It allows a Lisp user to sequentially examine a list and performs the desired computing on its elements. The following function *(!flex-chop)* is the complementary function to *(flex-chop)*, that is, it returns the original list used as input without some internal elements. Let us observe code 3-19:

```
;Code 3-19
(define (!flex-chop lst n1 n2, result-lst i lng)
    (setq result-lst '())
    (setq i 0)
    (while (< i n1)
            (begin
                    (setq result-lst (cons (nth i lst)
                                           result-lst))
                    (setq i (+ 1 i))
            );end begin
    );end while

    (setq lng (length lst))
    (setq i (- lng n2))

    (while (< i lng)
            (begin
                    (setq result-lst (cons (nth i lst)
                                           result-lst))
                    (setq i (+ 1 i))
            );end begin
    );end while
    (reverse result-lst)
)
```

The algorithm used for *(!flex-chop)* is rather easy: (a) initialize variables *result-lst* and *i*. (b) enter a loop and copy all the elements in the list until reaching the *n1* element in the original list. (c) calculate how many elements must be copied at the tail of the list, and (d) copy the rest of elements. Let us try this function at the Lisp prompt:

> (!flex-chop '(a b c d e f g h) 2 2)
: (a b g h)

It is interesting to observe a comparison of results between the functions *(flex-chop)* and *(!flex-chop)*, using, for example, the list *(a b c d e f g h)* as the first

Table 3.5 Results comparison between *(flex-chop) and (!flex-chop)*

Arguments n1, n2	List returned by *(flex-chop)*	List returned by *(!flex-chop)*
n1 = 0, n2 = 2	*(a b c d e f)*	*(g h)*
n1 = 2, n2 = 0	*(c d e f g h)*	*(a b)*
n1 = 2, n2 = 2	*(c d e f)*	*(a b g h)*
n1 = 4, n2 = 4	*()*	*(a b c d e f g h)*
n1 = 0, n2 = 0	*(a b c d e f g h)*	*()*

argument for both functions. Table 3.5 show some results for several values of *n1* and *n2* as the rest of arguments.

As desired, and under the point of view of classical sets theory, the union of *(flex-chop)* and its complementary *(!flex-chop)* over any list always produces the original list.

$$(flex\text{-}chop\ lst\ n1\ n2) \cup (!flex\text{-}chop\ lst\ n1\ n2) = lst$$

Very interestingly, NewLisp has some built-in functions related to sets. I would like to invite the reader to consult the NewLisp manual (from the Menu: Help → NewLisp Manual and Reference) in order to appreciate the functions *(intersect), (difference)* and *(unique)*.

Now it is time to re-visit Mondrian, the painter. In Chap. 2 we had promised to find a solution to the mess of shapes in Fig. 2.5 in order to obtain the correct position of compositions shown in Fig. 2.4. This implies to design a function for moving all the rectangles in the *Mondrian-style* list:

(setq Mondrian-style '(((5-4) (8 12)) ((0 0) (10 10)) ((2-1) (16 2)) ((6.5 1) (13.5 8)))))

Since the required movement is horizontal and towards left, we would only need to subtract the x values from the sub-lists representing points. However, we are going to seize the opportunity to design a more general function that allows to move the composition both in X and Y directions by a given amount *delta-x, delta-y*. The difficulty in this moment is that the elements of the list are not 2D points, but rectangles defined by their diagonals, that are in fact, lists of two 2D points. If we type at the Lisp prompt:

> **(first Mondrian-style)**
: *((5 -4) (8 12))*

For accessing the first x value of the first point in the first rectangle of the list we would need to type:

> **(first (first (first m)))**
: *5*

thus, it is not a surprise that the desired function becomes a bit complex. All the
code is shown in listing Code 3-20. After initializing some variables we enter in a
while loop in order to traverse the list given as argument, in this case
Mondrian-style. With the help of two simple expressions we extract points *p1* and
p2, representing the diagonal of every rectangle in the list. After this required step
comes the real list manipulation in order to move points *p1* and *p2* by the required
magnitudes delta-x, delta-y:

(setq p1 (cons (add (first p1) delta-x) (add (last p1) delta-y)))
(setq p2 (cons (add (first p2) delta-x) (add (last p2) delta-y)))

As can be seen, we obtain the more internal values by using the functions *(first)*
and *(last)*. Then we add the desired shift given by *delta-x*, *delta-y*, and after *consing*
them into a 2D point structure we re-assign the obtained points to *p1* and *p2*. Please
take the required time to understand these two lines of code. It is not cryptic but it
can take some time at first sight.

After this list manipulation the rest is easy: first we build a list composed by *p1*
and *p2* using the function *(list)* and then we iteratively *cons* the resulting list in
every loop pass. Just after the end of the loop all the processing is almost done and
we only need to *reverse* the obtained list in order to get the desired result, as can be
seen in code 3-20:

```
;code 3-20
(define (shift-xy lst delta-x delta-y, n i lst-out lst-temp
   p1 p2)
     (setq i 0)
     (setq n (length lst))
     (setq lst-out '())

     (while (< i n)
            (begin
                     (setq lst-temp (nth i lst))
                     (setq p1 (first lst-temp))
                     (setq p2 (last lst-temp))
                     (setq p1 (cons (add (first p1) delta-x)
                     (add (last p1) delta-y)))
                     (setq p2 (cons (add (first p2) delta-x)
                     (add (last p2) delta-y)))
                     (setq lst-temp (list p1 p2))
                     (setq lst-out (cons lst-temp lst-out))
                     (setq i (+ 1 i))
            );end begin
      );while end
      (reverse lst-out)
 );end function
```

Now we only need to use the function in order to get a new location for the rectangles that form the Mondrian-style composition. The required two dimensional shift is *delta-x = -18, delta-y = 0*:

> **(shift-xy Mondrian-style -18 0)**
: *(((-13 -4) (-10 12)) ((-18 0) (-8 10)) ((-16 -1) (-2 2)) ((-11.5 1) (-4.5 8)))*

3.5.2 Other Looping Structures

Aside the while looping structure, Lisp has four looping alternatives, named *for*, *do-until*, *do-while* and *until*.

The *for* looping structure is useful when we know, beforehand, how much iterations must happen in a loop. Code 3-21a shows a function, named *(calc-factorial)* that returns a list containing the factorial of the first *n* numbers, given by the following mathematical expression:

$$n! = n \cdot (n - 1) \cdot (n - 2)...3 \cdot 2 \cdot 1 \qquad (3\text{-}3)$$

```
;code 3-21a
(define (calc-factorial1 n, x last-result lst-out)
    (setq last-result 1)
    (setq lst-out '())
    (for (x 1 n)
            (begin
                    (setq last-result (mul last-result x))
                    (setq lst-out (cons last-result lst-
                    out))
            );end begin
    );for end
    (reverse lst-out)
)
```

As usually, let's try the *(calc-factorial1)* function at the Lisp prompt:

> **(calc-factorial1 10)**
: *(1 2 6 24 120 720 5040 40320 362880 3628800)*

We all know that the factorial function grows very quickly, but *(calc-factorial1)* help us to visualize how much "quickly" really means:

> (calc-factorial1 171)

: (1 2 6 24 120 720 5040 40320 362880 3628800 39916800 479001600			
6227020800	8.71782912e+10	1.307674368e+12	2.092278989e+13
3.556874281e+14	6.402373706e+15	1.216451004e+17	2.432902008e+18
5.109094217e+19	1.124000728e+21	2.585201674e+22	6.204484017e+23
1.551121004e+25	4.032914611e+26	1.088886945e+28	3.048883446e+29
8.841761994e+30	2.652528598e+32	8.222838654e+33	2.631308369e+35
8.683317619e+36	2.95232799e+38	1.033314797e+40	3.719933268e+41
1.376375309e+43	5.230226175e+44	2.039788208e+46	8.159152832e+47
3.345252661e+49	1.405006118e+51	6.041526306e+52	2.658271575e+54
1.196222209e+56	5.50262216e+57	2.586232415e+59	1.241391559e+61
6.08281864e+62	3.04140932e+64	1.551118753e+66	8.065817517e+67
4.274883284e+69	2.308436973e+71	1.269640335e+73	7.109985878e+74
4.05269195e+76	2.350561331e+78	1.386831185e+80	8.320987113e+81
5.075802139e+83	3.146997326e+85	1.982608315e+87	1.268869322e+89
8.247650592e+90	5.443449391e+92	3.647111092e+94	2.480035542e+96
1.711224524e+98	1.197857167e+100	8.504785886e+101	6.123445838e+103
4.470115462e+105	3.307885442e+107	2.480914081e+109	1.885494702e+111
1.45183092e+113	1.132428118e+115	8.946182131e+116	7.156945705e+118
5.797126021e+120	4.753643337e+122	3.94552397e+124	3.314240135e+126
2.817104114e+128	2.422709538e+130	2.107757298e+132	1.854826423e+134
1.650795516e+136	1.485715964e+138	1.352001528e+140	1.243841405e+142
1.156772507e+144	1.087366157e+146	1.032997849e+148	9.916779349e+149
9.619275968e+151	9.426890449e+153	9.332621544e+155	9.332621544e+157
9.42594776e+159	9.614466715e+161	9.902900716e+163	1.029901675e+166
1.081396758e+168	1.146280564e+170	1.226520203e+172	1.324641819e+174
1.443859583e+176	1.588245542e+178	1.762952551e+180	1.974506857e+182
2.231192749e+184	2.543559733e+186	2.925093693e+188	3.393108684e+190
3.969937161e+192	4.68452585e+194	5.574585761e+196	6.689502913e+198
8.094298525e+200	9.875044201e+202	1.214630437e+205	1.506141742e+207
1.882677177e+209	2.372173243e+211	3.012660018e+213	3.856204824e+215
4.974504222e+217	6.466855489e+219	8.471580691e+221	1.118248651e+224
1.487270706e+226	1.992942746e+228	2.690472707e+230	3.659042882e+232
5.012888748e+234	6.917786473e+236	9.615723197e+238	1.346201248e+241
1.898143759e+243	2.695364138e+245	3.854370717e+247	5.550293833e+249
8.047926057e+251	1.174997204e+254	1.72724589e+256	2.556323918e+258
3.808922638e+260	5.713383956e+262	8.627209774e+264	1.311335886e+267
2.006343905e+269	3.089769614e+271	4.789142901e+273	7.471062926e+275
1.172956879e+278	1.853271869e+280	2.946702272e+282	4.714723636e+284
7.590705054e+286	1.229694219e+289	2.004401577e+291	3.287218586e+293
5.423910666e+295	9.003691706e+297	1.503616515e+300	2.526075745e+302
4.269068009e+304	7.257415615e+306 inf)		

The last element in this long list is not a number, but the NewLisp representation of *infinite*. Calculating the factorial of 171 NewLisp reaches the bigger positive integer it can handle. For those interested in pushing a computer towards its limits while having a lot of fun in the reading I would suggest reference (Gray and Glynn 1991). Also, and as an exercise, I would recommend to the more mathematical oriented readers to write a function for creating a list of n elements, given by the following mathematical expression in order to compare the growing of *n!* and *a^n*:

$$f(n) = \frac{n!}{a^n} \qquad\qquad (3\text{-}4)$$

Again at the function *(calc-factorial1)* we must remark the anatomy of a *for* loop:

(for (symbol initial-value final-value)
     (begin
             (lisp-expression_i)
     ); end begin
); end for

After the keyword *for*, the name of a symbol must be specified. This symbol works like an index for running the loop exactly *(+ (- final-value initial-value) 1)* times. This means that *for* loops are used when we know beforehand how many times it must loop. However, every *for* loop can be expressed by means of a *while* loop, as we can see in Code 3-21b, where an alternate version for creating a list of the first n factorial numbers is shown:

```
;code 3-21b
(define (calc-factorial2 n, i last-result lst-out)
     (setq i 1)
     (setq last-result 1)
     (setq lst-out '())

     (while (<= i n)
             (begin
                     (setq last-result (mul last-result i))
                     (setq lst-out (cons last-result lst-
                     out))
                     (setq i (+ 1 i))
             );end begin
     );end while
     (reverse lst-out)
)
```

Trying it at the keyboard we can see that it produces exactly the same results as its *for*-based loop alternative:

> (calc-factorial2 5)
: (1 2 6 24 120)

The *do-until*, *do-while* and *until* loop structures are variations of the same theme. In order to save space, we shall write a function that sums the first *n* numbers using these loop structures, shown in Code 3-22a, 3-22b, and 3-22c, allowing the reader to discover the subtle differences:

```
; code 3-22a, using the do-until looping structure:
(define (sum-n1 n, i result)
     (setq i 0)
     (setq result 0)
     (do-until (> i n)
              (begin
                      (setq result (+ result i))
                      (setq i (+ i 1))
              )
     )
     result
)

; code 3-22b, using the do-while looping structure:
(define (sum-n2 n, i result)
     (setq i 0)
     (setq result 0)
     (do-while (>= n i)
              (begin
                      (setq result (+ result i))
                      (setq i (+ i 1))
              )
     )
     result
)

; code 3-22c, using the until looping structure:
(define (sum-n3 n, i result)
     (setq i 0)
     (setq result 0)
     (until (> i n)
              (begin
                      (setq result (+ result i))
                      (setq i (+ i 1))
              )
     )
     result
)
```

Testing the three functions at the same time we can confirm they produce the same result:

> **(list (sum-n1 5) (sum-n2 5) (sum-n3 5))**
: (15 15 15)

As a hint, it must be noted that the *while* and *for* loop structures are the most used ones in Lisp. The *while* structure is the more general and flexible one, existing today in practically any other programming language. The *for* structure is the better option when it is known in advance how many loop iterations are needed, but even so, it can be substituted by a *while* structure, as we have already seen. My practical advice would be to master the *while* structure in order to not get confused by these multiple looping programming options.

3.6 Recursion Is Based on Recursion

The first time we spoke about recursion in this book, intentionally without mentioning it, was in Chap. 2, when we defined what a list is:

> **"A list is a formal collection of elements enclosed by parenthesis whose
> elements are either atoms or lists"**

This definition is recursive because the word, or term, that we try to define is included into the definition itself. Let us take as an example the list *(a b c (1 2 3) d e)*. From a newcomer to Lisp, and reading from left to right (I already assume you are not a Lisp newcomer!), the first elements, *a*, *b*, *c* are a collection of atoms that are enclosed by parenthesis, so for now, it seems a list. Then, the fourth element appears as a new expression: *(1 2 3)*. It is not an atom, but then we call recursively the definition of list and then we "jump" inside the fourth element for discovering that elements *1*, *2* and *3* are atoms enclosed by parenthesis, so it is in itself a list, so the definition seems to work, at least for now. After exiting the fourth element we found elements *d* and *e*, so we finally conclude that it is certainly a list because it is a collection of elements enclosed by parenthesis. It only happens that one of its internal elements is also a list. The definition holds completely.

Another example or recursion in natural language is the famous Hofstadter's Law, the author of the book "Gödel, Escher, Bach: An Eternal Golden Braid": "It always takes longer than you expect, even when you take into account Hofstadter's Law" (Hofstadter 1999). This sentence helps to know, for example, how much time does it takes to write a book: "longer than you expect", but then the law calls itself recursively so you intuitively understand that it takes many more time than you expect. By the way, as you have correctly guessed, the title of this section is strongly recursive.

From its very first design, Lisp has included recursion. It is a programming structure where a function is able to call itself. Before introducing a true Lisp example of recursive function we are going to comment first some pseudo-code, shown in Code 3-23a, which will help us to understand the key concepts of recursion in the language. The example takes an imaginary function, written in Lisp syntax, for helping a Formula One driver to win a race in a circuit of *n* laps:

```
;Code 3-23a
(define (runF1GrandPrix n)
    (if (= n 0)
            (stop-the-car)
            (begin
                    (run-as-fast-as-you-can)
                    (runF1GrandPrix (- n 1))
            );end begin
    );end if
)
```

It does not matter in this moment what the functions *(stop-the-car)* and *(run-as-fast-as-you-can)* mean, probably a very complex set of cognitive instructions. The important idea is to understand the recursive structure of the function. Basically speaking, the F1 driver is instructed to run as fast as he can do until finishing the last lap of the Grand Prix. Let us observe the details: First, the function needs an argument *n* that describes the number of the laps of the Grand Prix. Just after entering the function, a conditional *if* appears, testing the number of laps to go. If *n* equals zero, that is, if the GP has ended, the pilot is instructed to stop the car. This programming structure is very important in recursion: we always need a way to exit the function, a way to stop it. Without an exit-point a recursive function would run forever, so if you are going to use recursive functions in your Lisp programs, the first thing you should design is the exit point of these type of functions.

The "else" part of the *if* structure inside the function contains two lisp expressions enclosed inside the *begin* block that will be called if any lap remains in the race. The first, *(run-as-fast-as-you-can)* has an obvious meaning in this pseudo-code. The second function call is the nuclear point in a recursive function: A new call to the function itself. The key-point in this moment is that the argument for the function has varied its value, decreasing its value in one lap by means of using the expression *(- n 1)*.

Since Code 3-23a is pseudo-code, we cannot test it at the Lisp prompt. However, it is rather close in its syntax to authentic Lisp code, so nothing forbids us to replace the functions *(stop-the-car)* and *(run-as-fast-as-you-can)* for something simple but convenient for our purposes of understanding recursion. For *(stop-the-car)* we shall use the Lisp expression *"Hey, this has been great!"*, and for the *(run-as-fast-as-you-can)* we shall employ the expression *(print "I'm running lap" n "as fast as I can do!\n")*. This is the first time we use the function *(print)* in this book, but suffice is to say for now that it prints string type data in the session area of the NewLisp environment. We shall introduce it in more detail in the next chapter. The resulting function runF1GrandPrix is shown in Code 3-23b:

```
;Code 3-23b
(define (runF1GrandPrix n)
    (if (= n 0)      ;the exit to the function is here
        "Hey, this has been great!"
        (begin
            (print "I'm running lap "n" as fast as I can do!
            \n")
            (runF1GrandPrix (- n 1))
            );end begin
    );end if
)
```

Now we certainly can test it at the Lisp prompt. In order to save space, we shall simulate a very short GP, with only five laps:

> **(runF1GrandPrix 5)**
: I'm running lap 5 as fast as I can do!
: I'm running lap 4 as fast as I can do!
: I'm running lap 3 as fast as I can do!
: I'm running lap 2 as fast as I can do!
: I'm running lap 1 as fast as I can do!
: "Hey, this has been great!"

In this example, the function starts with an *n* value equal to 5, so the *if* structure evaluates its relational expression to *nil* and enters the *begin* block. There, after printing the message "*I'm running lap 5 as fast as I can do!*" the function is called recursively: *(runF1GrandPrix 4)*. The value of *n* decreases in every function call until becoming zero and then the function reaches its exist-point, meeting the relational expression *(= n 0)* and then printing *"Hey, this has been great!"*.

Maybe one of the most known examples of recursion in Lisp is a function for calculating the factorial of *n*. Yes, we have just seen in the previous section how to calculate it using a *for* and then a *while* looping structure. Code 3-24 shows the recursive version. It does not produce a list with the first n factorial numbers. It returns only the factorial of *n*:

```
;Code 3-24
(define (r-factorial n)
    (if (= n 1) ;terminates the function
            1
            (mul n (r-factorial (- n 1))))
    )
)
```

The first thing we note about this recursive version is the small quantity of code it uses. Let us try it at the Lisp prompt:

> **(r-factorial 5)**
: 120

Now let us examine it meticulously. As you can immediately observe, the function *(r-factorial)* has a similar structure to the function *(runF1GrandPrix)*. The argument *n* is a natural number that decreases its value in recursive calls to the function until reaching a value of one. In that moment the function terminates, returning the value of all the multiplications performed in every function pass. Very interestingly, NewLisp incorporates a function named *(trace)* that allows us to observe the internal working of any function step by step. Let us try it for the function call *(r-factorial 4)*:

> **(trace true)**
: *true*
: 0> **(r-factorial 4)**
: ‒‒‒‒‒‒
: *(define (r-factorial n)*
: *#(if (= n 1)*
: *1*
: *(mul n (r-factorial (- n 1))))#)*
:
:*[-> 2] s|tep n|ext c|ont q|uit >*

You have now the options "step", "next", "cont" and "quit". Pressing "s" at the keyboard followed by the Enter key you will appreciate how every Lisp expression is evaluated, one at a time. The NewLisp environment marks every expression by means of a double sharp # enclosing symbol, showing the intermediate results until reaching the returning value: 6. For ending any trace session and exit to the standard Lisp prompt you only need to press the "q" key and then Enter. Pressing "s" and the Enter key several times, we arrive a point where *n* equals one. Now things start to become fascinating. We have remarked in bold the expression traced, enclosed between sharp marks:

: *(define (r-factorial n)*
: *(if (= n 1)*
: *1*
: *(mul n #(r-factorial (- n 1))#)))*
: *RESULT: 1*

That's correct. By definition in the function, when *n = 1* the function call becomes *(r-factorial 1)* → *1*. Now the most interesting thing is going to happen when we continue pressing "s" and the Enter key: After arriving at the minimum allowed value of *n*, Lisp start to increment *n* in every iteration automatically from its internal memory management system, performing and storing the partial calculations and values. Now *n* is two:

: *(define (r-factorial n)*
: *(if (= n 1)*
: *1*
: *#(mul n (r-factorial (- n 1)))#))*
: *RESULT: 2*

Now *n* is three:

: (define (r-factorial n)
: (if (= n 1)
: 1
: #(mul n (r-factorial (- n 1)))#))
: RESULT: 6

and finally *n* reaches four:

: (define (r-factorial n)
: (if (= n 1)
: 1
: #(mul n (r-factorial (- n 1)))#))
: RESULT: 24

In Code 3-25 we can see a recursive function that calculates the sum of the first *n* natural numbers:

```
;Code 3-25
(define (r-sum n)
    (if (= n 1) ;terminates the function
        1
        (+ n (r-sum (- n 1)))
    )
)
```

This function is similar to the previous one and now you should understand it perfectly well. After testing it at the Lisp prompt, it can help you to *(trace)* it:

> (r-sum 100)
: 5050

Aside recursion for numerical calculations, Lisp not only allows, but it is especially useful for traversing lists. Here we must remember the Fig. 2.1. The bottom area of the figure represents the realm of using while looping structures for traversing. The top area suggests the world of recursion in lists. The following function, shown in Code 3-26, prints the elements of a list

```
;Code 3-26
(define (r-show lst)
    (if (null? lst) ;exit condition to the funtion
            nil
            (begin
            (print (first lst) "\n")
            (r-show (rest lst)) ;the trick here is to use
            (rest)
            );end begin
    )
)
```

First, let us test the function:

> (r-show '(recursion is driving me crazy))
: recursion
is
driving
me
crazy
nil

The performed strategy in this function resides in the exquisite processing of data by means of the functions *(first)* and *(rest)*. The test *(if (null? lst)* allows the exit point of the function when *lst* is *nil*, that is, when we have traversed the entire list. Meanwhile, the function takes the first element, prints it, and immediately calls itself, recursively, using *(rest lst)* as its argument. The second iteration is equivalent to *(print (first (rest lst)))* and so on until all the elements in the list are printed.

Recursion is especially adequate to Lisp system programming, that is, to program kern Lisp functions. As an example, Code 3-27 shows a version of *(length)*, returning, as you can guess, the length of a list:

```
; Code 3-27
(define (r-length lst)
    (if (null? lst) ;exit condition to the function
            0
            (+ 1 (r-length (rest lst))))
    )
)
```

Testing it is straightforward:

> (r-length '(a b c))
: 3

The strategy for this function, aside the usual design for its exit point, is a mixture of using *(rest)* and a recursive call similar to the used in the function *(r-sum)*. Just compare the respective Lisp expressions used in both functions:

$$(+ \ n \ (r\text{-}sum \ (\text{-} \ n \ 1)))$$
$$(+ \ 1 \ (r\text{-}length \ (rest \ lst)))$$

The first expression adds numbers; the second one counts elements in a list. Another fundamental function in Lisp, as we already know is *(nth)*. In code 3-28 we can see an implementation of this function, named in this case *(r-nth)*, using recursion.

```
; code 3-28
(define (r-nth n lst)
   (if (null? lst)
      nil
      (if (zero? n)
              (first lst)
              (r-nth (- n 1) (rest lst))
      ) ;internal if
   ) ;external if
)
```

We can test it at the Lisp prompt comparing it with the original *(nth)* function:

> (cons (nth 0 '(a b c)) (r-nth 0 '(a b c)))
: (a a)

As usual, the first thing to do when designing recursive functions is to provide an exit point. In this case this function has two exit points. The first one checks if the list supplied as an argument is null: *(if (null? lst)* → *nil*, the second one checks if the *n* argument is zero, in such case, it return the first element of the list. If *n* is not zero, then a recursive call to the function is made, traversing the list using *(rest)* and decreasing *n* at the same time until *n* becomes zero. Please note the coordinate use of (first) and *(rest)*, ultimately allowing the extraction of the *nth* element in the list. Looking at the code is also clear why the *(nth)* function is zero indexed.

Through all these function examples exposed in this section of the chapter we have seen that recursion is just another way of looping. Iterative looping, based on *while* and alternative structures such as *for* or *until* is usually better understood than recursion from those readers that already have previous programming knowledge. Recursion can seem at first sight more obscure than standard iteration, but it really deserves attention. Recursive functions use few code, are elegant and are the most pure Lisp you can write. In any case I am not a purist. Feel free to choose the programming style that suits you best.

3.7   A Note on Lambda Expressions

Until now we have used the keyword *define* for creating functions. Since the Lisp environment is so special, where you can establish interactive sessions with the system in real time, you can use anonymous functions, that is, functions that have not a name but even so can be called directly from other functions. Such a type of Lisp structures are called lambda expressions.

Let us imagine, as an example, you have a lists of points *(x y)*, such as:

> (setq lst '((1 3) (4 5) (6 1)))
: ((1 3) (4 5) (6 1))

For swapping the values of x and y, that is, to build a list of *(y x)* pairs, you could interactively write at the Lisp prompt:

> (map (lambda (pair) (cons (last pair) (first pair))) lst)
: ((3 1) (5 4) (1 6))

We have already introduced the function *(map)* at the beginning of this chapter. It is a function that allows a Lisp user to successively apply another function on the elements of a list. In those paragraphs we used a function built by means of the keyword *define*, that is, a standard Lisp function. The novelty now is the following expression:

(lambda (pair) (cons (last pair) (first pair)))

This lambda expression is equivalent in its results to the function shown in code 3.29:

```
;Code 3-29
(define (swap-pair pair)
    (cons (last pair) (first pair))
)
```

The difference with lambda expressions is that the Lisp user can define the function on the fly, without even needing a name for it. Needless to say, after defining *(swap-pair)* you can use it with (map), too:

> (map swap-pair lst)
: ((3 1) (5 4) (1 6))

Alternatively, you could also give a name to a lambda expression and then use it as an argument for *(map)*. As an example:

> (setq my-exchange (lambda (pair) (cons (last pair) (first pair))))
: (lambda (pair) (cons (last pair) (first pair)))

Now the symbol my-exchange stores a lambda expression:

> my-exchange
: (lambda (pair) (cons (last pair) (first pair)))

And now, as you already suspect:

> (map my-exchange lst)
: ((3 1) (5 4) (1 6))

Lambda expressions are an extremely exciting feature of Lisp and they would deserve a complete chapter for them alone. However, and aside the fact we are short on space, they are a relatively advanced topic on Lisp. The interested reader can obtain further information on advances books on Lisp or even get the whole picture in specific texts (Michaelson 2011).

3.8 As a Summary

An important quantity of material has been exposed in this chapter. Together with the concepts of list structure and list processing, already seen in Chap. 2, these materials constitute the foundations of the Lisp language. The essentials of this chapter can be summarized in the following points:

- A user-defined function is a set of Lisp expressions that, working together, performs a programming task. As it is the case with the functions already included in the language, every user defined function has an input (available information), an output (the result of the function) and a body that, containing a set of Lisp expressions, is the responsible of processing the information obtained from the input until obtaining the desired function output.
- In the NewLisp environment all the functions are written in the programming area of the environment and can be tested in the session area, at the Lisp prompt.
- All the functions defined by user become Lisp functions, converting automatically themselves in an extension of the language in such a way that can be combined with any structure of the language. From the programmer's viewpoint there is no difference in use between a user defined function and a function originally embedded into Lisp.
- Comments are extremely important in Lisp. Every Lisp user should comment profusely the functions that he or she writes. A semicolon (;) is the sign used in Lisp to tell the language where a comment begins. Lisp treats all the words written after a semicolon in a line as a comment.
- Internal variables live only inside the functions where they are defined. This means that an internal variable cannot be accessed from any other function. They are usually needed for storing values from internal calculations in the functions, that is, to help to process information. On the other hand, global variables can be defined in a Lisp program in such a way that they can be accessed from any function in the program. Since any function can change the value of a global variable, these must be handled with care.
- After writing and testing a function it is common practice to try to optimize it. However, no function should be optimized until an extensive testing of the non-optimized version is performed. Function optimization usually produces some lack of legibility and human interpretability of the resulting Lisp code, so a delicate balance between optimization and legibility must be selected in every occasion.
- All types of Lisp expressions such as numbers, atoms or symbols can be compared in NewLisp by using the relational operators: Less than (<), bigger than (>), equal to (=), less or equal to (<=), bigger or equal to (>=), and different to (!=). Logical tests in Lisp can be performed by using function predicates and/or relational operators. A logical test returns only two possible outcomes: *true* or *nil*.
- A multiple logical test can be made by using the logical connectives "*and*" and "*or*". These connectives can combine function predicates and relational

operators for producing sophisticated logical expressions. The results of combining logical connectives are obtained from the truth table shown previously in Table 3.4.

- An if-then-else structure is one of the bases of conditional programming in Lisp. It begins with a logical test, and depending on its *true* or *nil* evaluation, an alternative set of Lisp expressions is evaluated. In order to group several Lisp expressions inside the same logical direction NewLisp uses the keyword *begin*. Special attention must be observed for matching parenthesis while using *begin* blocks.

- Indentation consists in an adequate placement of text farther to the right in some lines of code inside a function. Such a shift in code location helps to make it clear the design of the function, helping at the same time to avoid programming mistakes. The use of indentation, although not mandatory, is specially recommended in if-then-else conditional structures.

- A *cond* structure is an alternative to an if-then-else construction. After the *cond* keyword several logical tests are used until covering all the possible options in the range where the logical tests are defined. If all the logical tests are evaluated to nil, an optional last expression can be included as a lifeboat that works as an *else* statement.

- A while looping structure is formed by the *while* keyword followed by a logical test and then a block of Lisp expressions. This block of expressions is repeated until reaching a point where the logical test is evaluated to *nil*. The logical test is evaluated one time every loop pass. At least one of the Lisp expressions inside the block must be able to modify the value of the expression contained in the logical test. Otherwise a while loop would run forever.

- Aside the while looping structure, Lisp has four looping alternatives: *for, do-until, do-while* and *until*. The most important one of these is the *for* structure. It is specially suited when we know beforehand how many iterations must happen in a loop. The *while* and *for* structures are the most used ones in this book. Even so, every *for* construction can be expressed by means of a *while* looping structure.

- Recursion is a very special computing structure where a function calls itself from inside its own body. Although recursion can be used practically in every program language, it almost reaches the status of art in Lisp. The design of a recursive function must include an exit point to the function and an expression where the function calls itself. In every call, the argument changes. Recursion is an excellent computing strategy for traversing lists by means of the use of *(first)* and *(rest)*.

- Recursive functions are usually more difficult to read and understand that functions where looping structures are used. This is caused by traditional computer science inertia, where other computer languages, not naturally oriented towards recursion, establish a traditional way of thinking in the programmers. The NewLisp function *(trace)* helps the Lisp user to follow the inner workings of any function, being a useful tool to better understand recursion.

- Lambda expressions are a special type of functions. They have not a name, that is, they are anonymous, but can be called from other functions. Moreover, they can be defined and applied on the fly, being a first-class computing resource in interactive Lisp sessions.

In the next chapter we shall try to put in practice all we have learnt so far in the book. Until now we have played with small pieces of code in order to know the essentials of the language. The next step will consist in designing, aggregating and coordinating functions in order to build programs. We shall extensively use conditional structures, looping structures, and lists, of course. We shall even seize the opportunity to include new programming concepts.

References

Graham, P.: What made Lisp different. http://www.paulgraham.com/diff.html. Acceded Mar 2014

Gray, T., Glynn, J.: Exploring Mathematics with Mathematica. Addison-Wesley, Boston (1991)

Hofstadter, D.: Gödel, Escher, Bach: An Eternal Golden Braid. Basic Books, New York (1999)

Michaelson, G.: An Introduction to Functional Programming Through Lambda Calculus. Dover, 2011

National Heart, Lung, and Blood Institute.: Metric BMI Calculator. http://www.nhlbi.nih.gov/guidelines/obesity/BMI/bmi-m.htm. Acceded Mar 2014

Chapter 4
Lisp Programming

4.1 Introduction

After reading the three previous chapters, understanding the exposed material and playing with the examples at the keyboard I must confess that you already know to program in Lisp. All the heavy stuff of the language has been digested and you only need to put it all together. Yes, there are advanced programming techniques that have not been exposed, and in this point we are still far from a professional level of Lisp programming, but even so you have enough knowledge for creating powerful applications, and what is more important: You now have understood a programming paradigm that will serve you for exploring the realms of fuzzy sets and fuzzy logic with confidence in the next part of the book.

This chapter will add the last touches for creating programs with NewLisp. You will learn to implement simple user interfaces, that is, programming techniques that allows a human being to interact with a computer in real time. The devices that allow this interaction are the screen, the keyboard, and a storage device such a hard disk. Lisp incorporates functions that are designed to perform input/output (I/O) tasks with these devices and we are going to learn how to use them in this chapter.

For moving forwards in our Lisp experiences we shall develop two complete applications in this chapter. The first one is a rather complete French roulette simulation that shows how to structure a Lisp program into a set of functions and to design a simple user interface. The second application is a small library of functions for simple database managing. For experimenting with the library we shall use a database of deep-sky celestial objects, although the database functions can be used with any CSV (Comma Separated Values) format database, obtained, for example, from spreadsheet software applications such as Open Office, Excel, etc.

© Springer International Publishing Switzerland 2016   93
L. Argüelles Méndez, *A Practical Introduction to Fuzzy Logic using LISP*,
Studies in Fuzziness and Soft Computing 327,
DOI 10.1007/978-3-319-23186-0_4

4.2 From Montecarlo with Love

Being one of the most famous cities in the Mediterranean coast, Montecarlo always surprises the visitor with its good climate, smiling people, pure blue sea and glamorous Casino. Inside the historical building, a roulette works seven days a week, always moving tokens and money behind simple random mechanisms. In this section of this chapter we are going to develop a Lisp program that emulates a traditional French single zero roulette. After developing the application, the reader will be able to simulate a night of play, observing how much money can be won or lost. *Rien ne va plus*!

As it happens in every computing problem on simulation, the first step consists in getting a clear description of the real system. For each roulette run, a small ball is thrown towards the opposite direction of a spinning wheel (roulette) that contains 37 numbered pockets. After the system loses momentum, the ball ultimately falls into one pocket of the wheel, producing a resulting number and a corresponding set of play outcomes, corresponding to a variety of betting options. In our simulation, the following bets are allowed:

- Single number
- Red or black
- Even or Odd
- Manque (numbers from 1 to 18) or Passe (numbers from 19 to 36)
- Dozens: First, second, or third dozen
- Columns: First, second or third (see Fig. 4.1)

Fig. 4.1 Layout of a French single zero roulette

Table 4.1 Payouts for each result/bet in a single zero French roulette

Result	Payout
0	Nil, casino wins all
Single number	35:1
Red/black	1:1
Even/odd	1:1
Manque/passe	1:1
Dozen	2:1
Column	2:1

The payouts for every bet are given in Table 4.1.

On a real roulette, some more bets are allowed, as, for example "streets" (three horizontal numbers), "corners" (four adjoining numbers in a block), and so on. In order to maintain a suitable degree of complexity in the Lisp code, only the bets shown in Table 4.1 will be included in the simulation. We encourage the interested reader to develop a full implementation of this classic casino game.

4.2.1   Declaring Global Variables

In the previous chapter we learnt about local variables inside user-defined functions. Now it is the time to introduce global variables, that is, variables whose content can be used—and modified—by any function in a Lisp program. The advantage of such a data structure is, obviously, the easy availability of them from any part of a program. On the other hand, such easiness of using them brings as a consequence the generation of programs that can be confusing, hard to read and ultimately, prone to failures that can be hard to detect. The trivial solution resides in not using global variables, but from a practical point of view, the best advice for keeping a good balance between easy availability of data and minimizing the possibility of errors consists in using global variables scarcely and giving them a special spelling, surrounding them, for example, by asterisks. As an example, *payout* could be the name of a local variable, and *payout* would be the chosen spelling for a global variable.

Observing the roulette layout in Fig. 4.1 we can realize that the color assignment for numbers follows a strange pattern, so instead of generating an algorithm for describing them, the best is to declare all the number-color relationship inside a Lisp list. Columns follow a lot simpler pattern, but even so, we are going to declare them as global variables, too. The same will be made for the rewards/payouts, as can be seen in code 4-1:

```
;code 4-1
(setq *numbers* '((1 RED) (2 BLACK) (3 RED) (4 BLACK) (5 RED)
(6 BLACK) (7 RED) (8 BLACK) (9 RED) (10 BLACK) (11 BLACK) (12
```

RED) (13 BLACK) (14 RED) (15 BLACK) (16 RED) (17 BLACK) (18
RED) (19 RED) (20 BLACK) (21 RED) (22 BLACK) (23 RED) (24
BLACK) (25 RED) (26 BLACK) (27 RED) (28 BLACK) (29 BLACK) (30
RED) (31 BLACK) (32 RED) (33 BLACK) (34 RED) (35 BLACK) (36
RED)))

```
;These are the column patterns:
(setq *column1* '(1 4 7 10 13 16 19 22 25 28 31 34))
(setq *column2* '(2 5 8 11 14 17 20 23 26 29 32 35))
(setq *column3* '(3 6 9 12 15 18 21 24 27 30 33 36))
```

```
;These are the rewards for each type of winning bet:
(setq *rewards* '((NUMBER 35) (RED 1) (BLACK 1) (EVEN 1)
(ODD 1) (PASSE 1) (MANQUE 1) (FIRST-DOZEN 2) (SECOND-DOZEN 2)
(THIRD-DOZEN 2) (FIRST-COLUMN 2) (SECOND-COLUMN 2) (THIRD-
COLUMN 2)))
```

Now the global variable *numbers* contains a list where each sublist follows the
pattern *(number color)*. The number zero is not included in the list because its
outcome signals a game event where the casino wins all the bets on the table. The
local variables *column1*, *column2* and *column3* store in simple lists their
corresponding numbers. Finally, the global variable *rewards* contains a list of
sublists with the following structure *(type-of-bet payout)*. An alternative strategy for
storing this type of information could be to use some files for storing it, but since
the number of data is relatively small it is better to directly include all the information in the Lisp code itself. Please note that the program uses only five global
variables, a number small enough to maintain it under good programming practices.
Needless to say, code 4-1, as the rest of the code hereafter, must be put into the
programming area of the NewLisp interface (see Fig. 3.1).

4.2.2   Throwing the Ball and Checking Results

After a roulette spin, a number between 0 and 36 should be generated. We shall
design a dedicated function for this important event later in our roulette simulation.
Before arriving that programming step we are going to concentrate first in a set of
functions that check what betting results produces every roulette spin. That is, we
are going to design a set of functions that, receiving an integer number between 1
and 36 as an input, return the associated game result. For example, if the input
parameter is the integer 16, the goal is to get the associated results *RED, EVEN,
MANQUE, SECOND-DOZEN, FIRST-COLUMN*. The functions shown in Code
4-2 satisfy this goal. They do not check for number zero because in that case the
casino wins it all:

```
;code 4-2:
define (red-or-black number)
    (if (= (last (assoc number *numbers*)) 'RED)
            'RED
            'BLACK
    )
)

(define (even-or-odd number)
    (if (= (% number 2) 0)
            'EVEN
            'ODD
    )
)

(define (passe-or-manque number)
    (if (> number 18)
            'PASSE
            'MANQUE
    )
)

define (what-dozen number, result)
    (if (and (> number 0) (< number 13))
            (setq result 'FIRST-DOZEN)
    )
    (if (and (> number 12) (< number 25))
            (setq result 'SECOND-DOZEN)
    )
  (if (and (> number 24) (< number 37))
            (setq result 'THIRD-DOZEN)
    )
    result
)

(define (what-column number, result)
    (if (find number *column1*)
            (setq result 'FIRST-COLUMN)
    )

    (if (find number *column2*)
            (setq result 'SECOND-COLUMN)
    )
```

```
(if (find number *column3*)
        (setq result 'THIRD-COLUMN)
    )
    result
)
```

Trying them at the Lisp prompt, we can have a first idea of their working:

> (red-or-black 16)
: RED

> (what-column 16)
: FIRST-COLUMN

Let us type a more complex expression:

> (list (red-or-black 16) (even-or-odd 16) (passe-or-manque 16))
: (RED EVEN MANQUE)

This expression is very interesting because it gives us a hint for designing a function that returns all the associated results after a roulette spin. Such a function, named *(winning-results)* is shown in Code 4-3:

```
;code 4-3
(define (winning-results number)
    (list
        number (red-or-black number) (even-or-odd number)
        (passe-or-manque number) (what-dozen number) (what-
            column number)
    )
)
```

Trying it at the Lisp prompt we can observe that we are on the right track. We include the number given as input as the first element of the result of the function because we want a list that contains all the data from a wheel spin:

> (winning-results 16)
: (16 RED EVEN MANQUE SECOND-DOZEN FIRST-COLUMN)

In Chap. 2 we already started to explore random numbers and roulettes by means of the NewLisp function *(random)*. Now we shall design a function that randomly produces an integer between 0 and 36 and returns two possible outcomes: *nil*, if the generated number is zero, that is, if the casino wins, or a list of associated results after a roulette spin by means of a call to *(winning-results)*. The function, named *(spin-wheel)* is shown in Code 4-4:

```
;code 4-4
(define (spin-wheel, number)
```

```
        (setq number (integer (mul (random) 37)))
        (if (> number 0)
                (winning-results number)
                nil
        )
)
```

Trying this function at the Lisp prompt we can observe that the roulette is already working:

> **(spin-wheel)**
: (30 RED EVEN PASSE THIRD-DOZEN THIRD-COLUMN)

Please note that you will need several attempts until getting a *nil* result, but as it happens in the real world, it always appears. In fact when you bet real money your perception is that it appears with excessive frequency. Note also that pressing the "up" and down keys (↑↓) in the keyboard while at the Lisp prompt, NewLisp repeats the last expression entered by user.

4.2.3   Betting and Playing

For representing a bet, we are going to use a very simple strategy. We only need to use a list of two elements with the following structure *(type-of-bet money)*, where *type-of-bet* is either a number between 1 and 36 or a bet of the type, *RED/BLACK, EVEN/ODD*, etc. *Money* means, obviously, how powerful is our bet. As examples, we can type at the Lisp prompt:

> **(setq my-bet '(12 50))**
: (12 50)

Meaning that we bet 50 Euros or Dollars to the number 12. For betting 100 Euros or Dollars, or your favorite currency, to the THIRD-COLUMN, we would type:

> **(setq my-bet '(THIRD-COLUMN 100))**
: (THIRD-COLUMN 100)

In this point of program development we have on one side the function *(spin-wheel)* that simulates a casino roulette, and on the other side a simple data structure for representing a bet. Now what we need is a function that takes a bet and a roulette run as inputs and then produces the money (payout), if any, we have won with our bet:

$$f((type\text{-}of\text{-}bet\ money)\ (spin\text{-}wheel)) \rightarrow payout/money$$

Inside the desired function we are going to use some conditional structures in order to distinguish, (a) if the roulette spin has produced the number zero for checking if we directly have lost our bet (casino wins all bets), (b) we have got a hit on a number, and (c) we have got a hit on a different roulette bet. Expressing it with pseudo-code we can write:

```
(calculate-win bet spin-wheel)
      (if (spin-wheel) returns nil then return nil) ;case a
      (else
            (if the bet is on a number          ;case b
                  (get our number from the bet)
                  (get the associated payout)
                  (check if we have got a hit with the number)
                  (if yes calculate reward, else return zero)
            )
            (else if the bet is not on a number ;case c
                  (get the associated payout)
                  (check if we have got a hit with our bet)
                  (if yes calculate reward, else return zero)
            )
      )
)
```

the desired function is shown in Code 4-5:

```
;code 4-5
(define (calculate-win bet spin, n reward prize)
(if (= nil spin)
    nil; ball falls on 0: Casino wins
  (begin
      (if (number? (first bet))
      (begin
            (setq n (first bet))
            (setq reward (last (first *rewards*)))
            (if (= n (first spin)) ;we have got the number
              ;calculate reward
              (setq prize (mul reward (last bet)))
              (setq prize 0) ;we lose our bet
            )
      ) ;end-internal begin
      ;now the else begins, the bet is not a number:
      (begin
            (setq  reward  (last  (assoc  (first  bet)
            *rewards*)))
            (if (! = (find (first bet) spin) nil) ;we've got
            the bet
              (setq prize (mul reward (last bet)))
```

```
            (setq prize 0) ;we lose our bet
          ) ; end if
      ) ;end second begin
    ) ; if number? ends
    (cons prize spin) ; this is only to check prize & results
  ) ;end external begin
) ;end external if
)
```

as can be seen, this function is relatively complex and a bit long since we, intentionally, have not optimized it. It is important to remark that the input parameters
are in fact two lists, the first one is a bet, for example *(THIRD-COLUMN 100)* and
the second one is a list produced as the result of a call to the function *(spin-wheel)*,
such as, for example, *(31 BLACK ODD PASSE THIRD-DOZEN FIRST-
COLUMN)*. These lists are stored respectively by the symbols *bet* and *spin*.

The first conditional structure is given by *(if (= nil spin)*. If it evaluates to *nil*,
then *(calculate-win)* returns *nil* and terminates (casino wins it all). The second
conditional structure checks if we have made a bet on a number by using the
predicate *(number?)*. If it is the case, the function stores our number in the internal
variable *n* and takes the first element of the list stored on the global variable
rewards, storing the numerical payout in the internal variable *reward*. Now it is
the time for checking if we have got a hit with our number by means of the
conditional expression *(if (= n (first spin))*. If the expression evaluates to *true* then
the function calculates the payout/price and exits the *if* block. If the expression
evaluates to *nil*, the function sets zero as the resulting payoff, exiting the *if* block.

The third conditional structure is in fact an *else* structure from the expression *(if
(number? (first bet))*, that is, if that expression evaluates to *nil* then our bet is not on
a number. As you can see, the *(begin)* blocks are used appropriately for instructing
Lisp what our programming strategies really are. Hereafter the function works in a
similar way than in the previous case: The expression *(setq reward (last assoc (first
bet) *rewards*)))* extract the appropriate payout, principally, by using the function
(assoc). Then, please observe how the function *(find)* makes it easy to check if we
have got a hit with our bet in the conditional expression *(if (! = (find (first bet) spin)
nil)*. From this point the function works as in the previous case. The last step before
terminating is a *cons* given by the Lisp expression *(cons prize spin)*. It seems at first
a redundant expression, since we theoretically only need to know the prize/payout
we have got, but "*consing*" it with the list stored on the symbol *spin* gives us more
information in a Lisp session, at the prompt. Let us try it:

> (setq my-bet '(RED 50))
: (RED 50)

That is, we bet 50 units of your favorite currency to RED. Then, let's see what
happens after a roulette spin is made:

> **(calculate-win my-bet (spin-wheel))**
> *: (50 14 RED EVEN MANQUE SECOND-DOZEN SECOND-COLUMN)*

Well, in this case, the resulting number is 14, RED, so we have won our bet. Since the payment for the bet RED/BLACK is 1:1 (remember Table 4.1) we have won 50 currency units. This is exactly what appears in the first element of the list returned by the function *(calculate-win)*. Let us see another run, this time, for example, betting 50 currency units to the number 8. The expected payout in this case is 35:1, so if we win, the expected sum would be 1750 currency units:

> **(setq my-bet '(8 50))**
> *: (8 50)*

after placing our bet, let us see what happens:

> **(calculate-win my-bet (spin-wheel))**
> *: (0 28 BLACK EVEN PASSE THIRD-DOZEN FIRST-COLUMN)*

in this case the resulting number is 28. We have loose our bet. Inspecting the first element of the returned list we can see that the function reflect this facts showing a zero in that position.

4.2.4 Building a Simple User Interface

Everything runs fine in our roulette simulation. However, typing bets in lists and calling *(calculate-win)* at the Lisp prompt is not the best environment for playing along hours. Not a single casino in Las Vegas should use such a computing approach. Casinos need you feel comfortable for losing dollar after dollar. They need you play quickly.

User interfaces are the way a program interacts with an user. The NewLisp environment bases its interface on a menu bar and a main window where both the programming area and the session area, containing the Lisp prompt, are included. NewLisp allows to build sophisticated graphical user interfaces (GUI) by means of an extensive set of functions contained in the file "guiserver.lsp". This material is out of the scoop of this book, but you can find detailed information in the NewLisp's help menu.

Our aim is simpler. We are going to develop an user interface where the user will be guided interactively through simple questions by the program, usually expecting a yes/no answer, or better said, a y/n answer from the keyboard. In such type of user interfaces it is more than convenient to design a function that takes care of the management of questions and answers. Code 4-6 presents such a function

```
;code 4-6
(define (yes-no message)
    (print message)
```

```
(if (= "Y" (upper-case (read-line)))
        true
        nil
   )
)
```

Despite being a short function, it deserves several comments. First, every function call to *(yes-no)* expects a string of characters as the input parameter. Here is where questions in natural language are passed to the function, as, for example, "are you sure you want to disconnect me, Dave?", or "do you really want to sell your portfolio now?" Just after going inside the function, the message is shown in the session area of the Lisp environment by using the function *(print)*. Then an important conditional structure appears that must be understood perfectly well. The more internal Lisp expression is *(read-line)*. This function, by defect, read characters from the keyboard until the user press the Enter/Return key, returning the string of characters introduced by him. The returned string serves as argument to the function *(upper-case)* that converts every character in the returned string to uppercase. It does not matter if every original character is lower case or upper case, *(upper-case)* will return only strings in upper case.

Then, the resulting string is compared to the one-character string constant "Y". If they are the same string, the function *(yes-no)* returns *true*, otherwise it returns *nil*. Let us try the function at the Lisp prompt with a pair of examples:

> **(yes-no "Are you sure you want to disconnect me, Dave? (y/n): ")**
: *Are you sure you want to disconnect me, Dave? (y/n): y*
: *true*

> **(yes-no "Is it true that recursion is certainly recursive? (y/n): ")**
: *Is it true that recursion is certainly recursive? (y/n); I'm still not sure*
: *nil*

Two remarks must be made now: first, please note that we add the characters *(y/n):* at the end of the question, helping the user to know what to answer at the keyboard, and second, only if the user types *"Y"* or *"y"* at the keyboard the function *(yes-no)* will return *true*. In any other case it will return *nil*.

Now we are ready to design a function named *(bet)* for interactively building a bet for us, that is, it will return a list of the type *(type-of-bet money)*. This function, shown in Code 4-7, does not need any input argument:

```
;code 4-7
(define (bet, type-bet amount)
     (if (yes-no "Are you going to bet for a number? (y/n): ")
     (begin
            (print "Enter a number between 1-36: ")
            (setq type-bet (int (read-line))))
     )
```

```
(begin
       (print   "Enter   type   of   bet   (RED/BLACK,
       EVEN/ODD,.. etc.): ")
       (setq type-bet (sym (upper-case (read-line))))
)
);end if

(print "Enter an amount of money you bet: ")
(setq amount (int (read-line)))
(list type-bet amount)
)
```

Before explaining its working details, let us see how it works at the Lisp prompt in order to see it in perspective:

> (bet)
: Are you going to bet for a number? (y/n): y
: Enter a number between 1-36: 21
: Enter an amount of money you bet: 50
: (21 50)

> (bet)
: Are you going to bet for a number? (y/n): not this night
: Enter type of bet (RED/BLACK, EVEN/ODD,... etc.): red
: Enter an amount of money you bet: 50
: (RED 50)

Observing Code 4-7 we can realize that, basically, the function first checks if the user wishes to place a bet on either a number or another type of bet (red/black and so on) by means of an *if* conditional structure, then asks for the amount of money to bet and finally returns the two elements list containing the bet in the desired format. Now let us comment the details: The expression *(setq type-bet (int (read-line)))* first read characters from the keyboard until the user press the Enter/Return key, as we already know. Then the function *(int)* converts the string of characters into an integer. Here please note that *(int "3.1416")* → *3*, *(int "77")* → *77* and *(int "this can not be converted")* → *nil*.

The expression *(setq type-bet (sym (upper-case (read-line))))* is also interesting. It again takes a string of characters from the user, converts them to upper case and then converts it to a Lisp symbol by means of the function *(sym)*! This is a powerful function in Lisp because it allows us to go from the world of strings to the realm of symbols. For example: *(sym (upper-case "red"))* → *RED*. Needless to say, this conversion ultimately will allow us to find a match between the bet we enter at the keyboard in string format with the list of data stored in the global variable *rewards*. The rest of the function simply ask the user for the amount of money to bet, converts it to an integer and stores it in the symbol *amount*. The last line of the function builds and returns the bet as a list.

4.2.5   *Putting It All Together*

It is time to finish the user interface of our roulette simulator and add the last touches to the program. For this, we shall build a function named *(play)* that will put together the entire Lisp pieces built previously in this chapter. The pseudo-code for this function is as follows:

```
(play)
      (make some initializations)
      (while (you wish to continue playing) ;main event loop begins
        (place your bet)
        (calculate-win)

        (if (the casino wins)
                (inform you and update your win/lose balance)
        (else
                (show the results of the roulette run)
                (if (you lose your bet)
                        (inform you and update your win/lose balance)
                (if (you win your bet)
                        (inform you and update your win/lose balance)
      ));end else, end if (casino wins)
      (ask if you wish to continue playing)
  );end while –main loop-
)
```

Code 4-8 shows the function *(play)* in Lisp code:

```
;code 4-8. Main function of the program
(define (play, balance yn result your-bet)
   (seed (time-of-day)) ;initializes the internal random
   generator
   (setq balance 0)
   (setq continue true)

   (while (= continue true)
      (setq your-bet (bet))
      (setq result (calculate-win your-bet (spin-wheel)))

      (if (= result nil)
         (begin
            (print "Ball has fall on zero. Casino wins\n")
             (setq balance (sub balance (last your-bet)))
```

```
        (print "Your actual balance of money is: "
        balance "\n")
  );end begin

  (begin ;casino has not won
        (print "These are the results of this run: "
        (rest result) "\n")
        (if (= 0 (first result)) ;you loose your bet
        (begin
          (setq balance (sub balance (last your-bet)))
          (print "You lose. Your actual balance of money
        is: " balance "\n")
        );end begin
        (begin ;you won!
          (setq balance (add balance (first result)))
          (print "You win. Your actual balance of money
            is: "
          balance "\n")
        );end begin
        );end if (= 0 (first result))
    );end begin casino has not won
  );en if (= result nil)

  (if (yes-no "\nWould you like to continue playing?
    (y/n): ")
    (setq continue true)
    (setq continue nil) ;else part
  );end if
  );end while
)
```

As it is usual in the main function of a program, no arguments are needed for the function *(play)*. The first line of the function, *(seed (time-of-day))*, seems a bit tricky, so let us see it in detail: We must realize that a computer is anything but a random system, so random numbers used in any program are not strictly random under a pure mathematical viewpoint, so in computer science we use what it is known as "pseudo random numbers"; numbers that, not being purely random, behave almost like real random numbers, that is, they are good enough to develop applications that require true random numbers. In our roulette application, the use of the function *(random)* alone would return a sequence of random numbers that under an extensive set of runs an expert gambler would start to recognize (this phenomenon happens exactly the same on on-line casinos in the Internet). The function *(seed)* is in fact a seed for the internal random generator algorithm that takes an integer argument. Since the function *(time-of-day)* returns the number of milliseconds from the start of the current day at midnight, the aforementioned expression *(seed (time-of-day))* serve us for generating excellent sequences of pseudo random numbers.

Two important variables are initialized, too. The variable *balance* will represent the balance of money you are getting while playing roulette, while the variable *continue* is the key for managing the main event loop. "Main event loops" are the basis of any interactive program: A program starts and then you interact with it through its user-interface until you perform a menu selection, a mouse click or something that breaks the loop, and thus, terminates the program. Initially, the variable *continue* is set to *true*, so your roulette session will continue until this variable becomes *nil*. This is exactly the mission of the looping construction *(while (= continue true)*.

The expressions *(setq your-bet (bet))* and *(setq result (calculate-win your-bet (spin-wheel)))* are easy to understand: The former interactively ask us for our bet, assigning the resulting list to the variable *your-bet*, while the second one spins the roulette, takes the placed bet and assigns the produced list to the variable *result*. Now it is time to see if the result of spinning the wheel returns *nil*, meaning that the roulette ball has fallen on the zero pocket, that is, the casino wins. We check for this event with the expression *(if (= result nil)*. If it evaluates to true, casino wins and we lose our bet. Almost all the rest of the function is dedicated to an *else* situation, that is, the roulette ball has not fallen on the zero pocket, but the program must check if we have won our bet or not.

Now the first thing to do is to show the results of the roulette run by using the expression *(print "These are the results of this run: " (rest result) "\n")*. Note here the sub expression *"\n"*: it instructs NewLisp to be ready to print the next message on a new line. After printing this message, we check the first element in the list stored in the variable *result*. If it equals zero, then it means we have lost our bet. This is made with the expression *(if (= 0 (first result))*. If the expression evaluates to *true* we calculate the actual balance of money you are winning/losing by using the function *(sub)*. We use *(sub)* in order to cover the event you bet on a not integer quantity of money, say, 100.3 currency units. Yes, nobody does that, but it is another example on robust programming. If *(if (= 0 (first result))* evaluates to nil, it means you have won, so we update the balance by using *(add)* and show an adequate message.

All is almost done. The program now asks you if you wish to continue by means of the expression *(if (yes-no "\nWould you like to continue playing? (y/n): ")*. If it evaluates to true, then the program sets the variable *continue* to *true* and the main event loop continues. On the other hand, if it evaluates to nil, then the variable *continue* is set to *nil* and the main event loop and the program terminates because the *while* expression *(while (= continue true)* evaluates to *nil*. Let us try the program with a short casino session:

> **(play)**
Are you going to bet for a number? (y/n): n
Enter type of bet (RED/BLACK, EVEN/ODD,... etc.): red
Enter an amount of money you bet: 10

These are the results of this run: (25 RED ODD PASSE THIRD-DOZEN FIRST-COLUMN)
You win. Your actual balance of money is: 10

Would you like to continue playing? (y/n): y
Are you going to bet for a number? (y/n): n
Enter type of bet (RED/BLACK, EVEN/ODD,... etc.): red
Enter an amount of money you bet: 20
These are the results of this run: (31 BLACK ODD PASSE THIRD-DOZEN FIRST-COLUMN)
You lose. Your actual balance of money is: -10

Would you like to continue playing? (y/n): y
Are you going to bet for a number? (y/n): n
Enter type of bet (RED/BLACK, EVEN/ODD,... etc.): red
Enter an amount of money you bet: 40
These are the results of this run: (8 BLACK EVEN MANQUE FIRST-DOZEN SECOND-COLUMN)
You lose. Your actual balance of money is: -50

Would you like to continue playing? (y/n): y
Are you going to bet for a number? (y/n): n
Enter type of bet (RED/BLACK, EVEN/ODD,... etc.): red
Enter an amount of money you bet: 80
These are the results of this run: (17 BLACK ODD MANQUE SECOND-DOZEN SECOND-COLUMN)
You lose. Your actual balance of money is: -130

Would you like to continue playing? (y/n): y
Are you going to bet for a number? (y/n): n
Enter type of bet (RED/BLACK, EVEN/ODD,... etc.): red
Enter an amount of money you bet: 160
These are the results of this run: (36 RED EVEN PASSE THIRD-DOZEN THIRD-COLUMN)
You win. Your actual balance of money is: 30

Would you like to continue playing? (y/n): no more today
nil

With this simple roulette simulation you can try many playing strategies in order to check what would be the outcome of a night at the casino. There is only a winning strategy: playing with quasi-infinite quantities of money. This is the reason casinos admit bets up to a top fixed amount of money.

About the Lisp program itself, it must be noted how much resources it takes to implement a user interface. You only need to compare the function *(bet)* with simply writing a list at the Lisp prompt. If we moreover realize that we have barely implemented robust constructions in the code in order to save space in the book, you can imagine the time and quantity of code it takes for building a well-behaved, friendly program. In fact, a programmer always dedicates a good quantity of time in order to build a library for managing and design good user interfaces. In the next section of this chapter we shall develop a set of functions for simple database processing without an user interface. The goal is to create an extension of Lisp for managing CSV type databases.

4.3   Messier Was a French Astronomer

Back in century XVIII, one of the most important branches of astronomy was comet observing. For an astronomer, discovering a new comet was a guarantee of becoming famous, getting new incomes and obtaining a better position in his career. Charles Messier (1730–1817) was a French astronomer that soon realized that some celestial objects in the night skies seemed at first comets but in fact they were not, because these look-like comets remained fixed in the sky through time, while real comets showed a change of position month by month or even day by day in some cases. Messier decided to compile a catalogue of "no-comets" in order to help to save time to other astronomers, avoiding them wasted observing time. The compilation of such a catalogue, published in 1774, gave to Charles Messier a fixed site in the history of Astronomy.

The Messier Catalogue consists of a set of 110 celestial objects that can be observed in the Northern hemisphere. Since they are in a fact a collection of galaxies, nebulae, globular clusters and, in general, deep-sky objects, the apparent position of these celestial objects remains "fixed" and can be observed nowadays with a good pair of binoculars under dark skies. Here "dark skies" means an observing place without light pollution, an observing place where at night you are not able to see your own feet. Messier used, seen from the current state of our technology, archaic telescopes, but the unpolluted skies of Paris in mid 1700s was a terrific observing advantage.

The complete Messier Catalogue can be easily found on the Internet, and you can also download it from the book's web site (http://www.fuzzylisp.com) in CSV format. CSV is an acronym for "Comma Separated Values" and is a data file structure in text format used by many programs as a neutral way to exchange data with other programs. "In text format" means that the entire file is formed by text characters only, without binary elements. As a result, any CSV file can be opened with standard programs such as Notepad in Windows, or TextEdit in OSX. In every CSV file, each line represents a record of the database, with the first line (usually) describing the names of the fields.

The Messier.csv file is composed by 111 lines, that is, 110 records for the celestial objects plus one header line for the name of the fields. The Lisp functions we are going to design are not specific for the Messier Catalogue. They are valid for

Table 4.2 Name of fields for the CSV Messier database

Field's name	Meaning
Name	Name of the object in the Catalogue
NGC	Reference in the new general catalogue, NGC
Constellation	Constellation where the object is located
Class	Type of object
Right ascension	First celestial coordinate
Declination	Second celestial coordinate
Magnitude	Measure of apparent bright
Angular size	Apparent size in the sky
Burnham	Comment in Burnham's celestial guidebook
Remarks	Additional comments

any CSV file where its first line represents the name of the fields in the database. That means that if you keep any database in, for example, an Excel file, you will be able to use that information with the Lisp extension to the language we are going to implement. The name of the fields of the database stored on the file Messier.csv are shown in Table 4.2 (Dreyer 1888; Burnham 1978).

After having finished the Lisp functions for simple CSV database management, any database session at the Lisp prompt will be able to manage the following tasks:

- Opening a database from hard-disk or alternative source
- Load it into computer's memory
- Perform queries
- Add fields to the database
- Supply new info
- Perform database filtering
- Get simple statistics
- Save changes to disk

More specifically, the name of the functions and their intended use are given in Table 4.3.

Table 4.3 Functions designed for CSV database management

Function name	Use
(db-load)	Loads a database into memory
(db-fields)	Returns the name of the fields in a database
(db-tell)	Returns a record from the database
(db-tell2)	Returns the value of a field from a record
(db-new-field)	Adds a new field to a database
(db-update)	Updates a database with new info into a field from a record
(db-filter)	Filters a database
(db-stats)	Calculates simple statistics from a numerical field
(db-save)	Saves a database to a storage device

4.3.1   Opening and Loading Databases in CSV Format

Files usually reside in hard disks, CDs, DVDs, solid-state memory devices or in the Internet. In order to open them we shall have a dedicated function, named *(db-open)* that after opening the file will load its content into the computer's memory. Such a function is shown in Code 4-9:

```
;code 4-9
(define (db-load, f1 str lst)
     (setq lst '())
     (setq f1 (open "/Users/research/Desktop/Messier.csv"
     "read"))
     (while (setq str (read-line f1))
            (setq lst (cons(parse str ",") lst))
     );while
     (close f1)
     (reverse lst)
)
```

As can be seen, the first line creates a new empty list, *lst*, where we shall load the file content. The second line is more interesting because it uses the standard Lisp function for opening files: *(open)*. This function uses two arguments: the first one is the name of the file, including the directory (folder) where the file is located. In this example, the file resides in the Desktop, and the complete path is expressed by the string *"Users/research/Desktop/Messier.csv"*. The second argument means the mode we use to open a file. Since we are going to load the file into memory, that is, no writing action is needed, we use the mode *"read"*. NewLisp supports several access modes for opening a file, as shown in Table 4.4.

The function *(open)* returns either an integer if the opening process is successful, or *nil*, if it fails, for example, if the file does not exist or is located on another folder. This integer becomes hereafter a handler for subsequent reading or writing operations on the file. It is usual to employ the symbols *f1, f2, ..., fn* for naming these file handlers in any computer language, so we shall adhere to this convention in this book.

Table 4.4 Access modes for opening a file

File access mode	Meaning
"read" or "r"	Read only access
"write" or "w"	Write only access
"update" or "u"	Read/write access
"append" or "a"	Append read/write access

After opening the file, we use a looping structure in order to read every line in the file and load the content on the *lst* list. We already know the function *(read-line)* from its use in our user-defined function *(yes-no)*. There, we used the keyboard as input. Since now the input comes from a file we need and additional parameter for telling Lisp to pay attention to the computer's file system. This parameter is the file handler, in this case, *f1*. The expression *(setq str (read-line f1))* reads a line from the file and stores it into the variable *str*. When no more lines remain to be read the function *(read)* returns *nil*, so the while loop ends. The body of the while loop contains an expression that deserves a detailed explanation.

Let us pay attention to the more internal function inside the *while* loop, *(parse)*. As its name implies, this function parses a string of characters, returning the individual tokens contained in the string as a list of string elements. It takes two arguments: first the original string of characters and then the used delimiter for separating the elements. Needless to say, in a CSV file the used delimiter by defect is a comma sign. Let us see an example. The first line of the Messier.csv file, after being stored in the variable *str*, has the following appearance:

"Name, NGC, Constellation, Class, Right ascension,
Declination, Magnitude, Angular size, Burnham, Remarks"

Then, by means of the expression *(parse str ",")*, it converts into this Lisp expression:

("Name" "NGC" "Constellation" "Class" "Right ascension" "Declination"
"Magnitude" "Angular size" "Burnham" "Remarks")

And this list, by means of the function *(cons)* is additively stored on the list *lst*. After exiting the *while* loop, the entire list, now containing the complete database, is reversed before becoming the returned value of the function *(db-load)*. Let us call it from the Lisp prompt:

> (setq messier (db-load))
: (("Name" "NGC" "Constellation" "Class" "Right ascension" "Declination"
"Magnitude" "Angular size" "Burnham" "Remarks")

...
("M110" "205" "Andromeda" "Elliptical galaxy" "0h 40.4m" "41d 41m"
"9.4" "8.0x3.0" "" "E6 - companion of M31"))

The sublists corresponding from celestial objects M1 to M109 have been omitted in order to allow a shorter output in this text. The important detail now is that we have stored the complete database in the symbol *messier*. All the future calls to any of the functions designed for CSV database management will use this symbol as an argument. Now our simple Lisp database system converts into a question of list manipulation, so we should feel ourselves at home.

4.3.2 Querying the Database

The simplest query to our database comes from the already known Lisp function *(nth)*. If we wish to retrieve the record number indexed by *i* in the database we only need to type *(nth i database)* at the Lisp prompt. For M42, the famous Orion's Nebula, we only need to type:

> **> (nth 42 messier)**
> *: ("M42" "1976" "Orion" "Diffuse nebula" "5h 35.4m" "-5d 27m" "2.9" "66x60" "!!!" "Great Orion Nebula")*

Especially important is the first record in the database because it tells us the name of its fields:

> **> (nth 0 messier)**
> *: ("Name" "NGC" "Constellation" "Class" "Right ascension" "Declination" "Magnitude" "Angular size" "Burnham" "Remarks")*

In an interactive database session at the Lisp prompt, this last expression is of such importance that it deserves, for coherence with the name of the rest of functions, to create a simple function for better representing it. Such a function, named *(db-fields)*, is shown in Code 4-10:

```
;code 4-10
(define (db-fields lst)
    (nth 0 lst)
)
```

Needless to say:

> **> (db-fields messier)**
> *: ("Name" "NGC" "Constellation" "Class" "Right ascension" "Declination" "Magnitude" "Angular size" "Burnham" "Remarks")*

Things become a bit more complex when we wish to perform more sophisticated queries. In order to make things easier, we shall use the first field of the database as a key for the database. In our case, the key field is the name of the Messier object. For your personal databases, you could, for example, use the social security number of a person, driving number license, etc. as a key. The function *(db-tell)* allows us to perform simple queries to our database. Code 4-11 shows it:

```
;code 4-11
(define (db-tell db id, i l aux-lst result)
    (setq i 0)
    (setq l (length db))
    (while (< i l)
```

```
                  (setq aux-lst (nth i db))
                      (if (= id (first aux-lst))
                            (setq result aux-lst)
                      )
                      (setq i (+ 1 i))
      );end while
      result
)
```

After initializing the index *i*, the function calculates the length of the database by means of the expression *(setq l (length db))* and then it enters a *while* loop in order to traverse the entire database. For each loop pass, every record is assigned to the variable *aux-lst*. As can be seen, inside the *while* loop there is a conditional expression that compares the content of the first field in a record with the supplied argument *id*. If a match is found, then the actual record is assigned to the variable *result* and the matching record is returned. If no match is found the function returns *nil*. Note that the supplied argument *id* must be a unique key value in the database, else the last keyed record would be returned by the function. This is not a problem with well-designed databases, where the first field has always unique values, as for example, car plates in a car database, or in our example, the Messier object identifier.

Trying it at the Lisp prompt:

> **(db-tell messier "M42")**
 : ("M42" "1976" "Orion" "Diffuse nebula" "5h 35.4m" "-5d 27m" "2.9" "66x60" "!!!" "Great Orion Nebula")

The output is exactly the same as the produced previously with the expression *(nth 42 messier)*, but you must note that typing *(db-tell messier "M42")* you don't know beforehand the position of M42 in the database. Imagine again you have a CSV database of cars where the first field is, aside the car plate, a key field. Then typing, for example *(db-tell car-database "GWG0560")* would retrieve all the stored data for a car with that plate, without regards at its position in the database.

Another query function in our Lisp language extension for CSV database management, *(db-tell2)*, allows us to obtain the value corresponding to a particular record and a particular field at the same time. In other words, if records are ordered by lines and fields are ordered by columns (as it happens in spreadsheet programs), the result of the function *(db-tell2)* is the value stored in the intersection of a particular record and field. The function can be seen in Code 4-12:

```
;code 4-12
(define (db-tell2 db id a-field, i l aux-lst result position)
    (setq i 0)
    (setq l (length db))
```

```
(while (< i l)
      (setq aux-lst (nth i db))
              (if (= id (first aux-lst))
                   (setq result aux-lst)
              )
              (setq i (+ 1 i))
);end while
(setq position (find a-field (nth 0 db)))
(nth position result);returns the database value
)
```

The code starts exactly the same as the written one for *(db-tell)* until obtaining the desired record stored in the variable *result*. Then the expression *(setq position (find a-field (nth 0 db)))* returns the position of the argument *a-field* in the list of fields given by *(nth 0 db)*. After storing this value in the variable position, obtaining the desired value is straightforward using the expression *(nth position result)*. Let us see how does it work:

> **(db-tell2 messier "M45" "Constellation")**
: *"Taurus"*

> **(db-tell2 messier "M42" "Remarks")**
: *"Great Orion Nebula"*

4.3.3   Updating the Database

Aside querying a database, it is common to manipulate the data it contains. One of the most usual activities in database management is the addition of a new field. Let us imagine we wish to add the field *"My own observations"* to the Messier database. The function *(db-new-field)* performs this task by means of the Lisp expressions shown in Code 4-13:

```
;code 4-13
(define (db-new-field db name-field, i l lst-out)
     (setq lst-out '())
     (setq l (length db));number of records in lst

     ;in the following line we append the new field's name
     (setq lst-out (cons (append (nth 0 db) (list name-field))
     lst-out))
```

```
;then we copy the entire database, making space in
every record
;for the new field

(setq i 1)
(while (< i l)
  (setq lst-out (cons (append (nth i db) (list ""))
  lst-out))
  (setq i (+ 1 i))
);while
(reverse lst-out)
)
```

The used strategy here consists first in appending the name of the new field at the rightmost position and then to traverse the entire database, record by record, creating new space in each record. Since all the new values are empty by default, we only need to append an empty string *""* at the end of each record. The function starts initializing the list *lst-out* to an empty list. In this list we shall copy the actual contents of the database, adding the required new data to it. The expression *(setq lst-out (cons (append (nth 0 db) (list name-field)) lst-out)* seem complex at first but it is not. It takes the first record in the database, that is, the record where the name of all the fields of the database are stored *(nth 0 db)*, and then appends *name-field* to it. Since this function argument will receive a string, we convert it to a list for making it possible to use the function *(append)*. Then we *cons* it to the list *lst-out*.

Now we only need to traverse the entire database using a *while* looping structure where we append the empty data *""* to every record by using the expression *(setq lst-out (cons append (nth i db) (list "")) lst-out))*. After exiting the *while* loop we reverse *lst-out* and we are done. Let us try the function:

> **(setq messier (db-new-field messier "My observations"))**
  : *(("Name" "NGC" "Constellation" "Class" "Right ascension" "Declination" "Magnitude" "Angular size" "Burnham" "Remarks" "My observations")*
  .....
  ("M110" "205" "Andromeda" "Elliptical galaxy" "0h 40.4m" "41d 41m" "9.4" "8.0x3.0" "" "E6 - companion of M31" ""))

Please note that the field *"My observations"* has already been added. From the last record, see also how an adequate empty string has been added at the rightmost position in the list.

4.3.4 Modifying the Database

No joy can be had from a database manager software if we are not able to supply new information to it, or modify its content. The function *(db-update)* will take care of this need. This is the most complex function in this section of the chapter, so we shall use the well-known "divide and conquer" strategy for developing it in order to break the required tasks into auxiliary functions. We need two auxiliary functions named respectively *(db-get-row-number)* and *(db-update-record)*. The former is shown in Code 4-14a:

```
;code 4-14a
(define (db-get-row-number db key, i length1 row-number)
     (setq i 0)
     (setq length1 (length db))

     (while (< i length1)
              (if (= key (first (nth i db)))
                    (setq row-number i)
              )
              (setq i (+ 1 i))
     );while end
     row-number
)
```

This function takes two arguments: first the name of the database and then the first field value that will act as a key field. It will return a number representing the position of the record in the database. As previously, in our case the key field is the name of the Messier object. After initializing the variable *i* and calculating the number of records in the database, we traverse it with a *while* loop. Inside the loop we use the following conditional expression: *(if (= key (first (nth i db))) (setq row-number i))*. The expression *(nth i db)* extracts a record and then we obtain its first element. If it is equal to the supplied function argument *key*, then we mark its position in the variable *row-number*. The function returns this number. Trying it at the Lisp prompt:

> **(db-get-row-number messier "M42")**
: 42

Please note that if the key value were not unique, then *(db-get-row-number)* would return the number corresponding to the last record in the database matching that value. Needless to say, this can only happen if the database design is poor, with a key field containing multiple instances of a particular value.

The other auxiliary function has the name *(db-update-record)*, and, as its name implies, modifies a single record. It requires several arguments: the symbol for the database itself, the record to update, the name of the field, and the data that will be supplied for filling or updating the field. The expressions for the function are shown in Code 4-14b:

```
;code 4-14b.
;This is an auxiliary function to (db-update).
```

```
(define (db-update-record db record field new-data, position
                                    i l record-out)
    (setq record-out '())
    (setq i 0)
    (setq l (length record))
    ;gets the position of the field looking in the first
    record:
    (setq position (find field (nth 0 db)))
    ;replaces the data:

    (while (< i l)
      (if (= i position)
          ;if it evaluates to true
            (setq record-out (cons new-data record-out))
          ;else copy the element
          (setq  record-out  (cons  (nth  i  record)
          record-out))
      );end if
      (setq i (+ 1 i))
    )
    (reverse record-out)
)
```

The list *record-out* will be the list returned by the function, so it is initialized in the first line. After initializing also an indexing variable *i*, we calculate the number of elements (fields) in the record and assign it to the variable *l*. Then we get the position of the required field, looking for it in the first record of the database and then we go inside a *while* loop. There we traverse the record: when we are located on the desired field, we update it using the expression *(setq record-out (cons new-data record-out))*, else we simply copy the existing information using the Lisp expression *(setq record-out (cons (nth i record) record-out))*. After exiting the while loop, we reverse the list *record-out* and we are done. Let us try the function:

> (db-update-record messier (nth 42 messier) "Remarks" "great sight with the Zeiss telescope")

: ("M42" "1976" "Orion" "Diffuse nebula" "5h 35.4m" "-5d 27m" "2.9" "66x60" "!!!" "great sight with the Zeiss telescope")

After having ready for action the functions *(db-get-row-number)* and *(db-update-record)* we now can write the desired function *(db-update)* more easily, as shown in Code 4-14c:

```
;code 4-14c
(define (db-update db key field new-data, row-number i
length1 db-out)
```

```
(setq row-number (db-get-row-number db key)) ;get the
row index
(setq i 0)
(setq db-out '())
(setq length1 (length db)) ;number of records in db

;copy the first set of records
(while (< i row-number)
        (setq db-out (cons (nth i db) db-out))
        (setq i (+ 1 i))
)
;update the record:

(setq   db-out   (cons   (db-update-record   db   (nth
row-number db) field new-data) db-out))
(setq i (+ 1 i)); advances one row

;copy the rest of records
(while (< i length1)
        (setq db-out (cons (nth i db) db-out))
        (setq i (+ 1 i))
)
(reverse db-out)
)
```

We must realize that *(db-update)* needs four arguments: The symbol for the database, the key field value, that is, the value located in the first field of the record where we want to make modifications, the field's name for telling the function what value to change and finally, the new value. Let us see a call to the function first before commenting it. Our aim is to change the Remarks for M1, the Crab Nebula. The original value in the database can be seen easily:

> (db-tell2 messier "M1" "Remarks")
"SNR (1054) - Crab Nebula"

> (setq messier (db-update messier "M1" "Remarks" "A Supernova remnant"))
: (("Name" "NGC" "Constellation" "Class" "Right ascension" "Declination" "Magnitude" "Angular size" "Burnham" "Remarks")

...
("M110" "205" "Andromeda" "Elliptical galaxy" "0h 40.4m" "41d 41m" "9.4" "8.0x3.0" "" "E6 - companion of M31"))

Now, let us check the first record (M1):

> > (db-tell2 messier "M1" "Remarks")
: "A Supernova remnant"

The function *(db-update)* starts with a call to *(db-get-row-number)*, assigning its returning value to the variable *row-number*. After that, we initialize some variables and count the number of records in the database. Then we enter inside a while loop where we simply copy record by record all the records in the database until arriving to the record marked by *row-number*. In this moment we use the following Lisp expression:

(setq db-out (cons (db-update-record db (nth row-number db) field new-data) db-out))

As can be seen, the key point in the expression is a call to the auxiliary function *(db-update-record)*, explained some paragraphs above, *consing* its return value to *db-out*, the target list where we are copying the original database. After Lisp evaluates this expression, the new information, stored in *new-data*, updates the record.

Since the record is updated and not copied, we add an important programming detail in this moment: *(setq i (+ i 1))* that advances a record. Hereafter we enter a second *while* loop and continue copying records until the entire database is updated. Finally, we reverse *lst-out* and the function finishes.

4.3.5   Filtering the Database

Another usual feature of database managing systems is the ability to apply filters to the records in a database. The filter is in fact a logical expression that, applied to the database, filters all the records that match the logical expression. The records that do not match the logical condition are filtered out. The function *(db-filter)* is shown in Code 4-15:

```
;code 4-15
(define (db-filter db expr, i,
                length1 header record field-id lst-out str)
    (setq i 1)
    (setq lst-out '())
    (setq lst-out (cons (nth 0 db) lst-out))
    (setq length1 (length db))
    (setq header (first db));name of fields
    (setq field-id (find (nth 1 expr) header));field index

    (while (< i length1)
        (setq record (nth i db));current record
        (setq str (nth field-id record))
```

```
          ;if (last expr) is a number, convert str into
          number:
          (if (number? (last expr))
             (setq str (float str))
          );if

          (if (eval (list (eval '(first expr)) str (last
          expr)))
             (setq lst-out (cons record lst-out))
          );if
          (setq i (+ 1 i))
       );while
       (reverse lst-out)
    )
```

The structure of the function is similar to other *(db*)* functions already seen, so we shall comment only the essential expressions. The strategy here is to traverse the entire database. If a record matches the filter then it is copied to the output list, otherwise it is not. The most interesting argument in the function is *expr*. This symbol is used to pass a logical Lisp expression to the function. Examples of valid logical expressions are *(= "Constellation" "Orion")*, or *(<= "Magnitude" 4)*. Note that every logical expression has the following pattern *(relational-operator field value)*. Inside the function we shall need to evaluate them adequately. Let us see the processes in detail:

The field index is easily obtained by means of *(setq field-id (find (nth 1 expr) header))* where *header* is the first record in the database, containing the names of all the fields. After entering the *while* loop we get a record by using *(setq record (nth i db))* and then the value of the field of interest in the variable *str*. Now we examine the last element of the logical expression *expr*: If it is a number then we convert the string stored in *str* to a number in order to match data types. Now we are ready to evaluate the filter on the actual record with the following Lisp expression:

(eval (list (eval '(first expr)) str (last expr)))

Let us imagine we have passed the expression *'(= "Constellation" "Orion")* to the function. If the current record is, for example, the corresponding one to M42, then *(first expr)* → = , *str* → *"Orion"* and *(last expr)* → *"Orion"*, and *(list (eval '(first expr)) str (last expr))* produces the list *(= "Orion" "Orion")*. Here, both the *quote* sign and the function *(eval)* have worked with exquisite caution for protecting and evaluating symbols in such a way that we have transformed the original argument value *(= "Constellation" "Orion")* into the corresponding to the current record *(= "Orion" "Orion")*. Now we can evaluate this expression again with *(eval)*, thus producing either a *true* or a *nil* value that is passed to the conditional *if*. If the expression evaluates to *true*, then the record is filtered in and *consed* to *lst-out*. Now, as usual, the *while* structure terminates and the resulting filtered database is reversed before finishing the function.

And additional comment must be taken into account. The first expression in the function is *(setq i 1)* instead of the usual *(nth i 0)* initialization, and the third expression is *(setq lst-out (cons (nth 0 db) lst-out))*. This is due to the need of maintaining the first record in the resulting database, that is, the header of the database is never filtered out. Why? Because the list resulting from using the function *(db-filter)* is in fact a sub-database itself. Let us see it in action at the Lisp prompt:

> **(db-filter messier '(= "Constellation" "Orion"))**

: (("Name" "NGC" "Constellation" "Class" "Right ascension" "Declination" "Magnitude" "Angular size" "Burnham" "Remarks")

("M42" "1976" "Orion" "Diffuse nebula" "5h 35.4m" "-5d 27m" "2.9" "66x60" "!!!"

"Great Orion Nebula")

("M43" "1982" "Orion" "Diffuse nebula" "5h 35.6m" "-5d 16m" "6.9" "20x15" "" "detached portion of M42")*

("M78" "2068" "Orion" "Diffuse nebula" "5h 46.7m" "0d 3m" "10.5" "8x6" "" "spec: B8 - comet shaped"))*

This function call produces a sub-database containing all the Messier objects contained in the Orion constellation. Now let us filter the brighter objects in the Messier Catalogue, that is, those object with apparent magnitude equal or less than 3:

> **(db-filter messier '(<= "Magnitude" 3))**

: (("Name" "NGC" "Constellation" "Class" "Right ascension" "Declination" "Magnitude" "Angular size" "Burnham" "Remarks")

("M42" "1976" "Orion" "Diffuse nebula" "5h 35.4m" "-5d 27m" "2.9" "66x60" "!!!"

"Great Orion Nebula")

("M45" "" "Taurus" "Open cluster" "3h 47m" "24d 7m" "1.2" "110" "" "number: 130 - The Pleiades"))*

Effectively, the two brighter Messier objects are M45 in Taurus and M42 in Orion.

4.3.6   Performing Simple Statistics

Databases usually contain numeric fields and sometimes it is useful to get some basic statistics from these collections of data. The following function, *(db-stats)* uses the NewLisp function *(stats)* in order to obtain statistical data such as mean, standard, deviation, etc. It is assumed that the contents of the field are strings representing numbers. The Lisp expressions for building the function are shown in Code 4-16:

```
;code 4-16
(define (db-stats db field, header field-id i lst length1)
     (setq header (first db));get name of fields
     (setq field-id (find field header));find position of field
     (setq lst '())
     (setq i 1)
     (setq length1 (length db))

     (while (< i length1)
        (setq lst (cons (eval-string (nth field-id (nth i
        db))) lst))
        (setq i (+ 1 i))
     );while
     (stats (reverse lst)) ;return the statistics
)
```

The function takes two arguments, the symbol for the database and the name of
the field that contains the numerical data. As usual, the first lines of the function get
the name of the fields in the database, the position of the field and then a *while* loop
processes the data. Inside the function, the most interesting expression is the following one:

$$(setq\ lst\ (cons\ (eval\text{-}string\ (nth\ field\text{-}id\ (nth\ i\ db)))\ lst))$$

It first get a record, then gets the string value stored in the field, converts the
string to a number and then *conses* it in the list *lst*. After finishing the *while* loop
this list contains all the numerical data in the field. The last line reverses the list. Let
us try the function with the only field in the Messier database that represents
numerical values, "Magnitude":

> (db-stats messier "Magnitude")
 : *(110 7.534545455 1.540495868 1.914715413 3.666135113 -0.4639868813
0.3590037872)*

As we saw in Chap. 2 (see Table 2.1), the list returned by *(db-stats)* is composed
by the number of values, its mean, the average deviation from the mean value, the
standard deviation, the variance, the skew and the kurtosis.

4.3.7 Saving the Database

In a database session, an user will maybe perform only query actions, but if he or
she makes modifications to the database, sooner or later will wish to save the
modifications to a file in order to make the changes more or less permanent (no
storage device, as of this writing, is able to guarantee data survival for, let us say,

one hundred years). The function *(db-save)*, shown in Code 4-17, performs this task:

```
;code 4-17
(define (db-save db, i j length2 lenght2 record)
    (setq  f1   (open  "/Users/research/Desktop/Out.csv"
    "write"))
    (println "saving database...")
    (setq i 0)
    (setq length1 (length db)) ;number of records in db

    (while (< i length1)
            (setq record (nth i db)) ;loads a record
            (setq length2 (length record)) ;fields in record
            (setq j 0)

            ;now, for each record:
            (while (< j length2)
                    (write f1 (nth j record))
                            (if (< j (- length2 1))
                                    (write f1 ",")
                            );end if
                            (setq j (+ 1 j))
            );internal while

            (write f1 "\n")
            (setq i (+ 1 i))
    );end while i
    (close f1)
    (println "Database saved Ok")
    (println i " records successfully saved.")
    true
)
```

Initially, the function opens the file *Out.csv* in write mode in the computer's desktop. The parameter *"write"* means that the file is opened for writing procedures: If the file does not exist, the call to the function *(open)* will create it. If the file already exists, opening the file in this mode will overwrite it. After measuring the number of records in the database we enter a first *while* looping structure that retrieves a record in each loop pass by means of the expression *(setq record (nth i db))* and then measures the number of elements (fields) in the actual record.

The internal *while* structure processes the actual record in the following way: first writes an element to the file identified by the file handler *f1* by means of the Lisp expression *(write f1 (nth j record))*. Then, the most interesting point in the function happens: If we still have not reached the end of the record, then we write a comma sign to the output file and we continue writing elements. When we arrive at the end of the record, the internal while loop terminates and then we write a new-line code and a carriage-return code using the expression *(write f1 "\n")*. After all the records are saved, we close the file and print two messages for the user *"Database saved Ok"* and *"x records successfully saved"*, where *x* is the number of saved records in the database, including the first record containing the number of fields.

4.4   A Word on Function Design for This Chapter

In all the functions developed in this chapter we have not used the most pure Lisp style, that is, we have not been worried about generating the most academic, beautiful Lisp code. In fact we have not used some functions that would have made things easier for us because our aim has been to show how many things can be done with a relatively small number of kern Lisp functions and to give a helping hand to those persons used to other programming languages such as C or Pascal. As an example, we have not used the function *(assoc)*, which would have helped us a lot in shortening the code for the *(db*)* functions. As an example, code 4-18a and 4-18b shows an alternative design for the functions *(db-tell)* and *(db-tell2)*, respectively:

```
;code 4-18a, a shorter, more elegant version than code 4-11
(define (ddb-tell db id)
    (assoc id db)
)

;code 4-18b, a shorter, more elegant version than code 4-12
(define (ddb-tell2 db id a-field)
    (nth (find a-field (first db)) (assoc id db))
)
```

We have not used recursion, neither, and that would have been an excellent approach for designing our *(db*)* functions because almost all of them traverse lists, and with recursion and the functions *(first)* and *(rest)* we would have written pure Lisp poetry. We have opted instead for looping structures. Probably not a very romantic style of Lisp, but <u>practical</u> for those with previous knowledge in classical programming languages.

4.5   As a Summary

This chapter has shown two approaches for developing applications in Lisp. The first one, a French roulette simulation game, follows the traditional way of programming of other functional languages: It builds a set of functions and then a main function calls them adequately from the inside of a main event loop. The second one, a library of CSV database functions, shows the flexibility of Lisp: Every function is relatively independent and all of them can be called from the Lisp prompt in an interactive manner.

We have learnt how the functions *(read-line)* and *(print)* are the basic tools in Lisp for Input/Output activities in a program, either for interacting with the keyboard and computer screen or for reading and writing text-type files. The functions *(open)* and *(close)* allows the Lisp user to communicate to the computer's operating system for creating, accessing, and closing files.

We have developed a simple user interface for the Roulette application and we have observed how many lines of code must be devoted to it. We have realized also that the more sophisticated the user interface, the more time and resources it requires for developing it. The database library for managing CSV databases has not an user interface, but code 4-19 suggest a way to implement a simple one:

```
;code 4-19, a simple menu
(define (db-menu)
     (print "\nCSV database manager\n\n")
     (print "1. Load a database into memory\n")
     (print "2. Get the name of fields in a database\n")
     (print "3. Query the database\n")
     (print "4. Add a new field to the database\n")
     (print "5. Update the database\n")
     (print "6. Filter the database\n")
     (print "7. Calculate statistics\n")
     (print "8. Save the database\n")
     (print "\nChoose an option (1-8): ")
     (read-line)
)
```

trying the function *(db-menu)* at the Lisp prompt:

> (db-menu)
:CSV database manager

1. *Load a database into memory*
2. *Get the name of fields in a database*
3. *Query the database*
4. *Add a new field to the database*
5. *Update the database*
6. *Filter the database*
7. *Calculate statistics*
8. *Save the database*

Choose an option (1-8): 1

In this case, the function *(db-menu)* would return *"1"* and hereafter the Lisp user would only need to link the menu options to the developed *(db*)* functions, adding some interface-user touches here and there.

The set of database functions developed in the second part of this chapter is not extensive. Even so, it has the basic features of a database system. Since this set of functions is in fact an extension to the language itself, it offers lots of possibilities for managing data stored in CSV files because all the functions can be directly used from the language. For example, you can open several databases in a single Lisp session, you can get information from a database, transfer it to other, filter the obtained database with some criteria and then save the obtained new (sub)database to an storage device. You can do these things interactively at the Lisp prompt or you can write a new application using the *(db*)* functions as building blocks for even more sophisticated data processing applications.

FuzzyLisp follows the same philosophy of the *(db*)* library shown in this chapter. It is a set of Lisp functions that extends the language itself. We shall describe it in the next chapters of this book after exposing the basics on fuzzy-sets theory.

References

Burnham, R.: Burnham's Celestial Guidebook. Dover Publications (1978)
Dreyer, J.L.: A New General Catalogue of Nebulae and Clusters of Stars, being the Catalogue of the late Sir John F.W. Herschel, Bart., revised, corrected, and enlargued. Memoirs of the Royal Astronomical Society (1888)

Chapter 5
From Crisp Sets to Fuzzy Sets

5.1 Introduction

We are now starting the second part of this book. After the introduction to the Lisp programming language in part I, you are now ready to begin the phase of our travel that will require the most of you. Not because it is complex, but because it represents a shift of paradigm not only in the way we think about sets, but also in the form that we usually reason and analyze things. We shall start revisiting the essentials of classic sets theory, including the concepts of belonging to a set, union and intersection of sets, general properties of sets and the concepts of Cartesian Product and Relations between sets.

Hereafter, we shall leave the traditional way of thinking in sets theory and will start to introduce fuzzy sets. It will not be an abrupt quantum leap, but a smooth pathway towards the new paradigm. Here, the exposed material on crisp sets will be helpful for establishing a contrast between the two worlds, and the Lisp code from this chapter will help you not only as a pedagogical tool, but also to build your own fuzzy sets, that is, to experiment with Lisp expressions in your own area of expertise under the viewpoint of the new theory.

5.2 A Review of Crisp (Classical) Sets

A scientific theory does not appear suddenly. The development of a new scientific paradigm usually follows several phases. At a given time one or more scientists develop some ideas and soon they exchange them and derive new knowledge. It may happen that no clear, defined theory is immediately available, but the scientific community quickly recognizes that something new is in the air, creating an exciting scientific ambience. Later, more people is attracted by the fresh concepts and the theory gains momentum until arriving a point in time where the overall concepts are

© Springer International Publishing Switzerland 2016
L. Argüelles Méndez, *A Practical Introduction to Fuzzy Logic using LISP*,
Studies in Fuzziness and Soft Computing 327,
DOI 10.1007/978-3-319-23186-0_5

distilled into a new scientific frame. Sets theory is an example of this evolution in human thinking, appearing in the second half of century XIX in Germany, mainly from the works of Georg Cantor, Richard Dedekind and Ernst Zermelo, becoming a recognized branch of mathematical logic from 1915, approximately.

5.2.1   Definition of Sets and the Concept of Belonging

The definition of a set is in some way difficult because it is in itself an extremely simple concept. A set is a collection of objects, things, or put into a more formal mathematical language, elements. When put together, physically or conceptually, a collection of things becomes a set. As an example we can mention the set of odd numbers from a roulette, the set of Galilean Moons, the set of Messier objects, etc. In fact, we can describe a set by means of three ways:

1. By simple enumeration where you describe, one by one, the name of all the elements in a set. For example the set S of Galilean Moons can be mathematically represented as follows:

$$S = \{Callisto, Ganymede, Europa, Io\}$$

2. Conceptually, by means of an established condition:

$$S = \{x \mid x \text{ is one of the four biggest moons of Jupiter}\}$$

   This type of expressions can be read as "x such as x is … (condition)". In this case, (condition) equals to "one of the biggest moon of Jupiter".
3. By means of a Venn diagram:
   A Venn diagram, as shown in Fig. 5.1, consists in a circle or ellipse containing the elements of a set, becoming an excellent graphical method for representing them.

Fig. 5.1 A venn diagram representing the set S of Galilean Moons

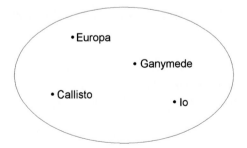

4. Additionally, in this book we are going also to represent sets by means of Lisp expressions. If lists are composed by elements, it seems a natural step to use lists for representing sets:

(setq S '(Callisto Ganymede Europa Io))

Independently from the way of representing sets, we shall always use capital letters for naming them.

Intrinsically related to the very definition of a set is the concept of *belonging*. In crisp, classical sets, a given element x either belongs or does not belong to a given set A. Formally:

$$x \in A \ (x \text{ belongs to set A}) \tag{5-1}$$

$$x \notin A (x \text{ does not belong to set A}) \tag{5-2}$$

In Lisp, we can use the user-defined predicate *(belongs?)*, shown in Code 5-1, for checking if a given element x belongs or not to a given set A:

```
;code 5-1
(define (belongs? x A)
    (if (or (intersect (list x) A) (= x '()))
            true
            nil
    )
)
```

Now, after typing *(setq S '(Callisto Ganymede Europa, Io))* at the Lisp prompt we can interrogate the system with, for example: *(belongs? 'Europe S)* → *true*, and, for example, *(belongs? 'Phobos S)* → *false*.

5.2.2 Subsets

If every element of a set A is also an element of a set B, we say that A is a subset of B. We can also say that A is included in B. Formally:

$$A \subset B \tag{5-3}$$

Conversely, if at least an element of set A is not an element of set B then we say that A is not a subset of B. We can also say that A is not included in B:

$$A \not\subset B \tag{5-4}$$

There is a peculiar set in Sets Theory, named "empty set" that is simply a set without any elements and is usually represented by the Greek letter ϕ or alternatively by {}. The empty set should not be strange for us because in Lisp we have extensively used the empty list () in many functions in the previous section of this book, proving its utility in real world applications. By definition, the empty set is included in every set. With these ideas on mind, we can write a Lisp predicate for checking if a set A is a subset of a set B, as shown in Code 5-2:

```
;code 5-2
(define (subset? A B)
    (if (or (= A (intersect A B)) (= A '()))
            true
            nil
    )
)
```

As an example, if we define the set U as the set of planets in the Solar System, U = {Mercury, Venus, Earth, Mars, Jupiter, Saturn, Uranus, Neptune}, H as the set of hard planets in the Solar System, H = {Mercury, Venus, Earth, Mars} and then G as the set of gaseous planets G = {Jupiter Saturn Uranus Neptune}, we can write:

$$H \subset U, G \subset U$$

Expressing these sets in Lisp notation: *(setq U '(Mercury Venus Earth Mars Jupiter Saturn Uranus Neptune)), (setq H '(Mercury, Venus, Earth, Mars)), (setq G '(Jupiter Saturn Uranus Neptune))*, then the following Lisp expressions hold: *(subset? H U)* → *true, (subset? G U)* → *true*. Please note that following the actual classification of the Astronomical Union, Pluto is not a planet of the Solar System.

Also by definition, every subset is included into itself, as we can see typing at the Lisp prompt: *(subset? G G)* → *true*.

Another peculiar set in sets theory is the Universal Set, usually represented by the capital letter U. The most trivial Universal Set would be the set of all things contained in the known universe, including atoms, quarks and any imaginable thing the reader can think in this moment. Needless to say, such a set can only be handled from a philosophical point of view, and from a mathematical and computer science perspective we define a Universal Set as the Set that contains all the objects under consideration, that is, the set of all the elements about a given subject matter. For example the Universal Set of numbers in a French roulette is the set U_1 of integer numbers from 0 to 36. The Universal Set of possible outcomes of throwing a dice is the set U_2 of integers from 1 to 6 and so on. Some example subsets of U_1 are, for example

$$S_1 = \{x \mid x \text{ is red}\}$$

$$S_2 = \{x \mid x \text{ belongs to the first column}\}$$

Remembering the Lisp code from the previous chapter, the set S_2 was already represented by the symbol *column1* as *(setq *column1* '(1 4 7 10 13 16 19 22 25 28 31 34))*. Needless to say, both S_1 and S_2 are included in U_1 and are subsets of it:

$$S_1 \subset U_1, S_2 \subset U_1$$

The imaginable set of "all things contained in the known universe" is, obviously uncountable, but uncountable sets can be represented mathematically, too. For example the set of real numbers between 1.0 and 10.0 can be expressed as:

$$S = \{x \mid x \geq 1.0 \text{ and } x \leq 10.0\}, x \in R$$

Theoretically, Lisp can handle uncountable sets, too. The previous expression can be put into Lisp code using the expression *(if (and (>= x 1.0) (<= x 10.0)) (lisp-expression))*. Here *(lisp-expression)* would be the action that Lisp would follow if x belongs to S, so Lisp can also conceptually describe uncountable sets. However, it is common, both in mathematics and Lisp, to use countable sets. In this case, the number of elements contained in a set S is called *cardinality*, and it is represented by the Greek letter η, or simply by $|S|$. For example: $\eta(S_1) = |S_1| = 18$, and $\eta(S_2) = |S_2| = 12$.

Cardinality in Lisp is trivially expressed by Code 5-3:

```
;code 5-3
(define (cardinality S)
      (length S)
)
```

Although this function is simply a call to the function *(length)* it serves us well for establishing a continuum between Sets Theory and Lisp. Now, for example, *(cardinality *column1*)* → 12. Please note that in Lisp we can also re-name any function at the Lisp prompt:

> **(setq cardinality length)**
: *length<1B3C6>*

And then, as before, *(cardinality *column1*)* → 12.
We define sets S_1 and S_2 as equivalent if their cardinality is the same, that is:

$$\eta(S_1) = \eta(S_2) \tag{5-5}$$

(= (cardinality S1) (cardinality S2)) → *true*

S_1 and S_2 are equal if the elements in both elements are exactly the same. We then write:

$$S_1 = S_2 \qquad\qquad (5\text{-}6)$$

Conversely, sets S_1 and S_2 are unequal if their cardinality is different or if their elements are not the same. Mathematically, we express it as follows:

$$S_1 \neq S_2 \qquad\qquad (5\text{-}7)$$

From a Lisp point of view, it cannot be easier for checking if two sets are equal or unequal. Taking as example, *(setq A '(a b c)), (setq B '(a b c d)), (setq C '(x y z)), (setq D '(a b c))*, we have that *(= A B)* → *nil, (= A D)* → *true, (= C D)* → *nil,* but *(= (cardinality A) (cardinality D))* → *true.* These expressions show us the road to write two simple Lisp predicates, shown in Code 5-3a and 5-3b:

```
;code 5-3a
(define (equivalent? A B)
    (if (= (cardinality A) (cardinality B)))
)

;code 5-3b
(define (equal? A B)
    (if (= A B))
)
```

The previous paragraph can be thus rewritten as: *(equal? A B)* → *nil, (equal? A D)* → *true, (equal? C D)* → *nil, (equivalent?A D)* → *true.*

5.2.3   Union, Intersection, Complement and Difference

The union $S_1 \cup S_2$ of two crisp sets S_1, S_2 is a set formed by all the elements of S_1 and all the elements of S_2 after eliminating any possible repeated element. Mathematically:

$$S_1 \cup S_2 = \{x \mid x \in S_1 \text{ or } x \in S_2\} \qquad\qquad (5\text{-}8)$$

As an example, if we have two sets A and B composed by some lower-case letters:

$$A = \{a, b, c, d, e\}, B = \{b, c, x, y, z\}$$

then,

$$A \cup B = \{a, b, c, d, e, x, y, z\}$$

NewLisp incorporates a function, not surprisingly named *(union)*, that returns the union of two sets as a list, that is, as another set. For example, if we create the

Fig. 5.2 A graphical representation of the union of two sets, A and B

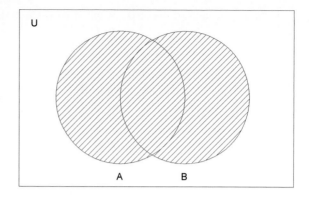

sets A and B by typing *(setq A '(a b c d e))* and *(setq B '(b c x y z))*, then *(union A B)* → *(a b c d e x y z)*. Figure 5.2 shows graphically the union of two sets.

The intersection $S_1 \cap S_2$ of two crisp sets S_1, S_2 is a set formed by all the common elements of S_1 and S_2. Mathematically:

$$S_1 \cap S_2 = \{x \mid x \in S_1 \text{ and } x \in S_2\} \tag{5-9}$$

Following the previous example with sets A and B, we have:

$$A \cap B = \{b, c\}$$

Although the reader should be able to write a Lisp function for obtaining the intersection of two sets, NewLisp already incorporates it in its library of functions, and we only need to type the following at the Lisp prompt: *(intersect A B)* → *(b c)*. By the way, please note how we have seized the opportunity to use the function *(intersect)* for creating the functions *(belong?)* and *(subset?)* as shown in Code 5-1 and 5-2, respectively. Figure 5.3 shows graphically the intersection of two sets.

Fig. 5.3 A graphical representation of the intersection of two sets, A and B

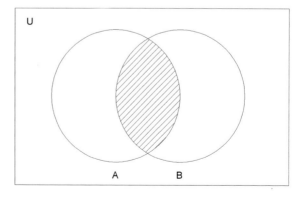

If the intersection of two sets, A and B results into the empty set, ϕ, that is;

$$A \cap B = \phi = \{\} \qquad (5\text{-}10)$$

we say that A and B are disjoint. For example, if A = {a, b, c}, B = {x, y, z} then A and B are disjoint because they do not share any element, that is, its intersection is null. Code 5-4 shows a simple Lisp predicate for testing if two sets are disjoint:

```
;code 5-4
(define (disjoint? A B)
    (if (= (intersect A B) '())))
)
```

Now, following the previous example, *(setq A '(a b c)), (setq B '(x y z)* then *(disjoint? A B)* → *true*. It is interesting to note that if two sets A and B are disjoint, then an obvious relationship does exist between the concepts of cardinality and union of sets:

$$\eta(A \cup B) = \eta(A) + \eta(B) \qquad (5\text{-}11)$$

Using the last example for sets A and B in Lisp for trying expression (5.11), we have:

(cardinality (union A B)) → *6*
(+ (cardinality A) (cardinality B)) → *6*

In general, any two sets A and B, despite they are disjoint or not, satisfy the following property:

$$\eta(A \cup B) = \eta(A) + \eta(B) - \eta(A \cap B) \qquad (5\text{-}12)$$

Again trying the two non-disjoint sets *(setq A '(a b c d e)), (setq B '(b c x y z)),* then we have: *(cardinality A)* → *5, (cardinality B)* → *5, (cardinality (union A B))* → *8, (cardinality (intersect A B))* → *2*. Expressed in only one line:

(= (cardinality (union A B)) (- (+ (cardinality A) (cardinality B)) (cardinality (intersect A B))) → *true*

After a simple manipulation of expression (5-12) we finally obtain:

$$\eta(A \cup B) + \eta(A \cap B) = \eta(A) + \eta(B) \qquad (5\text{-}13)$$

Fig. 5.4 A graphical
representation of the
complement of a set A, A'
with re-spect to an Universal
set U

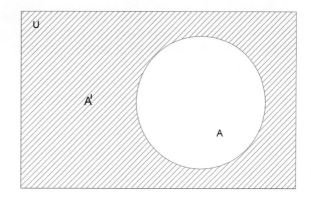

The complement of a set A with respect to a Universal set U is by definition the set composed by all the elements belonging to U that are not included in A. Formally:

$$A' = \{x \mid x \in U \text{ and } x \notin A\} \tag{5-14}$$

A Venn diagram, shown in Fig. 5.4, will help to visualize expression (5-14).

As an example, let us take again the set U as the set of planets in the Solar System, U = {Mercury, Venus, Earth, Mars, Jupiter, Saturn, Uranus, Neptune} and then H as the set of hard planets in the Solar System, H = {Mercury, Venus, Earth, Mars}. Then the complementary of H, H' is:

$$H' = \{\text{Jupiter Saturn Uranus Neptune}\}$$

Code 5-5 shows a Lisp function for obtaining the complementary set of a set A with respect to a universal set U:

```
;code 5-5
(define (complement A U, lU i set-out)
    (setq set-out '())
    (setq lU (cardinality U))
    (setq i 0)

    (while (< i lU)
            (if (! = (belongs? (nth i U) A) true)
                    (setq set-out (cons (nth i U) set-out))
            )
            (++ i);this is equivalent to (setq i (+ 1 i))
    );end while
    (reverse set-out)
);end function
```

Expressing again sets U and H in Lisp we have: *(setq U '(Mercury Venus Earth Mars Jupiter Saturn Uranus Neptune)), (setq H '(Mercury, Venus, Earth, Mars)), (setq G '(Jupiter Saturn Uranus Neptune)).* And now:

$$(complement\ H\ U) \rightarrow (Jupiter\ Saturn\ Uranus\ Neptune$$
$$(complement\ G\ U) \rightarrow (Mercury\ Venus\ Earth\ Mars)$$

In this case, G and U are disjoint sets and its union covers the complete Universal Set:

$$(intersect\ H\ G) \rightarrow ()$$
$$(union\ H\ G) \rightarrow (Mercury\ Venus\ Earth\ Mars\ Jupiter\ Saturn\ Uranus\ Neptune)$$

The difference between two sets A and B is another set whose elements belong to A but do not belong to B. Formally:

$$A - B = \{x \mid x \in A \text{ and } x \notin B\} \tag{5-15}$$

A graphic representation of the difference between sets can be seen in Fig. 5.5 with the help of Venn diagrams.

As can be inferred from Fig. 5.5, the difference between two sets A and B can be also described by the following expression:

$$A - B = A \cap B' \tag{5-16}$$

NewLisp incorporates a function named *(difference)* that automatically calculates the difference between two sets. As an example, let's take set A as the sets of satellites in the solar systems easily to observe with a small quality telescope: *(setq A '(Moon Callisto Ganymede Europa Io Titan))*, and B as the set of Galilean moons in Jupiter: *(setq B '(Callisto Ganymede Europa Io))*. Then, *(difference A B) → (Moon Titan)*. It is important to note that the difference between two sets is not commutative, that is: A − B ≠ B − A, for example: *(difference B A) → ().*

Fig. 5.5 A graphical representation of the difference between sets A and B

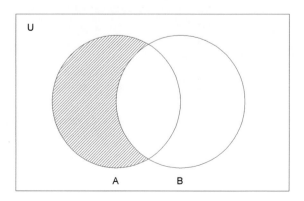

5.2.4   Set Properties

In this section we are going to expose the main properties of sets. Aside the formal description we shall give a simple Lisp example for each of the properties using the following expressions as sets: *(setq U '(0 1 2 3 4 5 6 7 8 9)), (setq A '(1 3 5 7)), (setq B '(5 6 7 8 9)), (setq C '(0 1 2 3 4))*. Each family of properties will be included into a single table as follows: Table 5.1 shows the Identity properties of sets, Table 5.2 the Idempotent ones, Table 5.3 the Complement ones, Table 5.4 the Associative ones, Table 5.5 the Commutative ones, Table 5.6 the Distributive ones and Table 5.7 shows the De Morgan's Laws.

Table 5.1 Identity properties of sets

$A \cup \phi = A$
(union A '()) $\rightarrow$ (1 3 5 7)
$A \cup U = U$
(union A U) $\rightarrow$ (1 3 5 7 0 2 4 6 8 9)
$A \cap U = A$
(intersect A U) $\rightarrow$ (1 3 5 7)
$A \cap \phi = \phi$
(intersect A '()) $\rightarrow$ ()

Table 5.2 Idempotent properties of sets

$A \cup A = A$
(union A A) $\rightarrow$ (1 3 5 7)
$A \cap A = A$
(intersect A A) $\rightarrow$ (1 3 5 7)
(intersect A '()) $\rightarrow$ ()

Table 5.3 Complement properties of sets

$A \cup A' = U$
(union A (complement A U)) $\rightarrow$ (1 3 5 7 0 2 4 6 8 9)
$A \cap A' = \phi$
(intersect A (complement A U)) $\rightarrow$ ()

Table 5.4 Associative properties of sets

$(A \cup B) \cup C = A \cup (B \cup C)$
(union (union A B) C) $\rightarrow$ (1 3 5 7 6 8 9 0 2 4) *(union A (union B C)) $\rightarrow$ (1 3 5 7 6 8 9 0 2 4)*
$(A \cap B) \cap C = A \cap (B \cap C)$
(intersect (intersect A B) C) $\rightarrow$ () *(intersect A (intersect B C)) $\rightarrow$ ()*

Table 5.5 Commutative properties of sets	$A \cup B = B \cup A$
	(union A B) → *(1 3 5 7 6 8 9)* *(union B A)* → *(1 3 5 7 6 8 9)*
	$A \cap B = B \cap A$
	(intersect A B) → *(5 7)* *(intersect B A)* → *(5 7)*

Table 5.6 Distributive properties of sets	$A \cup (B \cap C) = (A \cup B) \cap (A \cup C)$
	(union A (intersect B C)) → *(1 3 5 7)* *(intersect (union A B) (union A C))* → *(1 3 5 7)*
	$A \cap (B \cup C) = (A \cap B) \cup (A \cap C)$
	(intersect A (union B C)) → *(1 3 5 7)* *(union (intersect A B) (intersect A C))* → *(5 7 1 3)*

Table 5.7 De Morgan's Laws	$(A \cup B)' = A' \cap B'$
	(complement (union A B) U) → *(0 2 4)* *(intersect (complement A U) (complement B U))* → *(0 2 4)*
	$(A \cap B)' = A' \cup B'$
	(complement (intersect A B) U) → *(0 1 2 3 4 6 8 9)* *(union (complement A U) (complement B U))* → *(0 2 4 6 8 9 1 3)*

5.2.5   Cartesian Product and Relations

In sets theory, the Cartesian Product of two sets, A and B, denoted by A × B, is the set of all possible ordered pairs (x,y) whose first component x is a member of A and whose second component y is a member of B. Formally:

$$A \times B = \{(x,y) \mid x \in A \text{ and } x \in B\} \qquad (5\text{-}17)$$

As an example, if A = {a, b, c}, and B = {1, 2, 3, 4}, then:

A × B = {(a,1), (a,2), (a,3), (a,4), (b,1), (b,2), (b,3), (b,4), (c,1), (c,2), (c,3), (c,4)}

Needless to say, the Cartesian Product is not commutative, that is A × B ≠ B × A. In fact, the commutative property for the Cartesian Product between two sets A and B only holds when A = B. Using the same previous example with sets A and B:

B × A = {(1,a), (1,b), (1,c), (2,a), (2,b), (2,c), (3,a), (3,b), (3,c), (4,a), (4,b), (4,c)}

Neither Lisp nor the NewLisp dialect incorporate a function for calculating the Cartesian Product of two sets, but it is not a hard undertaking to write one. Code 5-6 shows a simple Lisp implementation of such a function:

```
;code 5-6
(define (cartesian-product A B, 1A 1B i j set-out)
   (setq 1A (cardinality A))
   (setq 1B (cardinality B))
   (setq i 0 j 0);initializes i and j at the same time to
      zero
   (setq set-out '())

   (while (< i 1A)
      (while (< j 1B)
            (setq set-out (cons (list (nth i A) (nth j B))
                            set-out))
                   (++ j)
      );end while j
      (++ i)
      (setq j 0);reinitializes j
   );end while i
   (reverse set-out)
)
```

Then, making *(setq A '(a b c))* and *(setq B '(1 2 3 4))*, we only need to write at the Lisp prompt: *(setq U1 (cartesian-product A B))* and Lisp will answer:

((a 1) (a 2) (a 3) (a 4) (b 1) (b 2) (b 3) (b 4) (c 1) (c 2) (c 3) (c 4))

Conversely, the Lisp expression *(setq U2 (cartesian-product B A))* produces, as expected:

((1 a) (1 b) (1 c) (2 a) (2 b) (2 c) (3 a) (3 b) (3 c) (4 a) (4 b) (4 c))

A Cartesian Product can be represented in two dimensions using a simple two-axis graphic. Figure 5.6a, b show A × B and B × A, respectively. Please note from the simple observation of the figures that A × B ≠ B × A, as previously stated:

The definition of a Relation between a set A and a set B is simple: A Relation between sets A and B (or from A to B) is any subset R of the Cartesian Product A × B. After a Relation is established we can say that a ∈ A and b ∈ B are related by R. Using the same previous sets A = {a, b, c} and B = {1, 2, 3, 4} a Relation R_1 can be, for example:

$$R_1 = \{(a,3), (a,4), (b,1), (b,2), (b,3), (b,4)\}$$

R_1 can be represented graphically as shown in Fig. 5.7.

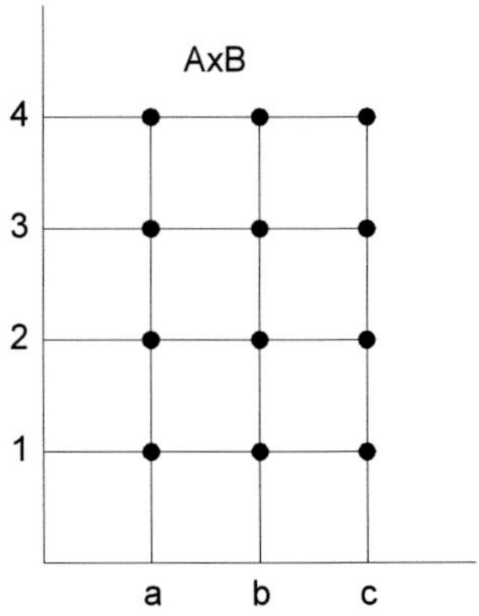

Fig. 5.6a A graphical
representation of the
Cartesian product A × B

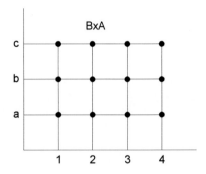

Fig. 5.6b A graphical
representation of the
Cartesian product B × A

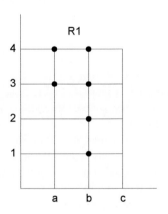

Fig. 5.7 A graphical
representation of relation R1

Expressing this into Lisp can not be simpler: *(setq R1 '((a 3) (a 4) (b1) (b 2) (b 3) (b 4)).* For testing if this is a Relation from A to B, we only need to type at the Lisp Prompt: *(subset? R1 U1) -> true*, and, as expected, *(subset? R1 U2) -> nil*, since R1 is not a relation from B to A.

5.3   Moving Towards Fuzzy Sets

In the previous sections of this chapter we have seen that probably the most important concept in sets theory is the concept of belonging or membership of an element to a set. In fact, without this concept it would be impossible to describe sets. Aside enumerating the elements of a set or expressing it conceptually, or by means of a Venn diagram, we can also express the membership of an element x to a set A using what is known as a membership function, $\mu_A(x)$ or characteristic function. Such a function can only take two values in classic sets theory: 0 if an element x does not belong to a set A, or 1 if an element x certainly belongs to a set A. Formally:

$$\mu_A(x) = 1 \text{ for } x \in A \tag{5-18a}$$

$$\mu_A(x) = 0 \text{ for } x \notin A \tag{5-18b}$$

As an example, let us take the set A as the set of French people, and a very small subset E of European citizens such as E = {Klaus, Jean, John, Paolo, Maurice, Juan}. Only two elements from E belong to A. Using expressions (5-18a) and (5-18b) we have:

$$\mu_A(\text{Klaus}) = 0, \ \mu_A(\text{Jean}) = 1, \ \mu_A(\text{John}) = 0$$
$$\mu_A(\text{Paolo}) = 0, \ \mu_A(\text{Maurice}) = 1, \ \mu_A(\text{Juan}) = 0$$

This type of situations is perfectly covered by using classic, crisp sets. Other examples could be the set of cars that use V12 engines, the set of rockets used in manned spaceflight missions and so on, where the membership value $\mu(x)$ to every element of their respective universal sets is either one or zero. However, there are also examples, in fact the most of things we observe in nature, that do not adhere to this formal framework of yes/no, 1 or 0 belonging.

Let us take the age of a human being as an example, ranging from 0 to 80 years (here we use 80 years old as a top limit since that figure expresses well an average of life duration in developed countries from occidental societies). The key question here is simple at first sight: How do we define the set of old people? That is, where do we establish a sharp separation between young and old people, at which age? After a while, you, dear reader, maybe would answer: "well, maybe it would be better to divide the universal set representing age from 0 to 80 years old in several subsets such as 'child, young, mature and old' in order to have a better representation of the concept of age". It is not a bad answer. However, if you reflect a bit about it you will soon realize that your answer put the question to sleep for a while, but it does not solve the problem because, for example, it leads to questions such us "how do we establish the dividing line between mature and old?". Pressed by my questions you maybe would tell me, probably with a challenging tone of voice:

"Well, we can not do infinite partitions in the set that represents age, right?" This is not a bad reply, either.

For exposing clearly the nature of the problem at hand let us say that there is an agreement if I say that an 80 years old person is an old person. If we subtract one year we obtain 79, and a 79 years old person is again an old person. Following the procedure we have the sequence of numbers 80, 79, 78, 77, 76, ... When do we stop enumerating the concept of old age? Since you are delighted with the classic set theory exposed until now in this chapter you decisively say: "ok, a decision must be made. Let us take 50 years old as the dividing line in such a way that we obtain a set A of young people and a set B of old people. Then for every man of woman on the entire Earth we can express formally:"

$$A = \{x \mid x \geq 0 \text{ and } x \leq 50\}$$
$$B = \{x \mid x > 50 \text{ and } x \leq 80\}$$

Sadly, this leads to another set of itching, irritating questions: Is it really a 51 years old person an old person?, or, do you really believe that a 49 years old person is a young person? Finally you exclaim "well Luis, at least you can not negate that a 75 years old person is older than a 55 years old person, and that a 23 years old person is younger than a 62 years old person!". No, I shall not negate that. In fact I think these are excellent observations.

5.3.1 The "Fuzzy Sets" Paper

The previous "age problem" is in fact an example of the well known "Sorites Paradox", a class of paradoxical arguments, based on little-by-little increments or decrements in quantity. A heap formed by sand grains is the primitive paradox (in fact "Sorites" means "heap" in ancient Greek) where decrementing the heap size grain by grain is impossible to establish when a heap turns into a no-heap. The same paradox arises when we try to describe the set of rich people, the set of ill patients in a hospital, the set of luminous galaxies, the set of beautiful women…, the number of instances of the Sorites paradox is quasi-infinite in the real world. These paradoxes can not be adequately solved by using classic sets theory.

Lofti Zadeh, the father of the fuzzy sets theory, was born in 1921 in Baku, Azerbaijan. Soon after graduating from the University of Tehran in electrical engineering in 1942 he emigrated to the United Sates, entering the Massachusetts Institute of Technology, MIT, in 1944 and getting an MS degree in electrical engineering in 1946. Not much later he moved again, this time to New York City, where he received his PhD degree in electrical engineering from the University of Columbia in 1949. After ten years of lecturing at Columbia he finally moved to Berkeley in 1959. While I am writing this book (2014) he still continues writing papers in the famous University on the eastern side of the San Francisco Bay.

Back in the summer of 1964, Zadeh was preparing a paper on pattern recognition for a conference to be held at the Wright-Patterson Air Force Base in Dayton. Ohio. The flight to Dayton made a stopover in New York, so Zadeh enjoyed an evening of free time, an evening free of academic and social encounters that conceded him the freedom of thinking at his best. In his own words:

I was by myself and so I started thinking about some of these issues (pattern recognition). And it was during that evening that the thought occurred to me that when you talk about patterns and things of this kind, ... that the thing to do is to use grades of membership. I remember that distinctly and once I got the idea, it became grow to be easy to develop it (Seising 2007)

As it usually happens, inspiration comes when you are working hard into a problem and you have a strong knowledge not only in the discipline where the problem to solve is defined, but in other more or less parallel and related disciplines. Under these conditions the human brain tends to establish new connections from patterns to patterns of neurons. This neurophysiological process is in fact what we call inspiration. In the case of Zadeh, the nucleus of his inspiration can be summarized in only five words: "to use grades of membership" and that realization materialized into what is probably the most famous paper in the history of fuzzy-logic. This paper, unambiguously titled "Fuzzy Sets", symbolized a shift of paradigm in the theory of sets. I cannot renounce to remember the first words of the abstract of such an important tour de force in the history of computer science and mathematics:

A fuzzy set is a class of objects with a continuum of grades of membership. Such a set is characterized by a membership (characteristic) function which assigns to each object a grade of membership ranging between zero and one (Zadeh 1965)

It is difficult to express the definition of a fuzzy set in a better way. Before presenting some mathematical and Lisp expressions for representing fuzzy sets, I think it is convenient to show a graphical representation of fuzzy sets. In Fig. 5.8a we can see a traditional Venn diagram showing a classic, crisp set A. It is a simple sketch drawn by hand with a pencil, but interesting enough for our discussion. For convenience we have shown it in black. If you wish so, you can imagine it

Fig. 5.8a A sketch of a crisp set, A

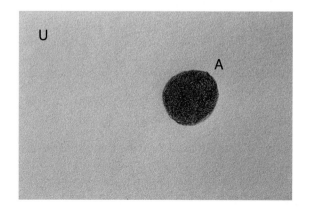

Fig. 5.8b A sketch of a fuzzy set, *fzA*. The *black* nucleus represents a whole membership degree to the fuzzy set

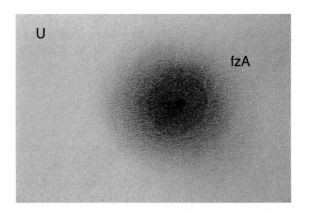

represents a set of old people from the traditional point of view of classic sets theory where every person x belonging to it satisfies the previous expression {x | x > 50 and x ≤ 80}. In this way, every person older than 50 years old would be "located" inside the black Venn diagram. Needless to say, every young person, less than 50 years old, would be located outside it.

Zadeh's shift of paradigm can be appreciated from the simple observation of Fig. 5.8b. There, it is easy to observe how the blackness decreases from its nucleus towards the exterior in a continuous way, representing different grades of membership. An element x_1 representing an 80 years old person would be located just in the centre, showing a whole membership degree to the fuzzy set fzA, that is $\mu_{fzA} = 1$. As the age of a person decreases, the location of its corresponding element x_i would move away from the centre, getting diminishing values of μ_{fzA}, that is, $\mu_{fzA} < 1$, such as, for example $\mu_{fzA} = 0.7$, $\mu_{fzA} = 0.45$ or $\mu_{fzA} = 0.2$ for decreasing values of age. Just note that now there is no need to define a threshold value for separating in a sharp way the set of old people from the set of young people.

Definition A fuzzy set *A* is defined by a characteristic function $\mu_A(x)$ that maps every element x belonging to A to the closed interval of real numbers [0,1]. Formally we can write:

$$A = \{(x, \mu_A(x)) \mid x \in A, \mu_A(x) \in [0,1]\} \qquad (5\text{-}19)$$

That is, we can create a fuzzy set by means of enumerating a collection of ordered pairs $(x_i, \mu_A(x_i))$ where $\mu_A(x_i)$ is the membership degree of an element x_i to the fuzzy set A. In general:

$$\mu_A : X \to [0,1] \qquad (5\text{-}20)$$

In this expression, the function μA completely defines the fuzzy set A (Klir and Yuan 1995). Following the example of age in human beings we can enumerate a precise, however subjective, characterization of the fuzzy set *A* of old people as, for example: $\mu_A(35) = 0.1$; $\mu_A(45) = 0.2$; $\mu_A(55) = 0.4$; $\mu_A(65) = 0.7$; $\mu_A(75) = 0.9$; $\mu_A(80) = 1$. Using the formal representation given by (5-19) we would have:

$$A = \{(35,0.1), (45,0.2), (55,0.4), (65,0.7), (75,0.9), (80,1.0)\}$$

Since we are dealing with persons, and admitting Paul is 35 years old, John is 45, Mary is 55, Klaus is 65, Juan is 75 and Agatha is 80, we can also write:

$$A_{names} = \{(Paul,0.1), (John,0.2), (Mary,0.4), (Klaus,0.7), (Juan,0.9), (Agatha,1.0)\}$$

As the reader has quickly realized, representing these fuzzy sets using Lisp expressions is straightforward:

(setq A '((35 0.1) (45 0.2) (55 0.4) (65 0.7) (75 0.9) (80 1.0)))
(setq A-names '((Paul 0.1) (John 0.2) (Mary 0.4) (Klaus 0.7) (Juan 0.9)
(Agatha 1.0)))

For graphically representing fuzzy sets, traditional Venn diagrams are not enough since they were designed for representing crisp sets. Inspired by the sketch shown in Fig. 5.8, we can use a radar type diagram as a sort of enhanced Venn diagram, as shown in Fig. 5.9. The inner the dot in the radar diagram, the higher is its membership degree to the set. Circles representing membership degrees are spaced every 0.2 units in the figure.

When elements from a fuzzy set are based on numbers, it is usually more convenient to use grid diagrams, as the one shown in Fig. 5.10. The vertical axis

Fig. 5.9 A radar-type diagram for representing a fuzzy set. *Inner circles* represent higher membership degrees to the set

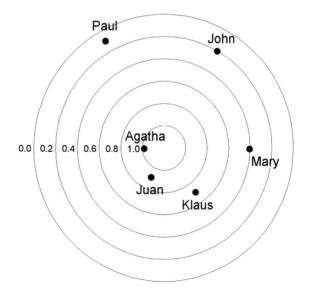

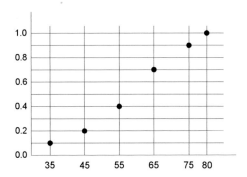

Fig. 5.10 A grid-type diagram for representing a fuzzy set. The vertical axis shows the membership degree to the set. The horizontal axis shows age in years

represents the membership degree to the set, while the horizontal axis shows the numerical elements. In this book we shall use the most suitable type of graphic for our needs. At the end of this chapter we shall make an intense use of grid-type diagrams.

It is especially important to remark that the subjective definition of a fuzzy set by means of its characteristic function is an advantage when modeling vague concepts because we can adapt, or better said, choose the most suitable one depending on context. Continuing with the fuzzy set of "old people" A, we have exposed a general example that suits well for a general population of human beings. However, if we are speaking about professional tennis players we can use, for example, the following fuzzy set T:

$$T = \{(\text{Dimitrov},0.2), (\text{Djokovic},0.5), (\text{Nadal},0.6), (\text{Federer},0.9), (\text{Borg},1.0)\}$$

where Borg is definitely an old professional tennis player, despite he would have a membership degree about 0.5 to the set of old people A in 2014. In fact you should take into account that the definition of set T is valid from the point in time I am writing this book. If you are reading this book in 2035 all the elements from T will have a membership degree of 1.0, that is, all of them will be old for playing tennis professionally.

As it was the case with crisp sets, and as the reader has probably suspected, the concept of belonging to a fuzzy set deserves a dedicated Lisp function. It is shown in Code 5-7:

```
;code 5-7
(define (fz-belongs? x A)
     (if (assoc x A)
             (last (assoc x A))
             nil
     )
)
```

As can be observed after reading the code, the function *(fz-belongs?)* returns *nil* if a given element x is not a member to the fuzzy set A, else, that is, if x belongs to A, then it returns its membership degree. Taking again the Lisp definition of the fuzzy set *A-names: (setq A-names '((Paul 0.1) (John 0.2) (Mary 0.4) (Klaus 0.7) (Juan 0.9) (Agatha,1.0)))*, then we would have, for example: *(fz-belongs? 'Klaus A-names)* → *0.7*, but *(fz-belongs? 'Paolo A-names)* → *nil*.

5.3.2 Union, Intersection and Complement of Fuzzy Sets

As we did in Sect. 5.2.3, we are going to explore now how the union, intersection and complement of fuzzy sets are defined. Basically the concepts remain the same, but the introduction of the idea of membership degree to a given fuzzy set adds important details that must be taken into account.

Definition The union $C = A \cup B$ of two fuzzy sets A and B, determined respectively by their characteristic functions $\mu_A(x)$, $\mu_B(x)$, is defined formally by the following expression:

$$C = A \cup B = \mu_C(x) = \max[\mu_A(x), \mu_B(x)] \qquad (5\text{-}21)$$

As an example, let us take the fuzzy sets A and B:

$$A = \{(1,0.7), (2,0.1), (3,0.3), (4,0.9), (5,0.2)\}$$
$$B = \{(1,0.1), (2,0.8), (3,0.9), (4,0.2), (5,1)\}$$

Figure 5.11a, b show a grid representation of these fuzzy sets: Then, the union $C = A \cup B$ is:

$$C = \{(1,0.7), (2,0.8), (3,0.9), (4,0.9), (5,1)\}$$

Fig. 5.11a Grid representation of fuzzy set A

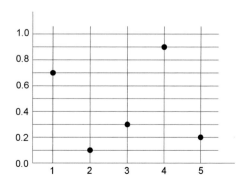

Fig. 5.11b Grid
representation of fuzzy set *B*

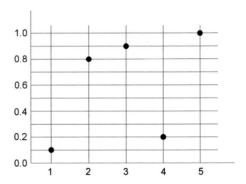

In order to have a Lisp function for obtaining the union of two fuzzy sets (and also for obtaining its intersection), it is convenient to have first an auxiliary function, named *(clean-md)* that given a fuzzy set *A* returns a related crisp set, that is, a set that conserves all the elements *x* in *A* but cleans all its membership values. Such a function is shown in Code 5-8:

```
;code 5-8.
(define (clean-md A, 1A i set-out)
     (setq 1A (cardinality A))
     (setq i 0)
     (setq set-out '())

     (while (< i 1A)
            (setq set-out (cons (first (nth i A)) set-out))
            (++ i)
     ); end while i
     (reverse set-out)
)
```

Then, for example, is we take *(setq A '((1 0.7) (2 0.1) (3 0.3) (4 0.9) (5 0.2)))* and *(setq B '((1 0.1) (2 0.8) (3 0.9) (4 0.2) (5 1.0)))*, then *(clean-md A)* → *(1 2 3 4 5)* and also *(clean-md B)* → *(1 2 3 4 5)*. Now, we can easily write the function *(fz-union)*, as shown in Code 5-9:

```
;code 5-9
(define (fz-union A B, temp 1A 1B 1t i element md-a md-b
set-out)
     (setq temp (union (clean-md A) (clean-md B)))
     (setq 1A (cardinality A))
     (setq 1B (cardinality B))
     (setq 1t (cardinality temp))
     (setq i 0)
     (setq set-out '())
```

```
(while (< i lt)
        (setq element (nth i temp))
        (setq md-a (assoc element A))
        (setq md-b (assoc element B))
        (if (> = md-a md-b)
                (setq set-out (cons md-a set-out));
                else:
                (setq set-out (cons md-b set-out))
        )
        (++ i)
); end while i
(reverse set-out)
)
```

Now, for testing the function, we only need to type at the Lisp prompt: *(fz-union A B)*, obtaining: *((1 0.7) (2 0.8) (3 0.9) (4 0.9) (5 1))*. Figure 5.12a, b show a graphic representation of the union of *A* and *B*.

Please observe how for both figures the membership degree of each element in the resulting union of sets get "higher" in its respective graphic. In the grid representation, this raising is clear, while in the radar-type graphic, the elements 1, 2, 3, 4 and 5 are closer to the centre. Imagining the curves representing membership degrees as contour lines in a terrain, we can appreciate that the resulting elements are closer to the top.

Definition The intersection $C = A \cap B$ of two fuzzy sets A and B, determined respectively by their characteristic functions $\mu_A(x)$, $\mu_B(x)$, is defined formally by the following expression:

$$C = A \cap B = \mu_C(x) = \min[\mu_A(x), \mu_B(x)] \qquad (5\text{-}22)$$

Taking again the same example sets A and B, we have the intersection $C = A \cap B$:

$$C = \{(1,0.1), (2,0.1), (3,0.3), (4,0.2), (5,0.2)\}$$

Fig. 5.12a A grid representation of $A \cup B$

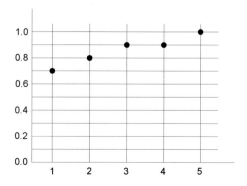

Fig. 5.12b A radar-type
representation of $A \cup B$

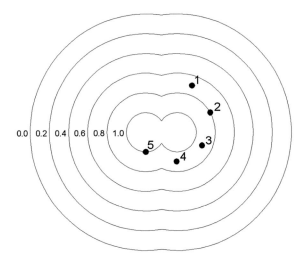

Translating this to Lisp results into the function *(fz-intersect)*, shown in Code 5-10:

```
;code5-10
(define (fz-intersect A B,
          temp lA lB lt i element md-a md-b set-out)
    (setq temp (intersect (clean-md A) (clean-md B)))
    (setq lA (cardinality A))
    (setq lB (cardinality B))
    (setq lt (cardinality temp))
    (setq i 0)
    (setq set-out '())
    (while (< i lt)
          (setq element (nth i temp))
          (setq md-a (assoc element A))
          (setq md-b (assoc element B))

          (if (<= md-a md-b)
                (setq set-out (cons md-a set-out));
                  else:
                (setq set-out (cons md-b set-out))
          )
          (++ i)
    ); end while i
    (reverse set-out)
)
```

Fig. 5.13a A grid representation of $A \cap B$

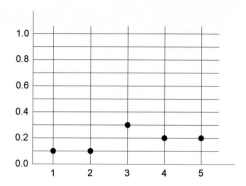

Fig. 5.13b A radar-type representation of $A \cap B$

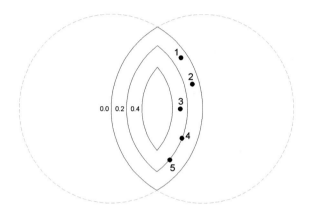

And testing the function at the Lisp prompt we get: *(fz-intersect A B)* → *((1 0.1) (2 0.1) (3 0.3) (4 0.2) (5 0.2))*. Figure 5.13a, b show a grid-type and a radar-type representation of the intersection of fuzzy sets A and B, respectively.

The dashed curves in Fig. 5.13b show the original external shapes (that is, 0.0 membership degrees) of the Venn diagrams corresponding to A and B. Since $A \cap B$ has resulted into a set with low membership degrees, you can visualize it as two radar-type Venn diagrams relatively separated with a common part (continuous lines). If the resulting fuzzy intersection had elements with high membership degrees, this fact would imply that the radar-type Venn diagrams would be closer. In fact, $A \cap A \rightarrow A$, and then the separation would be inexistent. Expressing it into Lisp: *(fz-intersect A A)* → *((1 0.7) (2 0.1) (3 0.3) (4 0.9) (5 0.2))*.

Definition The complement A' of a fuzzy set A determined by its characteristic function $\mu_A(x)$ is defined formally by the following expression:

$$A' = 1 - \mu_A(x) \qquad (5\text{-}23)$$

The translation of expression (5-23) is not especially complex, and it is shown in Code 5-11:

```
;code 5-11
(define (fz-complement A, lA i set-out element)
    (setq lA (cardinality A))
    (setq i 0)
  (setq set-out '())

    (while (< i lA)
            (setq element (nth i A))
            (setq set-out (cons (list (first element)
                    (sub 1.0 (last element))) set-out))
            (++ i)
    ); end while i
    (reverse set-out)
)
```

Remembering the fuzzy sets *A* and *A-names* in Sect. 5.3.1 of this chapter representing membership degrees to the concept of "old people": *(setq A '((35 0.1) (45 0.2) (55 0.4) (65 0.7) (75 0.9) (80 1.0))), (setq A-names '((Paul 0.1) (John 0.2) (Mary 0.4) (Klaus 0.7) (Juan 0.9) (Agatha 1.0)))*, then we have the following calls to the function *(fz-complement)*:

(fz-complement A) → *((35 0.9) (45 0.8) (55 0.6) (65 0.3) (75 0.1) (80 0))*
(fz-complement A-names) → *((Paul 0.9) (John 0.8) (Mary 0.6) (Klaus 0.3) (Juan 0.1) (Agatha 0))*

Now it is easy to realize that the fuzzy complements *A'* and *A-names'* represent the concept of "young people". For example, a 65 years old person had a 0.7 membership degree to the fuzzy set of old people and now has a 0.3 membership degree to its complement. Juan (75 years old) had a 0.9 membership degree to *A-names*, but only a 0.1 membership degree to its fuzzy complement and so on. Figure 5.14a, b show a grid and a radar-type representation of *A'* and *A-names'*, respectively.

It is interesting to compare these figure to Figs. 5.9 and 5.10, especially the radar-type one. Surprisingly at first sight, the geometrical positions of Agatha, Juan, Klaus, Mary, John and Paul are exactly the same in Figs. 5.9 and 5.14b. However, if you observe meticulously both figures you will soon realize that the values of the membership degrees associated to every contour line are reversed. In other words: while the radar-type diagram of Fig. 5.9 increases the values of its contour lines from the outside to the center of the diagram, in Fig. 5.14b the external line represents the maximum membership degree (1.0) and the inner the position in the diagram of an element the lesser its membership degree to the set. Any hypothetical

Fig. 5.14a Grid diagram for representing the complement fuzzy set A'

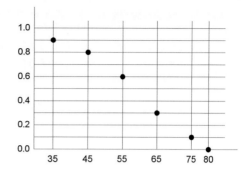

Fig. 5.14b Radar-type diagram for representing the complement fuzzy set *A-names'*

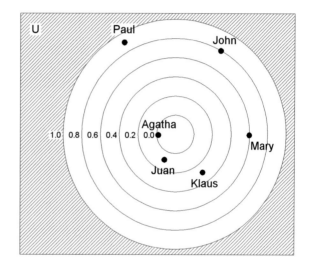

point located outside the external circle, that is, located on the shaded area, would have a 1.0 membership degree to *A-names'*.

5.3.3 Fuzzy Sets Properties

Fuzzy sets do not satisfy every property of classic sets as shown in Sect. 5.2.4. While the identity, idempotent, associative, commutative, De Morgan's Laws and distributive properties are perfectly satisfied by fuzzy sets, the complement properties are not. Especially interesting is the second expression from Table 5.3, named 'law of non contradiction' in logic (Trillas 2009):

$$A \cap A' = \phi$$

Let us take again the fuzzy set A as the set representing the concept of "old people", then: A → ((35 0.1) (45 0.2) (55 0.4) (65 0.7) (75 0.9) (80 1)) and (fz-complement A) → ((35 0.9) (45 0.8) (55 0.6) (65 0.3) (75 0.1) (80 0)). When we intersect these fuzzy sets, that is, after typing (fz-intersect A (fz-complement A)) at the Lisp prompt, we obtain the following fuzzy set:

$$((35\ 0.1)\ (45\ 0.2)\ (55\ 0.4)\ (65\ 0.3)\ (75\ 0.1)\ (80\ 0))$$

And this set is far from being equal to the empty set, φ. As we have already seen, since A represents the concept of "old people", its complement, A', represents the concept of "young people". If we were dealing with crisp sets, the intersection of A and A' would be the empty set, that is, every person is either young or old. However, when dealing with fuzzy sets we usually have:

$$A \cap A' \neq \phi$$

So, what is the meaning of this expression translated to normal language? If we take the example obtained from evaluating (fz-intersect A (fz-complement A)) at the Lisp prompt, it means that the elements belonging both to the fuzzy sets A and A' are at the same time young and old people! How can this be possible? Let us examine some elements of A ∩ A': For an 80 years old person her membership degree to the intersection is zero, so this person is not young and old at the same time, because from the definition of A she is "entirely" old. In a similar way, a 35 years old person has only a 0.1 membership degree to the intersection. We could say that he is mainly young, but some oldness has already started to appear in his physiology, in his organic development. Let us analyze now the element representing a 55 years old person. We can see that he has a 0.4 membership degree to A ∩ A', so he is clearly young and old at the same time. Is this statemen false? Under classical sets theory it certainly is. However, it is perfectly true when we use fuzzy sets, and what is especially interesting: it describes the real world in a very sensible way because a 55 years old person, while not young at all is still far from being old. He is transiting in time from young to old, and this transition is what the use of fuzzy sets represents perfectly well. At the beginning of Sect. 5.3 we defined the crisp set A as the set of young people and B as the crisp set of old people in the following way:

$$A = \{x \mid x \geq 0 \text{ and } x \leq 50\}$$
$$B = \{x \mid x > 50 \text{ and } x \leq 80\}$$

And as you can remember, these crisp definitions were generating serious irritating questions. Now, by means of using fuzzy sets the itching has disappeared. This key concept, the concept of transition between infinites shades of gray in the [0,1] interval of real numbers is what makes fuzzy sets theory so attractive for modelling systems from the real world. We use mathematics, physics, biology, astronomy and other sciences to model Nature, written here with capital letter at the beginning of the word to remark the broad meaning of the term, and we usually get a good representation of the systems we observe. However, when we introduce fuzzy sets in these sciences we usually get an even better representation of natural systems and in some cases we obtain a representation that is impossible to obtain by means of the simple use of crisp sets.

We are now immersed in the conceptual nucleus of this book, so it is a good time to remark another interesting matter that usually arises from using the [0,1] interval of real numbers for describing the membership degree on an element to a fuzzy set A, as already shown in expressions (5-19), (5-20): Membership degrees are not probabilities. As the reader already knows, probability is a mathematical concept for measuring the likeliness that an event E will happen. Formally:

$$p(E) \rightarrow [0,1]$$

When $p(E) = 0$ we say that an event E will not happen, for example, the probability that I will finish this book just tomorrow is 0 (I am now writing Chapter five), while when we write $p(E) = 1$ we are expressing that an event E will occur with absolute certainty. For example, the probability that the Sun will be a bright celestial object tomorrow is 1 (and it will continue being one well after the extinction of the human race). Any other value of p between 0 and 1 is an attempt to measure the likeliness that an event E will happen. Just observe the grammar I have used in this last sentence: As a general feature of probability, this branch of mathematics deals with things that can or cannot happen in the future. Membership degrees in fuzzy sets, on the other hand, do not need to deal with future events. They exclusively deal with actual facts, that is, with intrinsic features of an existing natural system. When we say, for example that Mary (55 years old) has a 0.4 membership degree to the fuzzy set of old persons we are expressing something that actually exists in reality without any mention to the future. However, if we say that the likelihood of Mary to live until 75 years old is $p = 0.85$ we are speaking about probabilities, since it is a measure of something that will or will not happen in the future and that actually we do not know for sure. Another example, taken from the famous potable drink problem by Bezdek (2013), will help to perfectly distinguish between membership degrees to a fuzzy set and probabilities.

Just imagine you have decided to spend your holidays in the Sahara desert. Don't ask me why, but after a few days and some discomforting adventures you are suddenly alone and lost in the hot African sands, and what is even worst: without water. After a while your luck seems to change a bit and you arrive to a small, very special oasis where you find two bottles with exactly the same shape placed over a table. Both are big and full of liquid, and the only difference between the two bottles is their labels: One of them, let us say, bottle A, says: "The liquid in this bottle has a 0.75 membership degree to the set of potable drinks". On the other hand, bottle B has a label that says: "The liquid in this bottle has a 0.9 probability of being potable". Now the important question is: Which bottle would you choose for drinking?

Bottle B seems attractive, especially if you do not like mathematics (we must recognize that since you are reading this book this is unlikely) because after all, 0.9 is a bigger number than 0.75 and everybody knows that the bigger the better. However, if p(B) = 0.9, it implies that p(B′) = 0.1, that is, it means that there is also a 0.1 probability value that the content of bottle B is not potable, being, for example, poisoned water. If you decide to drink from bottle B you are making a bet with two possible outcomes: To replenish your body with potable, pure water, with an associated probability p = 0.9 or to drink some poison and die, with an associated probability p = 0.1. Now, let us think about bottle A. The label tells us that it has a $\mu = 0.75$ membership degree to the set of potable drinks. Well, as any person versed in fuzzy sets theory, you know that the label of bottle A is informing you that its content is not pure water (that would have a 1.0 membership degree to the set of potable drinks). Maybe it is a commercial drink with lots of sugar, colorants and the like, but it is, intrinsically, a potable liquid so if you drink from bottle A you will have more time to find someone in the Sahara and eventually escape the desert. Even more: before drinking from bottle A you already know that you have find a suitable solution to your problems of thirst. If you choose to drink from bottle B you do not know beforehand what will happen. Especially interesting is what happens after, let us say, one hour after you drink: Bottle A will continue to have a $\mu = 0.75$ membership degree to the set of potable drinks because membership degrees are intrinsic to the features of a given element in a set. However, the probability value associated with bottle B has disappeared: Time has passed by and you now have a complete knowledge about its content if it was potable liquid. If it was not a potable liquid, say, poisoned water, then now you have not knowledge at all.

5.3.4 Fuzzy Relations

As we have seen in Sect. 5.2.4, the Cartesian Product A × B between two sets A and B composes a new set of ordered pairs (x,y) where the first component of each element belongs to A and the second component belongs to B, as shown by expression (5-17). Now, we can extend this idea to the realm of fuzzy sets by using expression (5-24):

$$R = \{((x,y), \mu_R(x,y))|(x,y) \in A \times B, \mu_R(x,y) \in [0,1]\} \qquad (5\text{-}24)$$

A fuzzy relation is a mapping from the Cartesian Product A × B to the closed interval [0,1]. The membership degree of the Relation is given by the function $\mu_R(x, y)$, that is, the value of $\mu_R(x, y)$ expresses the strength of the relation between the elements x and y of the pairs (x, y). Let us take, as an example, the following sets A and B:

$$A = \{a, b, c\}$$
$$B = \{x, y, z\}$$

Table 5.8 Basic operations on fuzzy relations

Operation	Expression
Union	$\mu_{R\cup S}(x,y) = \max(\mu_R(x,y), \mu_S(x,y))$
Intersection	$\mu_{R\cap S}(x,y) = \min(\mu_R(x,y), \mu_S(x,y))$
Complement	$(\mu_R(x,y))' = 1 - \mu_R(x,y)$

Then, a possible fuzzy relation, *R1*, between A and B could be:

$$R1 = \{(a,x,0.5), (a,z,0.8), (b,y,0.3), (b,z,1.0), (c,x,0.6)\}$$

Expressing fuzzy relations in Lisp is, needless to say, straightforward. *R1* can be represented as *(setq R1 '((a x 0.5) (a z 0.8) (b y 0.3) (b z 1.0) (c x 0.6)))*.

The basic operations on fuzzy relations defined in the Cartesian space A × B are given in Table 5.8.

5.3.4.1 Fuzzy Cartesian Product

Especially interesting is the situation where two sets *A* and *B* related by a fuzzy relation *R* are also fuzzy. In this case, every element from the pairing (x,y), x ∈ A, y ∈ B, already carries a membership degree. For obtaining the resulting ordered triples (x, y, $\mu_R(x,y)$) we use the Fuzzy Cartesian Product *A* × *B*, given by:

$$\mu_R(x,y) = \min(\mu_A(x), \mu_B(y)) \tag{5-25}$$

As an example, let us take the following two fuzzy sets *A* and *B*:

$$A = \{(x1,0.4), (x2,0.7)\}$$
$$B = \{(y1,0.8), (y2,0.6), (y3,0.4)\}$$

Then, the Fuzzy Cartesian Product *A* × *B* is:

$$A \times B = \{(x1,y1,0.4), (x1,y2,0.4), (x1,y3,0.4), (x2,y1,0.7), (x2,y2,0.6), (x2,y3,0.4)\}$$

In order to automatically calculating the Fuzzy Cartesian Product of two fuzzy sets, we only need to translate expression (5-25) into Lisp, as shown in Code 5-12:

```
;code 5-12
(define (fz-cartesian-product A B, lA lB i j set-out)
   (setq lA (cardinality A))
   (setq lB (cardinality B))
   (setq i 0 j 0)
   (setq set-out '())
```

```
(while (< i lA)
    (while (< j lB)
         (setq set-out (cons (list (first (nth i A))
         (first (nth j B)) (min (last (nth i A))
          (last (nth j B))))) set-out))

         (++ j)
     );end while j
     (++ i)
     (setq j 0);reinitializes j
  );end while i
  (reverse set-out)
)
```

Using the previous example as a test, we only need to type at the Lisp prompt: *(setq A '((x1 0.4) (x2 0.7)))*, *(setq B '((y1 0.8) (y2 0.6) (y3 0.4)))*, and then *(fz-cartesian-product A B)* → *((x1 y1 0.4) (x1 y2 0.4) (x1 y3 0.4) (x2 y1 0.7) (x2 y2 0.6) (x2 y3 0.4))*.

Since the concept of fuzzy relations and fuzzy Cartesian product is of pivotal importance for the future material in this book, we shall see another example. Let us say two women, Ana and Mary, belong to the fuzzy set A of highly communicating woman with membership degrees 0.9 and 0.1, respectively. On the other hand, John and Paul belong to the fuzzy set B of highly communicating men with membership degrees 0.8 and 0.4, respectively. The fuzzy Cartesian product between A and B will give us a fuzzy relation that expresses the strength of the possible communication links between all the members from A and B. Let us type at the Lisp prompt the following expressions: *(setq A '((Ana 0.9) (Mary 0.1)))*, *(setq B '((John 0.8) (Paul 0.4)))*. Now we obtain: *(fz-cartesian-product A B)* → *((Ana John 0.8) (Ana Paul 0.4) (Mary John 0.1) (Mary Paul 0.1))*. That is, Ana and John will be able to exchange a lot of ideas because their inherent communicating abilities, while Mary will not be able to communicate well nor with John neither with Paul. Figure 5.15 shows a graphical representation of $A \times B$:

The three dimensional appearance of the graphical representation of the fuzzy relation between A and B is important. Since the grade of complexity has grown from the relations in classic sets theory, a new (third) dimension is needed to correctly represent fuzzy relations. The vertical axis represents the membership degree, $\mu_R(x,y)$, of the relation. The other two axes represent the membership degrees $\mu_A(x)$ and $\mu_B(y)$ to the fuzzy sets A and B, respectively. In future chapters in this book we shall seize the opportunity of using 3D representations as a tool for a better understanding of new concepts.

Fig. 5.15 3D representation
of the fuzzy relation between
fuzzy sets *A* and *B*. Please
note that the membership
degree of the pair
"Mary-Paul" has not be
in-cluded in order to improve
the clarity of the image

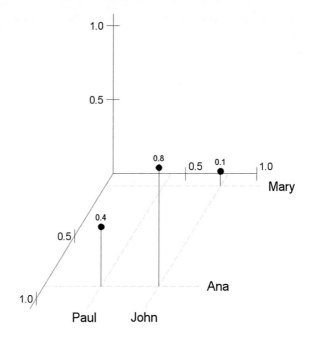

5.4   Membership Degrees: An Example Application in Medicine

Since this book has a strong practical vocation we are going to expose in this section a practical and complete use of the concept of membership degrees, in this case applied to the medical practice. We have already discussed the fuzzy set of old people. Now we can improve it in such a way that we are ready to speak about illness along the life of a person. For this we shall introduce the concept of Life Illness Curves, learning at the same time how a lot of phenomena in nature can be modelled by means of membership degrees and time.

We define a Life Illness Curve, LIC, as a graphical representation of the membership degree that a person has to the fuzzy set *I* of Illness over time, that is:

$$y = \mu_I(x), t \tag{5-26}$$

In these parametric graphics, the vertical axis represents the membership degree $\mu_I(x)$, defined as usually in the interval [0,1], while the horizontal axis represents time in years, from 0 to 80 years old. The value $\mu_I(x) = 0$ means an absolutely absence of illness that is experienced by the human being only at birth, when no congenital disorder is present. We emphasize the condition "only at birth" because cellular deterioration begins just with life, albeit usually extremely slowly. The value $\mu_I(x) = 1$ means an integral and definitive presence of illness that happens only at the individual's *exitus*. In the next paragraphs we present some examples of Life Illness Curves.

Figure 5.16 shows a LIC of a normal, healthy individual. The value $\mu_I(x)$ remains low for almost the entire life of the person and only in the last months of his/her life the organism looses its healthy state.

Figure 5.17 represents the evolution in time of a patient affected by amyotrophic lateral sclerosis (ALS), where the value $\mu_I(x)$ remains low and normal until the disease's debut that leads into a relatively quick outcome (Brown 2010). This figure represents exactly the case of the famous baseball player Lou Gehrig, which suffered this condition from 1938 to 1941. Interestingly, Fig. 5.18 shows the LIC of the same illness, this time representing the case of the known cosmologist Stephen Hawking. The curve shows the debut of the condition in his twenties, a tracheotomy

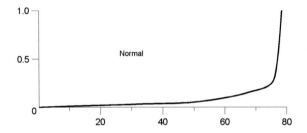

Fig. 5.16 LIC of a healthy person

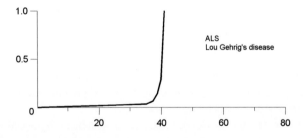

Fig. 5.17 LIC of a patient suffering Amyotrophic Lateral Sclerosis. Lou Gehrig's case

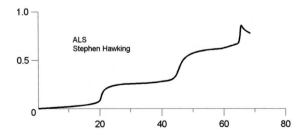

Fig. 5.18 LIC of a patient suffering Amyotrophic Lateral Sclerosis. Stephen Hawking's case

that resulted into a permanent aphonia in his forties and some severe infectious disorders at his sixties.

Figure 5.19 shows the LIC of a patient affected by type-1 diabetes mellitus that debuts at 15 years old, is correctly diagnosed and is treated adequately by means of insulin therapy, diet and exercise. As can be seen, under such circumstances the values $\mu_I(x), t$ remain relatively low through his life and only from his sixties he could start to suffer diabetes-related complications that spark the apparition of other conditions such as blindness, kidney failure and so on (Guyton 1996).

Figure 5.20 shows two possible evolutions of patients suffering acquired immune deficiency syndrome (AIDS): one of them living in a third world country (the one with the less favourable LIC) and another one living in a modern country. The difference in LICs is due to both a correct diagnostic and an appropriate treatment.

Figure 5.21 shows the LIC of a patient that has suffered a car accident in his thirties, resulting in permanent spine damage. Two main regions can be seen for the LIC: the one before the accident and the one after it. Despite the sharp increase in the $\mu_I(x)$ value resulting from the accident, the shape of each region resembles that of a normal life, like in Fig. 5.16.

We define a Life Quality Curve (LQC) as a graphical representation of the membership degree that a person has to the fuzzy set L of good quality of life over time. This curve is defined by the expression:

$$y = \mu_L(x) = (1 - \mu_I(x)), \, t \tag{5-27}$$

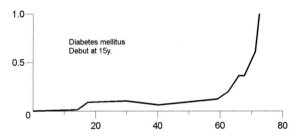

Fig. 5.19 LIC of a patient affected by diabetes with good diagnostic and treatment

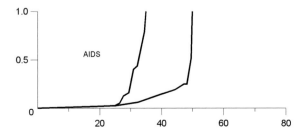

Fig. 5.20 LIC of a patient affected by AIDS. Two possible outcomes

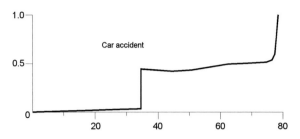

Fig. 5.21 LIC of a patient with permanent spine damage caused by an accident

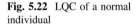
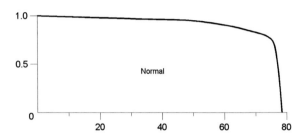

Fig. 5.22 LQC of a normal individual

As can be easily seen, such an expression generates a symmetrical curve from LICs whose axis of symmetry is $y = 0.5$. In other words, LQCs represent the fuzzy complement of LICs. Since the meaning of such type of curves is immediate after having exposed LICs, we shall offer only three examples of LQCs, shown in Figs. 5.22, 5.23 and 5.24:

At this point we must note an important remark: both LIC and LQC curves are fuzzy and are not carved in stone, as we can immediately realize from Figs. 5.17 and 5.18, where the same illness, ALS, show two dramatically different behaviours in two different patients, although we must concede that Hawking's case is certainly rare. In any case, it's really interesting to note that the shape of LIC and LQC curves are affected by the perception of the person who observes the condition. Let us take again as an example the LQC of Stephen Hawking as shown in Fig. 5.25: While we can interpret the bold curve as the perception of a neurologist, it is more than likely

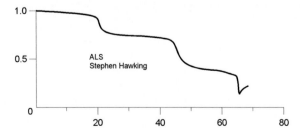

Fig. 5.23 LQC in Amyotrophic Lateral Sclerosis, Stephen Hawking case

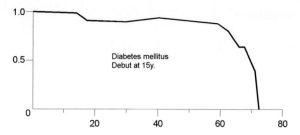

Fig. 5.24 LQC of a patient affected by diabetes with good diagnostic and treatment

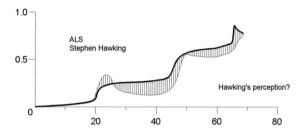

Fig. 5.25 Two possible LQC perceptions for ALS: Physician and patient

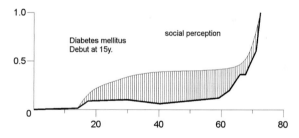

Fig. 5.26 Two possible LIC perceptions for diabetes: Patient and society

that the own patient's perception is different, expressed as an example by the thin curve in the graphic.

The difference in perception between patient and physician is not the only one at play. Society also usually perceives a given condition from a different point of view that the one from the affected person, as we can observe in Fig. 5.26 for diabetes, where the bold line shows a patient's possible own perception of the condition, used to daily subcutaneous insulin injections, while the fine line expresses a possible generalized external perception as a result of social lack of information about diabetes

Needless to say, the shape of Life Illness Curves and Life Quality Curves are not only dependent on the perception of the patient, physician or society, but also from the family environment, social class, economic scenario, politics, etc.

5.5   As a Summary

Lofti Zadeh was the man that back in 1965 realized that the membership of an element x to a set A can be expressed as a real number between 0 and 1. Since in mathematics, any closed interval of real numbers [a,b] contains infinite numbers, then this holds too for the closed interval [0,1] and hence, there are infinite membership degrees between 0 (meaning no membership to a set) and 1 (meaning a whole membership degree to a set). These types of sets are named fuzzy sets.

The seminal paper, written in late 1964 was published in 1965 and titled "Fuzzy Sets". The first words of its abstract say: "A fuzzy set is a class of objects with a continuum of grades of membership. Such a set is characterized by a membership (characteristic) function which assigns to each object a grade of membership ranging between zero and one".

Nowadays, a usual definition of a fuzzy set is the following one: A fuzzy set A is defined by a characteristic function $\mu_A(x)$ that maps every element x belonging to A to the closed interval of real numbers [0,1]. Formally:

$$A = \{(x, \mu_A(x)) \mid x \in A, \mu_A(x) \in [0,1]\}$$

That is, we can create a fuzzy set by means of enumerating a collection of ordered pairs $(x_i, \mu_A(x_i))$ where $\mu_A(x_i)$ is the membership degree of an element x_i to the fuzzy set A.

Geometry is an excellent tool for understanding fuzzy sets. In fact, grid and radar-type graphics are convenient and expressive tools for visualizing the meaning of fuzzy sets. When using grid-type graphics, the elements of the set are represented on the horizontal axis, while their respective membership degrees are shown on the vertical axis. Radar-type graphics are an enhancement of classic Venn diagrams where several contour lines show the different membership degrees. The external contour line usually shows the 0.0 membership degree, and then, the inner the circle, the bigger the membership degree until reaching a 1.0 value, located at the centre of the diagram.

In general, fuzzy sets satisfies all the properties of crisp sets, that is, the identity, idempotent, associative, commutative, De Morgan's Laws and distributive properties are perfectly satisfied by fuzzy sets. However, the complement properties are not satisfied in fuzzy sets theory. Especially significant is the so named 'law of non contradiction'. In classic sets theory it holds that: $A \cap A' = \phi$, but when operating with fuzzy sets, generally it holds that $A \cap A' \neq \phi$. That is, some elements of a fuzzy set A belong both to A and to its complement A'. As examples, some men can be young and old at the same time, some cars can be fast and slow, some houses can be expensive and cheap at the same time, etc. Since we use membership degrees in fuzzy sets there is no contradiction in these statements. In fact these types of statements reflect many times just the things we observe in nature. We can remember in this moment the famous words by A. Einstein: "as far as the laws of mathematics refer to reality, they are not certain, and as far they are certain, they do not refer to reality".

Another important point in this chapter is the fact that membership degrees are not probabilities. If I say that an element x has a membership degree of 0.7 to a fuzzy set A, that is, $\mu_A(x) = 0.7$, I'm affirming something that actually exists or is intrinsic to an existing system. I'm in fact describing in a meaningful way a property of x with respect to the fuzzy set A. On the other hand, if I say that an event x has an associated probability 0.7, that is, $p(x) = 0.7$, I'm giving a measure of the likelihood that this event will happen in the future. As an example: let us imagine a black bag containing seven red balls and three black balls. The probability of extracting a red ball is $p(x) = 0.7$, but this value only exists before extracting the ball. While in probability there is always a random substratum, fuzzy sets theory deals with descriptions of existing, observable features of reality.

Fuzzy relations are an extension of classic, crisp relations between sets. While the classic theory of sets tells us if an element x of a set X is or is not related to an element y in a set Y, fuzzy relations inform us of the strength of the relation between elements, expressed in the closed interval [0,1]. Fuzzy relations can be established between elements from crisp sets or from elements belonging to fuzzy sets. This latter type of relation is the most interesting one for us in this book.

An engaging application of membership degrees in fuzzy sets is the construction of parametric curves of the type $y = \mu(x)$, t where t represents time along the horizontal axis and $\mu(x)$ represents membership degrees along the vertical axis. That is, this type of curves show the membership degree of an element x to a fuzzy set A along time. In this chapter we have seen as an example how the Life Illness Curves, ILC, and Life Quality Curves, LQC, can be implemented in such a way.

In the following chapter we shall continue exposing material from the fuzzy sets theory. I am sure many questions have flourished in the reader's mind after reading this chapter. I hope at least some of them will find its answer in the next one.

References

Bezdek, J.: The Parable of Zoltan. In: On Fuzziness, vol I, pp. 39–46. Springer, Berlin (2013)

Brown, R.: Amyotrophic lateral sclerosis and other motor neurone diseases. In: Harrison's Neurology in Clinical Medicine. McGraw-Hill (2010)

Guyton, A.: Textbook of Medical Physiology. W B Saunders Co (1996)

Klir, G., Yuan, B.: Fuzzy Sets and Fuzzy Logic: Theory and Applications. Prentice Hall (1995)

Seising, R.: The Fuzzification of Systems. Springer, Berlin (2007)

Trillas, E.: Non contradiction, excluded middle, and fuzzy sets. In: Fuzzy Logic and Applications, vol. 5571, pp. 1–11. Springer, Berlin (2009)

Zadeh, L.A.: Fuzzy Sets. Inf. Control **8**, 338–353 (1965)

Chapter 6
From Fuzzy Sets to Linguistic Variables

6.1 Introduction

This chapter is a large one. Not only because it introduces lots of additional material about the theory of fuzzy sets with respect to the previous chapter but because it includes too a big share of the Lisp functions that make up FuzzyLisp, so you will maybe find yourself a bit desperate trying to finish the chapter. Please don't feel so. Try to enjoy it at your own pace; always remembering that in these moments you are just at the core of the book.

FuzzyLisp is, simply put, a collection of Lisp functions that allows to explore the world of Fuzzy Logic theories and develop Fuzzy Logic applications with relative little effort. You can think of it as a small and compact metalanguage that permits you to concentrate in the construction of fuzzy models while still retaining full control of all the Lisp features. Its basic philosophy is the same as the one we discovered in Chap. 4 with the set of functions for managing CSV databases: To offer a set of Lisp building blocks for accomplishing a mission. In this case, to be fluent in fuzzy modeling.

All the functions that make up FuzzyLisp can be downloaded from the book's web site http://www.fuzzylisp.com as a single file named *fuzzylisp.lsp*. This means that you don't need to type code for getting them. However I would like to invite you to read the code in the following pages with attention, and especially, to develop your own examples while playing with the functions. A complete glossary of the metalanguage is placed in Appendix II for quick reference.

In this chapter we shall start to introduce triangular and trapezoidal membership functions with the aid of some geometry, having into account that in this chapter, and in the rest of the book, the expressions "membership function" and "characteristic function" have the same meaning since they are equivalent expressions in fuzzy set theory. Geometry will also help us to understand the concepts of *support, nucleus and alpha-cuts.*

© Springer International Publishing Switzerland 2016
L. Argüelles Méndez, *A Practical Introduction to Fuzzy Logic using LISP*,
Studies in Fuzziness and Soft Computing 327,
DOI 10.1007/978-3-319-23186-0_6

Later, fuzzy sets with discrete characteristic functions are introduced before revisiting the concepts of complement, union and intersection of fuzzy sets. Then an important section is introduced to the reader about fuzzy numbers, including the notion of intervals and interval arithmetic for presenting fuzzy numbers arithmetic and then fuzzy averaging together with a first view on defuzzification.

After all this material is exposed we finally arrive to the concept of linguistic variables, considering those formed by fuzzy sets with triangular or trapezoidal membership functions and also by fuzzy sets with discrete characteristic functions. Before closing the chapter, and from the point of view of a practical application, fuzzy databases are introduced.

6.2   Towards Geometrical Characteristic Functions

In Sect. 5.3.1 from the previous chapter we introduced an example of a fuzzy set A representing old people with membership degrees $\mu_A(35) = 0.1$; $\mu_A(45) = 0.2$; $\mu_A(55) = 0.4$; $\mu_A(65) = 0.7$; $\mu_A(75) = 0.9$; $\mu_A(80) = 1$, that is:

$$A = \{(35, 0.1), (45, 0.2), (55, 0.4), (65, 0.7), (75, 0.9), (80, 1.0)\}$$

where a 35 years-old person has a 0.1 membership degree value to the fuzzy set A, a 45 years old one has a 0.2 membership degree value and so on. However, some questions quickly arise after reflecting a bit on the definition of A, such as: Given A, what is the membership degree of, say, a fifty years-old person, or a seventy years-old one? As smart readers, we can answer that $\mu_A(50)$ will be bigger than 0.2 and smaller than 0.4, while $\mu_A(70)$ will be bigger than 0.7 and smaller than 0.9, that is:

$$0.2 < \mu_A(50) < 0.4$$
$$0.7 < \mu_A(70) < 0.9$$

However, from the definition of A and only from that definition, we cannot know, under a strictly mathematical point of view, the membership degrees $\mu_A(50)$ or $\mu_A(70)$. A workaround for solving this question seems at first to declare the fuzzy set A with bigger resolution, giving, for example, a paired, increasing membership degree for every age from 35 to 80 years old, such as, for example:

$$A = \{(35, 0.1), (36, \mu_A(36)), (37, \mu_A(37)), \ldots, (79, \mu_A(79)), (80, 1.0)\}$$

Now, a grid representation of A could be the one given in Fig. 6.1.

Now for every integer number from 35 to 80 years old we have associated an increasing membership degree. There is no doubt that the simple comparison of Fig. 6.1 against Fig. 5.10 shows that the new, enhanced fuzzy set A offers now a substantially richer description of the concept of old people.

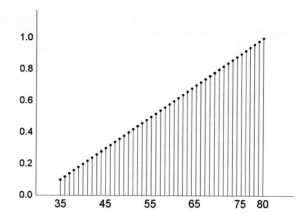

Fig. 6.1 A grid-type diagram for representing the fuzzy set *A* of old people with bigger resolution. See Fig. 5.10 for comparison

Nevertheless, this strategy has introduced a disadvantage. The new inconvenient is the now long enumeration of *A*, not only in a mathematical expression, but in a computational one, too: Expressed in Lisp, the new enhanced set *A* requires a list of 45 sublists representing pairs *(age membership-degree)*, far to be comfortable while declaring it in a Lisp program. Moreover, another question still arises, as can be easily observed with only one question: What is the membership degree of a 68.5 years-old person? Yes, the enhanced version of the fuzzy set *A* covers all the ages represented with integers from 35 to 80, but nothing can be said if we need to know the membership degree, for example, of a 68 years and six months old person. We shall solve these problems later in this chapter, but meanwhile we must say that the strategy of increasing the resolution in the fuzzy set *A* is not bad at all in its very concept. In fact, if we push it further until reaching "infinite resolution" we can arrive to a new representation of fuzzy sets that offer us, in many cases, a more practical approach to describe these types of sets.

"Infinite resolutions" can be easily implemented by means of continuous membership functions, that is, functions of the type $x \rightarrow f(x)$ where x is a point on the real axis and $f(x)$ returns a real number always defined in the closed interval [0,1]. Some examples of this type of functions are given in Table 6.1 (Klir and Yuan 1995).

Table 6.1 Some families of membership functions

	Membership function	x defined in
$\mu_1(x)$	$a(x - b) + 1$	$[b - 1/a, b]$
	$a(b - x) + 1$	$[b, b + 1/a]$
$\mu_2(x)$	$1/(1 + a(x - b)^2)$	$(-\infty, +\infty)$
$\mu_3(x)$	$e^{-\|a(x-b)\|}$	$(-\infty, +\infty)$
$\mu_4(x)$	$(1 + cos(a\pi(x - b)))/2$	$[b - 1/a, b + 1/a]$
	0	Rest of cases

The parameter b represents a real number that adjusts $f(x)$ in such a way that the image $f(x)$ has a value exactly of 1.0 for $x = b$, while the parameter a determines how fast the function increases from the left and decreases from the right with respect to its axis of symmetry. For getting a visual representation of these families of membership functions $\mu_1(x)$, $\mu_2(x)$, $\mu_3(x)$, $\mu_4(x)$, we can have a look at Fig. 6.2 with given parameters $a = 2$ and $b = 2$.

Experience has shown that many practical fuzzy logic based applications can be developed with the family of membership functions given by $\mu_1(x)$, and usually there are no sensible, substantial differences between models built with the triangular family of membership functions and the other ones, $\mu_2(x)$, $\mu_3(x)$ and $\mu_4(x)$, not to mention the difference in complexity that results from computational processes between them. In some applications of fuzzy-logic, such as for example fuzzy control, the algorithms used for computing results from the input variables that represent the status of a system are in some cases critical from the point of view of computational speed, and the family of membership functions expressed by $\mu_1(x)$ satisfy these requirements.

Triangular membership functions can be enhanced by some minor transformations until obtaining trapezoidal functions. This new family of membership functions $\mu_T(x)$ can be expressed by the following set of Eq. (6-1):

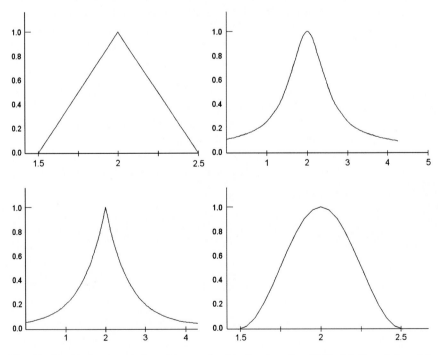

Fig. 6.2 Graphical representation of the membership functions from Table 6.1. From *left to right* and *top to bottom*, $\mu_1(x)$, $\mu_2(x)$, $\mu_3(x)$, $\mu_4(x)$

$$\begin{array}{ll} a(x-b)+1, & \forall x \in [b-(1/a),b] \\ 1, & \forall x \in (b,c) \\ a(c-x)+1, & \forall x \in [c,c+(1/a)] \end{array} \qquad (6\text{-}1)$$

These expressions have four <u>singular points</u> of capital importance for our understanding of fuzzy sets and its implications when representing them by means of a computer language:

$$x_1 = b-(1/a), \quad x_2 = b, \quad x_3 = c, \quad x_4 = c+(1/a) \qquad (6\text{-}2)$$

Needless to say, $f(x) = a(x-b) + 1$ is less than 0.0 when $x < b - (1/a)$ and bigger than 1.0 when $x > b$. The same applies in a symmetrical way to $f(x) = a(c - x) + 1$. Thus, we define adequately the intervals of existence of the function as shown in Eq. (6-1) in order to guarantee that the values returned by the characteristic function is always bounded between 0.0 and 1.0. In other words, the equations defined by (6-1) represent perfectly well trapezoidal shaped fuzzy sets and also triangular shaped fuzzy sets when $b = c$.

As an example, when $a = 2.0$, $b = 2.0$, and $c = 3.0$ the characteristic function, or membership function results as follows:

$$\begin{array}{ll} f(x) = 2x - 3, & \forall x \in [1.5,2] \\ f(x) = 1, & \forall x \in (2,3) \\ f(x) = 7 - 2x, & \forall x \in [3,3.5] \end{array}$$

and substituting the a, b and c values into expression (6-2), we have:

$$x_1 = 1.5, \quad x_2 = 2.0, \quad x_3 = 3.0, \quad x_4 = 3.5$$

Gathering all these ideas into a graphical representation we obtain Fig. 6.3a.

When $a = 2.0$, $b = 2.0$ and $c = 3.0$, the resulting trapezoidal characteristic function starts at $x1 = 1.5$, where its membership degree is zero. It increases from

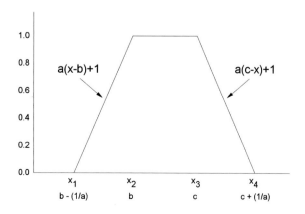

Fig. 6.3a Characteristic functions for trapezoidal shaped fuzzy sets

Fig. 6.3b Ttrapezoidal
characteristic function when *a*
= *2.0*, *b* = *2.0* and *c* = *3.0*

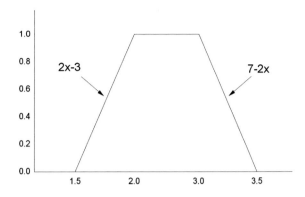

$x_1 = 1.5$ to $x_2 = 2.0$ until getting a membership degree equal to one. Then it maintains this value from $x_2 = 2.0$ to $x_3 = 3.0$, and finally it starts to decrease from $x_3 = 3.0$ until reaching $x_4 = 3.5$ where the membership degree is again zero. Figure 6.3b shows these values.

6.3   From Geometry to FuzzyLisp

In Sect. 2.5 from Chap. 2 of this book the reader enjoyed (I hope so) how well suited the Lisp language is to represent geometrical shapes. I would like to invite the reader to observe again Fig. 2.6. There, the list named *simple-composition* stored the value *((2 (4 4) (6 4) 8) (6 (8 4) 10) (8 (10 4) (12 4) 14))*, and Fig. 2.6 graphically represented that list. Was it a case of fuzzy-sets? Not exactly, but not far from it! No mention was made then, but now that we are immersed in the theory I must confess that in some way I was trying to guide your mind from the start of the book. We were learning Lisp, of course, but geometry has a tight relationship with fuzzy sets and fuzzy logic, so it was a great opportunity to pave the way towards undiscovered (by then) ideas and theories.

The expressions shown in (6-2) give us the key for representing fuzzy sets in Lisp. The following Lisp expression constitutes what we are going to name "FuzzyLisp Standard Set Representation", FLSSR:

$$(fuzzy\text{-}set\text{-}name\ x1\ x2\ x3\ x4) \tag{6-3}$$

As can be seen, the structure of a list representing a fuzzy set is composed by a name, expressed as a symbol, and then four real numbers x_1, x_2, x_3 *and* x_4, representing the singular points from (6-2). The FuzzyLisp Standard Set Representation, assumes two nuclear, basic points:

Table 6.2a Combination of singular points in trapezoidal membership functions

Combination	Meaning	Lisp example
$x_1 \neq x_2 \neq x_3 \neq x_4$	Normal trapezium	*(normal-tp 1.5 2 3 3.5)*
$x_1 = x_2 \neq x_3 \neq x_4$	Right trapezium by-left	*(left-tp 1.5 1.5 3 3.5)*
$x_1 \neq x_2 \neq x_3 = x_4$	Right trapezium by-right	*(right-tp 1.5 2 3.5 3.5)*

(a) All the fuzzy sets in the FuzzyLisp Standard Set Representation can be expressed by means of a trapezoidal or triangular shaped characteristic function.

(b) All the fuzzy sets in the FuzzyLisp Standard Set Representation have at least a point x_i on the real axis such as $\mu(x_i) = 1$. That is, all the fuzzy sets created with this representation must have at least one point whose membership degree equals to one.

The first point can be seen at first as restrictive, but aside helping to develop an efficient, compact and elegant layer of Lisp functions for creating fuzzy logic based applications it still offers a high level of flexibility resulting from several arrangements and combinations of the singular points from trapezoidal and triangular membership functions. Table 6.2a shows all the possible combinations for trapezoidal characteristic functions.

The graphical representation of these combinations can be seen in Fig. 6.4a.

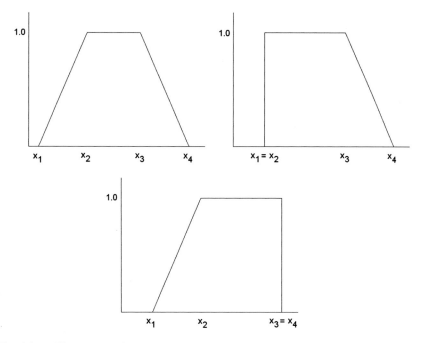

Fig. 6.4a Different types of trapezoidal shaped fuzzy sets

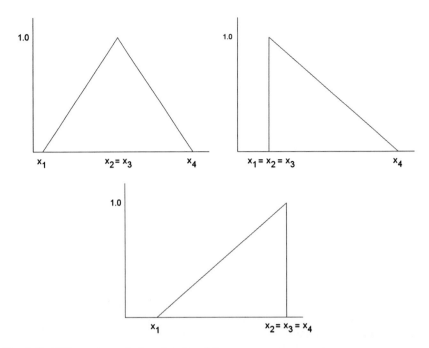

Fig. 6.4b Different types of triangular shaped fuzzy sets

Conversely, Table 6.2b shows now all the possible combinations for triangular characteristic functions.

The graphical representation of these combinations can be seen in Fig. 6.4b.

After the exposed material, declaring fuzzy sets at the Lisp prompt is straightforward:

> (setq mature-age '(mature 35.0 45.0 55.0 75.0))
: *(mature 35 45 55 75)*

Here the symbol *mature-age* stores a trapezoidal shaped fuzzy set named *mature* where $x_1 = 35$, $x_2 = 45$, $x_3 = 55$ and $x_4 = 75$. Let us see some other examples for representing temperatures. The Universe of discourse, that is, the range of values over the real axis, X, where the fuzzy sets are defined, is bounded in this example between 0 and 100 °C. First, let us use a right trapezium by-left for expressing the meaning of cold temperature:

Table 6.2b Combination of singular points in triangular membership functions

Combination	Meaning	Lisp example
$x_1 \neq x_2 = x_3 \neq x_4$	Normal triangle	*(normal-tr 1.5 2.5 2.5 3.5)*
$x_1 = x_2 = x_3 \neq x_4$	Right triangle by-left	*(left-tr 1.5 1.5 1.5 3.5)*
$x_1 \neq x_2 = x_3 = x_4$	Right triangle by-right	*(right-tr 1.5 2 3.5 3.5)*

> **(setq cold-temperature '(cold 0.0 0.0 10.0 20.0))**
: (cold 0 0 10 20)

Now the same goes for comfortable temperature. We shall use a symmetrical, normal trapezium:

> **(setq comfortable-temperature '(medium 10.0 20.0 30.0 40.0))**
: (medium 10 20 30 40)

And finally, let us see an example of use of a right trapezium by-right for expressing hot temperatures:

> **(setq hot-temperature '(hot 30.0 40.0 100.0 100.0))**
: (30 40 100 100)

The previously stated point (b) is also important: The FuzzyLisp Standard Set Representation only supports fuzzy sets that have at least one point defined on the universe of discourse whose image $f(x)$, that is, its membership degree, equals to one. In fuzzy sets theory, this type of sets are named <u>Normal fuzzy sets</u>. Formally:

$$\exists x \in [x_1, x_4]/f(x) = 1.0 \qquad (6\text{-}4)$$

Conversely, those fuzzy sets whose membership degree is always less than 1.0 are called Subnormal Fuzzy Sets, that is:

$$\forall x \in [x_1, x_4], f(x) < 1.0 \qquad (6\text{-}5)$$

A graphical example of normal and subnormal fuzzy set is shown in Fig. 6.5.

In the figure is easy to see how the trapezoidal fuzzy set reaches a membership degree $f(x) = 1.0$, so it is a normal fuzzy set. On the other hand, the triangular shaped fuzzy set has a maximum membership degree $f(x) = 0.8$, so it is a subnormal fuzzy set.

In order to know if a real number x belongs to a fuzzy set, either triangular or trapezoidal shaped, we can write a simple FuzzyLisp predicate. In fact we only need to know if x is included inside the interval $[x_1, x_4]$. If it certainly is, the predicate returns true, else it returns nil. Code 6-1 shows it:

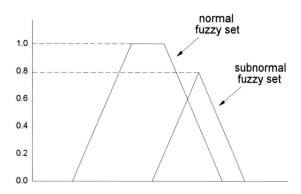

Fig. 6.5 An example of normal and subnormal fuzzy sets

```
;code 6-1
(define (fl-belongs? fset x)
    (if (and (>= x (nth 1 fset)) (<= x (nth 4 fset)))
            true
            nil
    )
)
```

For testing this predicate, and having into account the previous definitions of the fuzzy sets *cold*, *medium* and *hot*, we can try, for example: *(fl-belongs? cold-temperature 7.0)* → *true*, *(fl-belongs? comfortable-temperature 23.0)* → *true*, but as expected: *(fl-belongs? hot-temperature 12.0)* → *nil*.

A step further in computational complexity arises when we are interested not only in knowing if an element *x* belongs or not to a fuzzy set *A*, but also in knowing its membership degree if it indeed belongs to *A*. The function *(fl-set-membership?)* returns the membership degree of a given element *x* defined on its correspondent universe of discourse to a given fuzzy set *A*. Under a computational point of view this function plays a key role in FuzzyLisp because it is in fact a characteristic function in disguise, so we shall call it extensively from many other FuzzyLisp functions. Code 6-2 shows it:

```
;code 6-2
(define (fl-set-membership? fset x,
            name x1 x2 x3 x4 membership-degree)
    (setq name (nth 0 fset)) ;fuzzy set name
    (setq x1 (nth 1 fset)) ;support starts
    (setq x2 (nth 2 fset)) ;nucleus starts
    (setq x3 (nth 3 fset)) ;nucleus finishes
    (setq x4 (nth 4 fset)) ;support ends

    ;x <= x1 | x >= x4 => membership-degree = 0.0
    (if (or (<= x x1) (>= x x4))
      (setq membership-degree 0.0)
    )

    ;x >x1 && x < x2 => membership-degree increasing
    (if (and (> x x1) (< x x2))
      (setq membership-degree (div (mul (sub x x1) 1.0)
                          (sub x2 x1)))
    )

    ;nucleus membership degree is always 1.0
    (if (and (> = x x2) (<= x x3))
      (setq membership-degree 1.0)
    )
```

```
;x > x3 && x < x4 => membership-degree decreasing
(if (and (> x x3) (< x x4))
   (setq membership-degree (div (mul (sub x4 x) 1.0)
                                 (sub x4 x3))))
)

(list name membership-degree) ;gives results as a list
)
```

The design of the function is easy to follow: On the first lines the name of the fuzzy-set and the singular values x_i are stored in some variables and then it checks if x is inside x_1 and x_4. Later, the fuzzy set is scanned from left to right, calculating the membership values using simple analytic geometry. First it scans the increasing part of the characteristic function, then it tests if $f(x)$ equals to one in the "normal" section of the fuzzy set, and finally it scans the decreasing part of the characteristic function. Let us try the function at the Lisp prompt for the already defined fuzzy sets for temperature:

> (fl-set-membership? comfortable-temperature 25)
: (medium 1)

Now, let us try another interesting value of temperature, $x = 35\ ^\circ C$:

> (fl-set-membership? comfortable-temperature 35)
: (medium 0.5)

> (fl-set-membership? hot-temperature 35)
: (hot 0.5)

As we learned in Sect. 5.3.3 of this book, the "law of non contradiction" does not hold in fuzzy sets theory. From our definitions of temperatures in a range from 0.0 to 100.0 $^\circ C$, a temperature $x = 35$ belongs both to the fuzzy sets "comfortable-temperature" and "hot-temperature". Hereafter we assume that the reader is already familiar with this important concept in the theory and we shall not make additional comments to it.

As can be seen, and because a consideration of design, the function *(fl-set-membership?)* returns a list containing both the name of the fuzzy set and the calculated membership degree. For obtaining only the membership degree, we only would need to type the following expression at the Lisp prompt:

> (last (fl-set-membership? hot-temperature 35))
: 0.5

Code 6-1 and Code 6-2 could be unified as follows in Code 6-3:

```
;Code 6-3
(define (fl-belongs2? fset x)
   (if (and (>= x (nth 1 fset)) (<= x (nth 4 fset))))
```

```
      (fl-set-membership? fset x)
      nil;else returns nil
   )
)
```

And then, for example, *(fl-belongs2? comfortable-temperature 22)* → *(medium 1)*, but *(fl-belongs2? comfortable-temperature 42)* → *nil*.

The reader should note that *(fl-set-membership?)* and other related FuzzyLisp functions ending in a quotation mark are not Lisp predicates, that is, functions that return either *true* or *nil*, but functions that are used to interrogate a fuzzy model about something, usually the membership of a crisp value *x*. This has been unavoidable because the English word "set", aside "collection of things", also means "to adjust", as the reader well knows. The question mark at the end of the function intends to clarify the true meaning of the functions where it applies. As an example, the FuzzyLisp function *(fl-set-membership?)* should be translated into plain English as "What is the membership degree of x in a fuzzy set *A*?"

6.4 Support, Nucleus and Alpha-Cuts

As usually happens in science, terminology is important because technical terms help to explain concepts and to avoid vagueness when handling them. In this section we are going to explain three basic terms directly related with fuzzy sets.

The support, *s*, of a fuzzy set *A* is, conceptually, the horizontal segment of the X axis where a fuzzy set rests, that is, the horizontal segment bounded by the closed interval $[x_1, x_4]$. Formally:

$$s = \mathrm{x}/\mathrm{x} \in [x_1, x_4] \tag{6-6}$$

In other words, *s* is the geometrical projection of the entire membership function of a fuzzy set over the X axis.

The nucleus, *k*, of a fuzzy set *A* is the geometrical projection over the real X axis of the part of the membership function of a fuzzy set whose membership degree equals to 1.0, that is:

$$k = \mathrm{x}/\forall x \in [x_2, x_3], f(x) = 1.0 \tag{6-7}$$

From this definition is easy to follow that:

- Subnormal fuzzy sets have not a nucleus.
- The nucleus of a triangular shaped, normal fuzzy set is only a point over the X real axis.

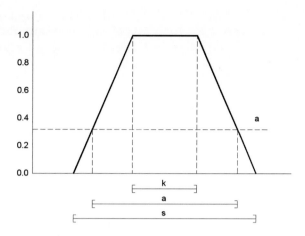

Fig. 6.6 Nucleus (k), support (s) and an alpha-cut (a) in a fuzzy set

$$\exists x \in [x_1, x_4]/f(x) = 1.0 \qquad (6\text{-}4)$$

Figure 6.6 will help the reader to better understand these terms.

An alpha-cut is also a segment, or closed interval, formed by a set of points over the X axis, but in this case it is produced by the projection of a cut given to the characteristic function of a fuzzy set by a horizontal line $y = \alpha$, where $\alpha \in [0,1]$. Formally:

$$^{\alpha}a = x/f(x) \geq \alpha \qquad (6\text{-}8a)$$

A strong alpha-cut is another crisp set of points that satisfy the following expression:

$$^{\alpha+}a = x/f(x) > \alpha \qquad (6\text{-}8b)$$

That is, an alpha-cut produces a closed interval, while a strong alpha-cut results into an open interval.

Alpha-cuts management is important in fuzzy-sets theory because they represent a nexus between membership functions and intervals defined on the real axis. Due to this we need to have in our FuzzyLisp toolbox a function for calculating alpha-cuts. Code 6-4 shows the function *(fl-alpha-cut)*:

```
;code 6-4
(define (fl-alpha-cut fset alpha,
                name x1 x2 x3 x4 extrem_left
                extrem_right tan_phi1 tan_phi2
                fraction numerator)

    setq name (nth 0 fset)) ;fuzzy set name
    (setq x1 (nth 1 fset)) ;support starts
```

```
(setq x2 (nth 2 fset)) ;nucleus starts
(setq x3 (nth 3 fset)) ; nucleus finishes
(setq x4 (nth 4 fset)) ;support ends

;left extrem of alpha-cut begins, vertical mf case:
(if (= x1 x2)
   (setq extrem_left x1)
)

;left extrem of alpha-cut begins, ascending mf case:
(if (! = x1 x2)
   (begin
      (setq tan_phi1 (div 1.0 (sub x2 x1)))
      (setq fraction (mul tan_phi1 x2))
      (setq numerator (sub (add fraction alpha) 1.0))
      (setq extrem_left (div numerator tan_phi1))
   )
)

;right extrem of alpha-cut begins, vertical mf case:
(if (= x3 x4)
   (setq extrem_right x4)
)

;right extrem of alpha-cut begins, descending mf case:
(if (! = x3 x4)
   (begin
      (setq tan_phi2 (div 1.0 (sub x4 x3)))
      (setq fraction (mul tan_phi2 x3))
      (setq numerator (sub (add fraction 1.0) alpha))
      (setq extrem_right (div numerator tan_phi2))
   )
)
;returns the alpha-cut as a list
(list name extrem_left extrem_right)
)
```

The inner workings of this function are not hard to follow. It scans a trapezoidal or triangular membership function from left to right and after some basic analytic geometry calculations it returns the obtained alpha-cut as a list, including the name of the original fuzzy set. Let us try this function at the Lisp prompt:

> **(fl-alpha-cut '(B1 7 10 12 15) 0.7)**
: *(B1 9.1 12.9)*

For a value of $\alpha = 0.7$, the trapezoidal shaped fuzzy set *(B1 7 10 12 15)* produces an alpha-cut represented by the interval $a = [9.1, 12.9]$. Please note that the support of this set is the interval $s = [7, 15]$, and its nucleus is the interval $k = [10, 12]$. The following relationship always holds:

$$k \leq a \leq s \qquad (6\text{-}9)$$

Let us try two other interesting values of α, $\alpha = 0.0$ and $\alpha = 1.0$:

> **(fl-alpha-cut '(B1 7 10 12 15) 0.0)**
: *(B1 7 15)*

> **(fl-alpha-cut '(B1 7 10 12 15) 1.0)**
: *(B1 10 12)*

After experimenting with these values of alpha, we can expose two important corollaries:

- The support of a fuzzy set is equal to its $\alpha = 0$ strong alpha-cut, ^{0+}a.
- The nucleus of a fuzzy set is equal to its $\alpha = 1$ alpha-cut, ^{1}a.

If we observe Fig. 6.6 with attention, and having these two corollaries in mind, we soon shall realize we can define a triangular or trapezoidal membership function if two alpha-cuts are given. If both the 0-strong-alpha-cut and 1-alpha-cut from a triangular or trapezoidal membership function are known the explanation is immediate: from ^{0+}a we obtain the x_1, x_4 singular values of the fuzzy set and from $^{1.0}a$ we obtain the x_2, x_3 singular values of the fuzzy set. For any other alpha-cuts, for example, $^{0.25}a$ and $^{0.75}a$, we shall need to compute some analytic geometry. Figure 6.7 gives an extra help in understanding this notion.

The FuzzyLisp function *(fl-def-set)* translates these ideas into Lisp code. It has three parameters: a name for naming the resulting fuzzy set and then two alpha cuts expressed as lists in the following form: *(extreme-left extreme-right alpha-cut-value)*. The function automatically identifies if the resulting fuzzy set will have a triangular or trapezoidal shape. The only required condition for using it is that the

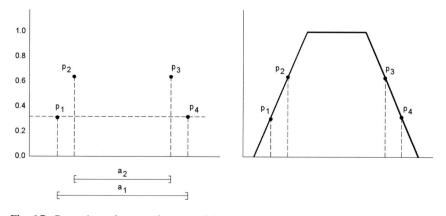

Fig. 6.7 Generating a fuzzy set from two alpha-cuts

first supplied value for the alpha-cut must be less than the second one, that is, $\alpha 1 < \alpha 2$. Code 6-5a shows the function. Although the function is a bit long it is not especially complex, especially if you read the comments included in the code. It must be also mentioned that the internal variables $m1$ and $m2$ are used for calculating the slopes of the lines that represent the resulting triangular or trapezoidal membership functions. For these calculation a dedicated function, named *(fl-aux-calculate-m)* is shown in Code 6-5b:

```
;Code 6-5a
(define (fl-def-set name a-cut1 a-cut2,
        triangle x1-a x1-b x2-a x2-b alpha-1 alpha-2 m1 m2
        base-1 base-2 base-3 base-4 temp-x temp-y)

    ;initially we assume it is not a triangular mf:
    (setq triangle nil)
    (setq x1-a (nth 0 a-cut1))
    (setq x1-b (nth 1 a-cut1))
    (setq alpha-1 (nth 2 a-cut1))
    (setq x2-a (nth 0 a-cut2))
    (setq x2-b (nth 1 a-cut2))
    (setq alpha-2 (nth 2 a-cut2))

    (if (< (abs (sub x1-a x2-a)) 0.000001)
        (setq m1 1e + 300) ;slope's tangent equals to infinity
        (setq m1 (fl-aux-calculate-m x1-a x2-a alpha-1
        alpha-2))
    )

    (if (< (abs (sub x1-b x2-b)) 0.000001)
        (setq m2 1e + 300)
        (setq m2 (fl-aux-calculate-m x1-b x2-b alpha-1
        alpha-2))
    )

    ;calculation of X axis intersections:
    ;base-1 and base_4 are the extremes of the set's support
    (setq base-1 (sub x1-a (div alpha-1 m1)))
    (if (< m2 0.0)
        (setq m2 (mul m2 -1.0)) ;absolute value
    )
    (setq base-4 (add (div alpha-1 m2) x1-b))
```

```
;base-2 and base-3 represent the extremes of the set's
nucleus
(setq base-2 (div (add 1 (mul m1 base-1)) m1))
(setq base-3 (div (sub (mul m2 base-4) 1.0) m2))

;check if the set will have a triangular membership
function
(if (> = base-2 base-3)
(begin
    (setq triangle true) ,
    (setq temp-x
            (div (add (mul m1 base-1) (mul m2 base-4))
                        (add m1 m2)))
    (setq temp-y (mul m1 (sub temp-x base-1))))
) ;end begin
) ;end if

(if (= triangle true)
    (list name base-1 temp-x temp-x base-4);it's a
    triangle
    (list name base-1 base-2 base-3 base-4);it's a
    trapezium
) ;end if
);end function

;Code 6-5b
(define (fl-aux-calculate-m x1 x2 y1 y2)
        (div (sub y2 y1) (sub x2 x1))
)
```

Let's us try the function with two simple alpha cuts expressed by the lists *(15.0 35.0 0)* and *(25.0 25.0 1.0)*, where we want to create a fuzzy set representing young age. From the first list we can observe that the support will extend from 15 to 35 years old, while from the second list is easy to see that the nucleus will extend from 25 to 25 years old (yes, it is a triangular shaped fuzzy set!):

> (fl-def-set 'young '(15.0 35.0 0) '(25.0 25.0 1.0))
: (young 15 25 25 35)

Let us now try a call for obtaining a trapezoidal shaped fuzzy set:

> (fl-def-set 'mature '(35.0 75.0 0) '(45.0 55.0 1.0))
: (mature 35 45 55 75)

In this moment, is probable that you would like to stop me for a moment and formulate an interesting question: Wouldn't be easier to simple type the following expressions at the Lisp prompt: *(setq age1 '(young 15 25 25 35))* and *(setq age2 '(mature 35 45 55 75))*?. The answer is simple: "horses for courses", as they usually say. Depending on the intended use while programming, *(fl-def-set)* will be the natural election, as we shall see later in this chapter.

6.5   Fuzzy Sets with Discrete Characteristic Functions

The FuzzyLisp standard set representation has strong practical advantages derived from its simplicity and computational compactness for creating fuzzy-logic applications. However we cannot easily represent subnormal fuzzy sets or other fuzzy sets where the characteristic function has not a triangular or trapezoidal shape. For solving this situation we introduce the FuzzyLisp Discrete Set Representation, FLDSR, where we express a fuzzy set by means of a discrete characteristic function defined by a Lisp list with the following structure:

$$(fuzzy\text{-}set\text{-}name\ (x_1 \mu(x_1))\ (x_2 \mu(x_2))\ldots(x_n \mu(x_n))) \qquad (6\text{-}10)$$

where again *fuzzy-set-name* is an identifier and each sublist $(x_i\ \mu\ (x_i))$ is formed by a point x on the real axis and its corresponding membership value $\mu\ (x_i)$. Figure 6.1 shows an example of a fuzzy set with a discrete characteristic function.

As stated in Sect. 6.2, the main inconvenient for using a FuzzyLisp discrete representation is its long manual enumeration when writing a Lisp program that uses it. However, we can use the language itself for helping to build fuzzy sets with discrete characteristic functions. The following function, named *(fl-discretize)* takes a trapezoidal or triangular FuzzyLisp standard set representation and discretizes it into a FuzzyLisp discrete representation as shown in Code 6-6:

```
;Code 6-6
(define (fl-discretize fset steps,
          name x1 x2 x3 x4 i resolution list-out x
          trapezium)
   (setq name (nth 0 fset))
   (setq x1 (nth 1 fset) x2 (nth 2 fset) x3 (nth 3 fset)
          x4 (nth 4 fset))
   (setq list-out (list name))
   (setq trapezium true)

   ;discretize from x1 to x2:
   (setq resolution (div (sub x2 x1) steps))
   (setq i 0)
   (setq x x1)
   (while (< i steps)
        (setq list-out
```

```
              (cons (list x (last (fl-set-membership? fset x))))
                        list-out))
       (setq x (add x resolution))
       (++ i)
   ); end while

   ;discretize from x2 to x3
   (if (< (sub x3 x2) 0.0000001) ;testing if fset is a
   triangle
   (begin
       (setq list-out
           (cons (list x2 1.0) list-out)) ;if it is only one
           point
       (setq trapezium nil)
   )
   ;else it is a trapezium:
   (begin
       (setq resolution (div (sub x3 x2) steps))
       (setq i 0)
       (setq x x2)
       (while (< i steps)
           (setq list-out (cons (list x 1.0) list-out))
           (setq x (add x resolution))
           (++ i)
       );end while
   );end begin
   );end if

   ;finally, discretize from x3 to x4:
   (setq resolution (div (sub x4 x3) steps))
   (setq i 0)
   (if (= trapezium true)
       (setq x x3)
       (setq x (add x3 resolution)) ;it's a triangle
   );if end
   (while (< i steps)
       (setq list-out
           (cons (list x (last (fl-set-membership? fset x))))
                        list-out))
       (setq x (add x resolution))
   (++ i)
   ); end while
```

```
;add the last element corresponding to x4
;if fset is a trapezium:
(if (= trapezium true)
    (setq list-out (cons (list x 0.0) list-out))
)
(reverse list-out)
)
)
```

This function takes as arguments a trapezoidal or triangular shaped membership function and the desired number of steps for discretizing the function. Then, every line that composes the trapezium or triangle is divided in n steps, and for each division the membership degree is calculated. For instance, in the trapezium case we first discretize the segment of the membership function corresponding to the interval x_1, x_2, then for the interval x_2, x_3 and finally for the interval x_3, x_4. Albeit being a long function, the Lisp expressions used inside are not especially complex. One of them, however, deserves our attention:

```
(cons (list x (last (fl-set-membership? fset x))) list-out))
```

The key point is the call to the function *(fl-set-membership?)* that calculates the membership degree of x to *fset*, being x a changing value that scans the complete support of the function *fset* step by step from left to right. Then the function *(list)* merges both the value of x and its obtained membership degree into a list. Finally the function *(cons)* adds the sublist to *list-out*. Another line of code seems initially intriguing:

$$(if (< (sub x3 x2) 0.0000001)$$

This conditional structure allows to automatically check if the membership function to discretize has a triangular or trapezoidal shape. It examines the x_2 and x_3 values by means of a simple subtraction. Just note that we don't use the expected expression *(=x3 x2)* because in technical programming it is not safe to compare two real numbers! Not only in Lisp but also in any computer language. Only integer numbers should be compared in such a way.

Now let us test the function. Under normal use, it is enough to use three or four steps because the functions to discretize, either triangular or trapezoidal, are linear. For discretizing the trapezoidal fuzzy set expressed by (B1 7 10 12 15): using four steps we only need to type:

> (setq dA (fl-discretize '(B1 7 10 12 15) 4))
: (B1 (7 0) (7.75 0.25) (8.5 0.5) (9.25 0.75) (10 1) (10.5 1) (11 1) (11.5 1) (12 1) (12.75 0.75) (13.5 0.5) (14.25 0.25) (15 0))

As expected, FuzzyLisp has divided the three segments of the trapezium from $x = 7$ to $x = 10$, from $x = 10$ to $x = 12$ and from $x = 12$ to $x = 15$ in four parts, respectively. It has calculated every associated membership degree and finally has

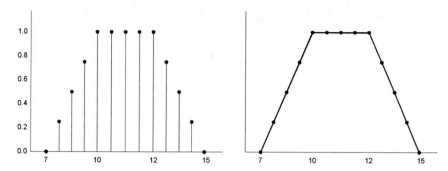

Fig. 6.8 Two versions of the same trapezoidal shaped fuzzy set: continuous and discrete characteristic functions

produced a FuzzyLisp discrete set representation stored in *dA*. Figure 6.8 shows both characteristic functions, discrete and continuous.

Now we need an additional FuzzyLisp function. A function that, given a FuzzyLisp discrete set representation and any value *x* from its support, returns its interpolated membership degree, $\mu(x)$. In other words: we need a function equivalent to *(fl-set-membership?)* for dealing with discrete membership functions. Such a function is shown in Code 6-7a and 6-7b:

```
;Code 6-7a
(define (fl-dset-membership? dfset x, i n pair-a pair-b
result)
     (setq result (list (first dfset) 0.0))
     (setq i 1) ;we are not interested in dfset's name anymore
     (setq n (length dfset))

     (while (< i (- n 1)) ;traverse list taking care at the end
       (setq pair-a (nth i dfset))
       (setq pair-b (nth (+ 1 i) dfset))

       ;if x is bounded:
       (if (and (<= (first pair-a) x) (>= (first pair-b) x))
         ;we pass the shift at right from the left value:
         (setq result (list (first dfset)
           (interpolation pair-a pair-b (sub x (first
           pair-a)))))
       ); end if
       (++ i)
     ); end while
     result
   )
```

For simplicity of design, *(fl-dset-membership?)* uses an auxiliary function for interpolation as shown in Code 6-7b:

```
;code 6-7b
(define (interpolation pair-a pair-b p, a b y1 y2)
    ;extract values:
    (setq a (first pair-a) b (first pair-b))
    (setq y1 (last pair-a) y2 (last pair-b))
    ;calculate the interpolation using simple analytic
    geometry
    (add y1 (div (mul p (sub y2 y1)) (sub b a))))
)
```

The function *(fl-dset-membership?)* is rather simple. It traverses the list representing a discrete fuzzy set until finding two bounding sublists *pair-a*, *pair-b*, geometrically located at left and right respectively from the *x* value supplied as an argument to the function. After both sublists are identified, the function makes a call to *(interpolation)*, thus obtaining the desired interpolated membership degree. Figure 6.9 shows the used magnitudes for the interpolation algorithm used in Code 6-5b.

From the simple observation of Fig. 6.9 we can easily derive the value of *q*:

$$q = p(y_2 - y_1)/(b - a) \tag{6-11}$$

Only a few paragraphs above we discretized the trapezoidal fuzzy set expressed by the list *(B1 7 10 12 15)*, obtaining its FuzzyLisp discrete set representation counterpart and storing it in the Lisp symbol *dA*. Now, for testing the function, when *x = 8.2*, we can type the following expression at the Lisp prompt.

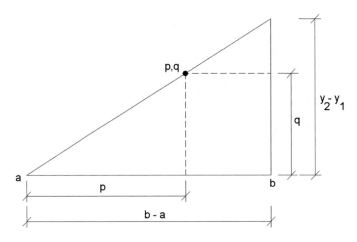

Fig. 6.9 Analytic geometry for interpolation

> **(fl-dset-membership? dA 8.2)**
: *(B1 0.4)*

After obtaining 0.4 as the membership degree from the discretized version of the fuzzy set I am sure the reader is really interested in knowing the membership degree corresponding to $x = 8.2$ when directly using the FuzzyLisp standard set representation *(B1 7 10 12 15)*. For testing this, we are going to use the appropriate FuzzyLisp function, *(fl-set-membership?)*, as follows:

> **(fl-set-membership? '(B1 7 10 12 15) 8.2)**
: *(B1 0.4)*

The reader can play with different values of x and different fuzzy sets using both a FuzzyLisp standard and discrete set representation in order to understand the differences and also the equivalences of using both forms of representation. The functions *(fl-discretize)*, *(fl-dset-membership?)* and *(fl-set-membership?)* open an entire universe of possibilities for fuzzy set modeling as we are going to discover.

The use of discrete characteristic functions suggest us the opportunity to design a FuzzyLisp function named *(fl-discretize-fx)* for discretizing any continuous function $y = f(x)$. In this way, we shall be able to use any of the membership functions shown in Fig. 6.1. Code 6-8 shows the way to do it:

```
;Code 6-8
(define (fl-discretize-fx name fx steps a b,
          x resolution list-out)
   ;the first element is the associated name
   (setq list-out (list name))
   (setq resolution (div (sub b a) steps))
   (setq x a);start from left ro right
   (while (<= x (add b 0.00001))
       ;eval fx, make a list (x fx) and cons it to list-out
       (setq list-out (cons (list x (eval fx)) list-out))
       (setq x (add x resolution))
   ); while end
   (reverse list-out)
)
```

The function takes five arguments: The first one is the name for the expected discretized fuzzy set, then it takes the $y = f(x)$ function to discretize expressed into Lisp form, and then the number of required steps between the start point $x = a$ and end point $x = b$. This FuzzyLisp function is rather simple: First it initializes the resulting list, that is, the desired discretized fuzzy-set, calculates the required resolution and then it evaluates the supplied function inside the while loop with the following Lisp expression:

```
(setq list-out (cons (list x (eval fx)) list-out))
```

As the reader already recognizes, it is a consing expression where the sublist formed by the scanning value x and its image $f(x)$ is added to *list-out*, the expected return of this FuzzyLisp function. Now let us try it using the membership function $\mu_4(x)$ from Fig. 6.2. Adjusting the parameters a and b to $a = 2$ and $b = 2$, respectively, we obtain the following expression:

$$y = (1 + cos(2\pi(x - 2)))/2, \ \forall x \in [1.5, 2.5]$$

Translating it into a Lisp expression and assigning it to the symbol f, we have:

$$(setq f \ `(div \ (add \ 1.0 \ (cos \ (mul \ 2.0 \ pi \ (sub \ x \ 2.0))))) \ 2.0))$$

Since it contains a trigonometric expression we must remember to include the value of Pi before calling the function by typing *(setq pi 3.1415926535)* at the Lisp Prompt. Then we can finally call the function:

> (setq dBell (fl-discretize-fx 'Bell f 10 1.5 2.5))
(Bell (1.5 0) (1.6 0.09549150283) (1.7 0.3454915028) (1.8 0.6545084972) (1.9 0.9045084972) (2 1) (2.1 0.9045084972) (2.2 0.6545084972) (2.3 0.345491502) (2.4 0.09549150283) (2.5 0))

The resulting fuzzy set *Bell* can be seen graphically in Fig. 6.10.

Aside making the call to the function *(fl-discretize-fx)* we have also assigned the resulting fuzzy set to the symbol *dBell*. Now if we wish to obtain the membership value of any other $x \in [1.5, 2.5]$, let us say, for example, $x = 1.75$, we only need to type the following at the Lisp prompt:

> (fl-dset-membership? dBell 1.75)
: (Bell 0.5)

This is extremely important because as already stated with the function *(fl-set-membership?)*, *(fl-dset-membership?)* is also in itself a membership function in

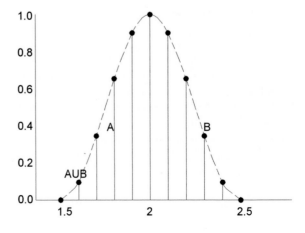

Fig. 6.10 Graphical representation of the discretized characteristic function y = (1 + cos (2π(x − 2)))/2

disguise, because both functions determine not only if a value $x \in R$ belongs to a given fuzzy set A, but also provides us its membership degree to A.

Some words must be said about resolution before closing this section: If we wish to discretize a triangular or trapezoidal shaped fuzzy set, the required number of steps does not need to be big and it can be as low as 2. Let us experiment it:

> **(fl-discretize '(B1 7 10 12 15) 2)**
: (B1 (7 0) (8.5 0.5) (10 1) (11 1) (12 1) (13.5 0.5) (15 0))

Pushing it at the minimum resolution, with only one step, we have:

> **(fl-discretize '(B1 7 10 12 15) 1)**
: (B1 (7 0) (10 1) (12 1) (15 0))

And this is enough to obtain any interpolated $\mu(x)$ value. Again:

> **(fl-dset-membership? (fl-discretize '(B1 7 10 12 15) 1) 8.2)**
: (B1 0.4)

The applied extreme discretization shows the big similarity between the discrete and continuous representation of a trapezoidal or triangular shaped fuzzy set:

$$(B1\ 7\ 10\ 12\ 15)$$
$$(B1(7\ 0)(10\ 1)(12\ 1)(15\ 0))$$

However, when we wish to discretize continuous $f(x)$ functions by means of *(fl-discretize-fx)*, the resolution must be increased accordingly. For the bell shaped function shown in Fig. 6.10 we used 10 steps for the sake of simplicity in the exposition, but in practical applications it is better to take 15 or 20 steps and thus, still maintaining a good level of precision with respect to the original $f(x)$ function.

6.6 Revisiting Complement, Union and Intersection of Fuzzy Sets

As we saw in Sect. 5.3.2 from the previous chapter, simple mathematical expressions provide us formal definitions for the complement, union and intersection of fuzzy sets. When handling fuzzy sets with continuous characteristic functions, these formal expressions must be slightly rewritten in order to maintain the mathematical rigor.

Complement:

$$A' = 1 - \mu_A(x)/x \in [x_1, x_4] \tag{6-12a}$$

$$A' = 1 \quad \forall x \notin [x_1, x_4], x \in U \tag{6-12b}$$

That is, for every value x included in the universe of discourse U, the membership degree of the complement A' of a fuzzy set A is given by the expression $1 - \mu_A(x)$ when x is contained inside the support of A, and it is equal to 1 when x is outside the support of A but is still contained into U. Graphically we can see it in Fig. 6.11, where A' is shown in thick line.

Union: The union $A \cup B$ of two fuzzy sets A and B is formally obtained by the following expression:

$$A \cup B = max[\mu_A(x), \mu_B(x)]/x \in [x_{1A}, x_{4B}] \qquad (6\text{-}13)$$

That is, the union of two fuzzy sets A, B is obtained by the maximum membership degree of either $\mu_A(x)$ or $\mu_B(x)$ when x is contained in the closed interval formed by the extreme left of the support of A and the extreme right of the support of B. Figure 6.12 Graphically shows in a thick line the union of two fuzzy sets A and B:

Fig. 6.11 Complement of fuzzy set A

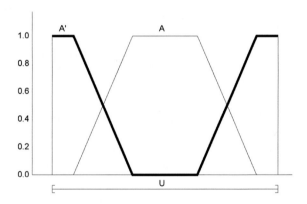

Fig. 6.12 Union of two fuzzy sets, A and B

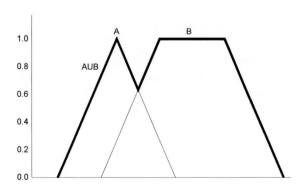

Intersection: The intersection $A \cap B$ of two fuzzy sets A and B is formally obtained by the following expression:

$$A \cap B = min[\mu_A(x), \mu_B(x)]/x \in [x_{1A}, x_{4B}] \qquad (6\text{-}14)$$

Similarly as with the case of the union, the intersection of two fuzzy sets A, B is obtained by the minimum membership degree of either $\mu_A(x)$ or $\mu_B(x)$ when x is contained in the closed interval formed by the extreme left of the support of A and the extreme right of the support of B. Figure 6.13 shows the intersection of A and B by means of a thick line.

Observing the Figs. 6.11, 6.12 and 6.13 we immediately realize that in general the FuzzyLisp standard set representation is not able to model neither the complement of a fuzzy set nor the union or intersection of two fuzzy sets. Only in the particular case of two trapezoidal shaped fuzzy sets sharing part of their nuclei we could obtain a representable union or intersection by the FuzzyLisp Standard Set Representation.

Nevertheless, we can develop some simple functions for testing if an element x belongs to the complement of a fuzzy set A, the union of two fuzzy sets, and the intersection of two fuzzy sets A, B getting at the same time its membership degree, if any. Code 6-9 shows the function *(fl-set-complement-membership?)* for obtaining the membership degree of the complement of a fuzzy set:

```
;code 6-9
(define (fl-set-complement-membership? fset x)
   (list
        (first (fl-set-membership? fset x))
        (sub 1.0 (last (fl-set-membership? fset x)))
   )
)
```

The function takes two arguments: *fset*, in FuzzyLisp Standard Set Representation and then x, the value for which we want to know its complement membership degree. Let us test the function with the fuzzy set A: *(B1 7 10 12 15)*.

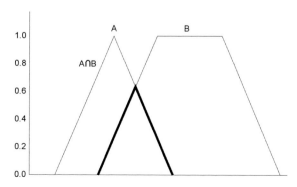

Fig. 6.13 Intersection of two fuzzy sets, A and B

When $x = 9$ its membership degree $\mu_A(x)$ is *0.6666666666*. Let us observe the resulting membership degree of its complement:

> **(fl-set-complement-membership? '(B1 7 10 12 15) 9)**
: *(B1 0.3333333333)*

For obtaining the union of two fuzzy sets we can design the function *(fl-set-union-membership?)* as shown in Code 6-10:

```
;code 6-10
(define (fl-set-union-membership? name fset1 fset2 x, mu1 mu2)
  (setq mu1 (last (fl-set-membership? fset1 x)))
  (setq mu2 (last (fl-set-membership? fset2 x)))
  (list name (max mu1 mu2))
)
```

The function takes four arguments: *name* is used for associating a name to the resulting list. *fset1* and *fset2* are two fuzzy sets using FuzzyLisp Standard Set Representation and finally, x is the value for which we want to know its $A \cup B$ membership degree. For testing it, let us take, for example, two fuzzy sets typing the following at the Lisp prompt: *(setq A '(Triangle 0 5 5 10))* and *(setq B '(Trapezium 5 10 15 20))*. Then for obtaining the membership degree of $A \cup B$ when x = 7.5 we can type:

> **(fl-set-union-membership? 'AuB A B 7.5)**
: *(AuB 0.5)*

Finally we shall write a function for getting the membership degree of the intersection of two fuzzy sets *A* and *B*, as shown in Code 6-11:

```
;code 6-11
(define (fl-set-intersect-membership? name fset1 fset2 x,
mu1 mu2)
   (setq mu1 (last (fl-set-membership? fset1 x)))
   (setq mu2 (last (fl-set-membership? fset2 x)))
   (list name (min mu1 mu2))
)
```

As it happened with the function *(fl-set-union-membership?)*, *(fl-set-intersect-membership?)* takes four arguments, too; again *name* is used for associating a name to the resulting list. *fset1* and *fset2* are two fuzzy sets using FuzzyLisp Standard Set Representation and finally, x is the value for which we want to know its $A \cap B$ membership degree. Using the same sets *A: (Triangle 0 5 5 10))* and *B: (Trapezium 5 10 15 20)* from the previous example, we can type:

> **(fl-set-intersect-membership? 'AintB A B 8)**
: *(AintB 0.4)*

In the design of these three functions we would like to remark again the importance of the function *(fl-set-membership?)* as a building block for producing new FuzzyLisp functions. The reader is thus invited to understand its code (Code 6-2) perfectly well.

6.7 Fuzzy Numbers

Human beings use fuzzy numbers everyday in common language and only a few examples will be enough to realize how useful they are for processing information and to transmit it to other persons: "I have parked my car about half a meter from the wall", "the distance from the Louvre Museum to Concorde Square is a bit more that one kilometer", "the distance between our Milky Way and the Andromeda galaxy is about two million light years", "the temperature inside my home in summer is 24 °C, more or less", "it takes me about twenty to thirty minutes to go by car to the train station". Even in science, when talking between colleagues (that is, when exchanging information) it is usual to use fuzzy numbers: No doctor will tell a colleague: "patient from room 122 in the hospital has a platelet count equal to 39,724", but instead he or she will say: "the patient from room 122 has a very low platelet count". If more "precision" is required the colleague will then probably ask: "how low is her platelet count?" and a possible reply could be "less than 40,000". This is enough for both specialists to know that patient from room 122 has developed a high risk of bleeding (Guyton and Hall 2005). "Less than 40,000" is a fuzzy number, as also are "about half a meter", "a bit more than one kilometer", "about twenty to thirty minutes" and so on.

Under a practical point of view, fuzzy numbers are the most important type of fuzzy sets because they have application in important areas of research such as fuzzy control, approximate reasoning or decision making, to name only a few of them. This is not strange because both in applied science and technology nothing can be defined without measuring it and then express the obtained measure with numbers. As Thomas William, better known as Lord Kelvin, wrote:

> … when you can measure what you are speaking about and express it in numbers, you know something about it; but when you cannot express it in numbers, your knowledge is of a meagre and unsatisfactory kind; it may be the beginning of knowledge, but you have scarcely in your thoughts advanced to the state of science, whatever the matter may be (Gray 2014)

We can update Lord Kelvin's citation replacing the word "number" with "numbers that are close to a real number" or "numbers that are around a given interval of real numbers". Let us try it: "when you can measure what you are speaking about and express it in numbers that are close to a real number then you know something about it". The reader can complete the rest of the citation, realizing, maybe with surprise, that the resulting set of sentences describes even better the original meaning of the English author.

For fixing ideas, every fuzzy number is a fuzzy set that satisfies the following properties:

- A fuzzy number must be a normal fuzzy set, that is, there exists at least a value x belonging to the support of the fuzzy set whose image $\mu(x)$ equals one, as we saw in (6-4).
- A fuzzy number must be a convex fuzzy set.
- The support of a fuzzy number must be bounded.

The first property seems to be the easier one to understand. First, because we have already seen what a normal fuzzy set is, and second because it seems obvious (an certainly it is) that if we are speaking about a fuzzy number n, then it will reach its maximum membership degree $\mu(x) = 1$ just when $x = n$. For example, if we define the fuzzy set "around three", it will reach its maximum membership degree just at x = 3, and then $\mu(3) = 1.0$.

For defining a convex fuzzy set we need again a helping hand from alpha-cuts. A convex fuzzy set is a fuzzy set where every alpha-cut on it is represented by one and only one closed interval on the real axis. For better understanding this concept it is usually convenient to show what is not a convex fuzzy set: In Fig. 6.12 we represented a fuzzy set $C = A \cup B$ that is not a convex set. While A and B are both convex fuzzy sets, the union C is not because there is at least one alpha-cut formed by more than one closed interval, as can be seen in Fig. 6.14 where taking for example $\alpha = 0.8$, its alpha-cut $^{0.8}a$ is represented by two closed intervals a_1, a_2. By the way, from the simple observation of Fig. 6.11 it can be said that in general, the complementary P' of a fuzzy number P is not a fuzzy number.

Finally, the support of a fuzzy number, that is, its strong alpha-cut ^{0+}a, must be bounded, that is, it must be a closed interval. Expressed with other words, the strong alpha cut of a fuzzy number is represented by the closed interval $[x_1, x_4]$.

6.7.1   Fuzzy Numbers Arithmetic

The existence of only one interval for every possible alpha-cut in a fuzzy number opens the door to performing arithmetic calculations on fuzzy numbers because it

Fig. 6.14 Example of a non-convex fuzzy set: $A \cup B$

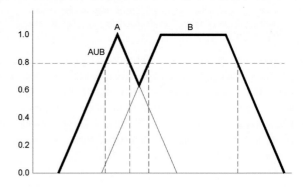

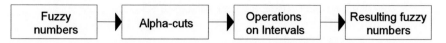

Fig. 6.15 Computational strategy for performing fuzzy number arithmetic

translates to perform operations on closed intervals, that is, to perform interval arithmetic, a branch of mathematics well established since the 50s and 60s last century. The computational strategy, shown schematically in Fig. 6.15 is simple: Take the fuzzy numbers involved in the desired arithmetic, obtain some alpha-cuts of them (thus obtaining closed intervals), perform the required arithmetic operation on the intervals, and finally, from the obtained intervals, build the resulting fuzzy number.

We define any arithmetic operator * on two closed intervals I_1, I_2 as the following expression:

$$I_1 * I_2 = \{z \,|\, \text{there is some } x \text{ in } I_1 \text{ and some } y \text{ in } I_2, \text{ such as } z = x * y\} \quad (6\text{-}15)$$

Then, for the addition, subtraction, multiplication and division of two closed intervals [a, b], [c, d], and having into account that both [a, b] and [c, d] are subsets of the real line, expressed in interval form as $[-\infty, \infty]$, we have the following expressions (6-16):

- Addition: *[a,b] + [c,d] = [a + c, b + d]*
- Subtraction: *[a,b] − [c,d] = [a − d, b − c]*
- Multiplication: *[a,b] · [c,d] = [min(ac, ad, bc, bd), max(ac, ad, bc, bd)]*
- Division: *[a,b]/[c,d] = [min(a/c, a/d, b/c, b/d), max(a/c, a/d, b/c, b/d)]*

Needless to say, the division *[a,b]/[c,d]* of two intervals is defined only in the case that the value *x = 0* is not contained on the interval *[c,d]*. With the help of the Lisp language is easy to define functions that perform interval arithmetic automatically. Code 6-12a through 6-12d shows these functions:

```
;code 6-12a
(define (fl-intv-add x1 x2 x3 x4)
    (list (add x1 x3) (add x2 x4))
)

;code 6-12b
(define (fl-intv-sub x1 x2 x3 x4)
    (list (sub x1 x4) (sub x2 x3))
)

;code 6-12c
(define (fl-intv-mult x1 x2 x3 x4, extrm-left extrm-right)
    (setq extrm-left (min
```

```
            (mul x1 x3)  (mul x1 x4)  (mul x2 x3)  (mul x2 x4))
   )
   (setq extrm-right (max
            (mul x1 x3)  (mul x1 x4)  (mul x2 x3)  (mul x2 x4))
   )
   (list extrm-left extrm-right)
)

;code 6-12d
(define (fl-intv-div a b d e, extrm-left extrm-right)
   (setq extrm-left (min
            (div a d)  (div a e)  (div b d)  (div b e))
   )
   (setq extrm-right (max
            (div a d)  (div a e)  (div b d)  (div b e))
   )
   (list extrm-left extrm-right)
)
```

Since the Lisp code from these four functions is a direct translation from the formulas for interval arithmetic we shall not make any additional comment on it. Let us quickly test these functions with the intervals $I_1 = [2,4]$ and $I_2 = [1,3]$ at the Lisp prompt:

> (fl-intv-add 2 4 1 3)
: (3 7)

> (fl-intv-sub 2 4 1 3)
: (-1 3)

> (fl-intv-mult 2 4 1 3)
: (2 12)

> (fl-intv-div 2 4 1 3)
: (0.6666666667 4)

Now, let us apply the algorithm shown in Fig. 6.15 for obtaining the sum of two fuzzy numbers A, B given by the following Lisp expressions: *(setq A '(around-2 1.75 2 2 2.25))* and *(setq B '(around-5 4.8 5 5 5.2))*. Then, for each of them let us obtain their 0.25 and 0.75 alpha-cuts, $^{0.25}a$, $^{0.75}a$, $^{0.25}b$ and $^{0.75}b$, respectively as shown in Table 6.3 by means of several calls to the function *(fl-alpha-cut)*:

So we have the following intervals/alpha-cuts: $^{0.25}a \rightarrow [1.8125\ 2.1875]$, $^{0.75}a \rightarrow [1.9375\ 2.0625]$, $^{0.25}b \rightarrow [4.85\ 5.15]$, $^{0.75}b \rightarrow [4.95\ 5.05]$. Now, adding the 0.25 and 0.75 alpha-cuts we have: *(fl-intv-add 1.8125 2.1875 4.85 5.15)* $\rightarrow$ *(6.6625 7.3375)*, and: *(fl-intv-add 1.9375 2.0625 4.95 5.05)* $\rightarrow$ *(6.8875 7.1125)*. That is:

Table 6.3 Some alpha cuts performed on the fuzzy numbers A, B

Alpha-cut	Function call	Function result
$^{0.25}a$	(fl-alpha-cut A 0.25)	(around-2 1.8125 2.1875)
$^{0.75}a$	(fl-alpha-cut A 0.75)	(around-2 1.9375 2.0625)
$^{0.25}b$	(fl-alpha-cut B 0.25)	(around-5 4.85 5.15)
$^{0.75}b$	(fl-alpha-cut B 0.75)	(around-5 4.95 5.05)

$$^{0.25}a + b \rightarrow [6.6625 7.3375]$$
$$^{0.75}a + b \rightarrow [6.8875 7.1125]$$

for obtaining the fuzzy number $A + B$ we only need to call the function *(fl-def-set)* from the Lisp prompt:

> **(fl-def-set 'A + B '(6.6625 7.3375 0.25) '(6.8875 7.1125 0.75))**
: *(A + B 6.55 7 7 7.45)*

Aside obtaining the addition of two fuzzy numbers, the last expression serves us well for demonstrating the utility of the function *(fl-def-set)* when dealing with fuzzy numbers arithmetic. Now, after calculating $A + B$ in a manual way, it would be interesting to have some functions for automatically calculating some fuzzy arithmetic. Code 6-13a and 6-13b show two FuzzyLisp functions for calculating the addition and subtraction of two fuzzy numbers, respectively:

```
;code 6-13a
(define (fl-fuzzy-add name A B, cut1A cut1B cut2A cut2B sum1
sum2)
    (setq cut1A (fl-alpha-cut A 0.25))
    (setq cut1B (fl-alpha-cut B 0.25))
    (setq cut2A (fl-alpha-cut A 0.75))
    (setq cut2B (fl-alpha-cut B 0.75))

    ;eliminate first element from every cut
    (pop cut1A) (pop cut1B)
    (pop cut2A) (pop cut2B)

    (setq sum1 (fl-intv-add
        (nth 0 cut1A) (nth 1 cut1A)
        (nth 0 cut1B) (nth 1 cut1B))
    )

    (setq sum2 (fl-intv-add
        (nth 0 cut2A) (nth 1 cut2A)
        (nth 0 cut2B) (nth 1 cut2B))
    )
```

```
;add alpha-cut value at the last position in the list

(push '0.25 sum1 2)
(push '0.75 sum2 2)
(fl-def-set name sum1 sum2)
)
```

The function for subtracting two fuzzy sets, *(fl-fuzzy-sub)* has the same structure:

```
;code 6-13b
(define (fl-fuzzy-sub name A B, cut1A cut1B cut2A cut2B sum1
sum2)
    (setq cut1A (fl-alpha-cut A 0.25))
    (setq cut1B (fl-alpha-cut B 0.25))
    (setq cut2A (fl-alpha-cut A 0.75))
    (setq cut2B (fl-alpha-cut B 0.75))

    ;eliminate first element from every cut
    (pop cut1A) (pop cut1B)
    (pop cut2A) (pop cut2B)

    (setq sum1 (fl-intv-sub
      (nth 0 cut1A) (nth 1 cut1A)
      (nth 0 cut1B) (nth 1 cut1B))
    )

    (setq sum2 (fl-intv-sub
      (nth 0 cut2A) (nth 1 cut2A)
      (nth 0 cut2B) (nth 1 cut2B))
    )

    ;add alpha-cut value at the last position in the list

    (push '0.25 sum1 2)
    (push '0.75 sum2 2)
    (fl-def-set name sum1 sum2)
)
```

Now we can test these functions simply typing the following at the Lisp prompt:

> (fl-fuzzy-add 'A + B A B)
: (A + B 6.55 7 7 7.45)

> (fl-fuzzy-sub 'A-B A B)
: (A-B -3.45 -3 -3 -2.55)

> **(fl-fuzzy-sub 'B-A B A)**
: *(B-A 2.55 3 3 3.45)*

And very interestingly:

> **(fl-fuzzy-sub 'A-A A A)**
: *(A-A -0.5 0 0 0.5)*

Figure 6.16 is a graphical representation of the fuzzy sets A, B, $A + B$ and $B - A$.
The addition or subtraction of two fuzzy numbers whose membership function
has a triangular or trapezoidal shape always generates a fuzzy number that also has
a triangular or trapezoidal membership function. However, the fuzzy multiplication
and fuzzy division of two fuzzy numbers does not produce, in general, a triangular
or trapezoidal fuzzy number, so FuzzyLisp cannot represent them directly. Anyway,
let us perform manually the calculations for obtaining $A.B$ and A/B. Table 6.4a is an
enhanced version of Table 6.3.

On the other hand, Table 6.4b shows all the intervals obtained after multiplying
the respective ${}^{\alpha}a \cdot {}^{\alpha}b$ and ${}^{\alpha}b/{}^{\alpha}a$.

This example helps to understand that any fuzzy set can be also defined by a set
of its own alpha-cuts. All the previous discussion about resolution applies here in
the same way. Figure 6.17 shows a discretized representation of the fuzzy sets
$A \times B$ and A/B. The hidden horizontal lines represent the 0, 0.25, 0.5, 0.75 and 1.0
membership degrees.

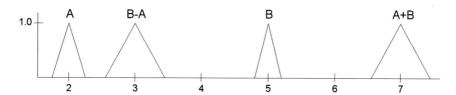

Fig. 6.16 Addition and subtraction of fuzzy sets: from *left to right*: A, B-A, B, A + B

Table 6.4a 0, 0.25, 0.5, 0.75 and 1.0 alpha cuts performed on the fuzzy numbers A, B

Alpha-cut	Function call	Function result
${}^{0}a$	*(fl-alpha-cut A 0)*	*(around-2 1.75 2.25)*
${}^{0.25}a$	*(fl-alpha-cut A 0.25)*	*(around-2 1.8125 2.1875)*
${}^{0.5}a$	*(fl-alpha-cut A 0.5)*	*(around-2 1.875 2.125)*
${}^{0.75}a$	*(fl-alpha-cut A 0.75)*	*(around-2 1.9375 2.0625)*
${}^{1.0}a$	*(fl-alpha-cut A 1)*	*(around-2 2 2)*
${}^{0}b$	*(fl-alpha-cut B 0)*	*(around-5 4.8 5.2)*
${}^{0.25}b$	*(fl-alpha-cut B 0.25)*	*(around-5 4.85 5.15)*
${}^{0.5}b$	*(fl-alpha-cut B 0.5)*	*(around-5 4.9 5.1)*
${}^{0.75}b$	*(fl-alpha-cut B 0.75)*	*(around-5 4.95 5.05)*
${}^{1.0}b$	*(fl-alpha-cut B 1)*	*(around-5 5 5)*

Table 6.4b Interval multiplication and division for *AxB* and *A/B*

Alpha value	Interval multiplication	Interval division
0	*[8.4 11.7]*	*[2.133333333 2.971428571]*
0.25	*[8.790625 11.265625]*	*[2.217142857 2.84137931]*
0.5	*[9.1875 10.8375]*	*[2.305882353 2.72]*
0.75	*[9.590625 10.415625]*	*[2.4 2.606451613]*
1.0	*[10 10]*	*[2.5 2.5]*

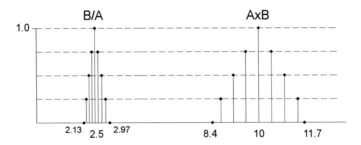

Fig. 6.17 Division and multiplication of two fuzzy sets: *B/A, AxB*

As the reader can observe, this is a tedious procedure, so two FuzzyLisp functions should be developed for calculating the multiplication and division of fuzzy numbers, respectively. For the inner working of the functions several alpha-cuts are obtained from every fuzzy set involved in the operation, then the adequate interval calculations are made and the final result is a discretized fuzzy number. Code 6-14 shows the function *(fl-fuzzy-mult)* for obtaining the multiplication of two fuzzy numbers *A, B*:

```
;code 6-14
(define (fl-fuzzy-mult name A B n,
          i alpha cutA cutB mult head tail interval)
   (setq head '() tail '())
   (setq interval (div 1.0 n))
   (setq alpha 0.0 i 0)

   (while (<= i n)
      (setq cutA (rest (fl-alpha-cut A alpha)))
      (setq cutB (rest (fl-alpha-cut B alpha)))
      ;perform  the  multiplication  of  alpha-cuts
      (intervals):
      (setq mult
         (fl-intv-mult
            (first cutA) (last cutA) (first cutB) (last cutB)
         )
      ) ;setq mult
```

```
                    ;carefuly construct the head and tail of the
                    ;discretized resulting set:
                (setq head (cons (append (list (first mult))
                                   (list alpha)) head))
                (setq tail (cons (append (list (last mult))
                                   (list alpha)) tail))

                (setq alpha (add interval alpha))
                (++ i)
            );while end
            (append (list name) (reverse head) (rest tail))
        )
```

Testing the function for sets *A*, *B*, we have:

> (fl-fuzzy-mult 'AxB A B 5)
*: (AxB (8.4 0) (8.712 0.2) (9.028 0.4) (9.348 0.6) (9.672 0.8) (10 1) (10.332 0.8)
(10.668 0.6) (11.008 0.4) (11.352 0.2) (11.7 0))*

In a similar way, Code 6-15 shows the FuzzyLisp function *(fl-fuzzy-div)* for
obtaining the division of two fuzzy numbers *A*, *B*:

```
;Code 6-15
(define (fl-fuzzy-div name A B n,
                  i alpha cutA cutB mult head tail interval)
    (setq head '() tail '())
    (setq interval (div 1.0 n))
    (setq alpha 0.0 i 0)

    (while (<= i n)
        (setq cutA (rest (fl-alpha-cut A alpha)))
        (setq cutB (rest (fl-alpha-cut B alpha)))
        ;perform the division of alpha-cuts (intervals):
        (setq mult
            (fl-intv-div
                (first cutA) (last cutA) (first cutB) (last cutB)
            )
        );setq mult

        ;carefuly construct the head and tail of the
        ;discretized resulting set:
        (setq head (cons (append (list (first mult))
                           (list alpha)) head))
        (setq tail (cons (append (list (last mult))
                           (list alpha)) tail))
```

```
        (setq alpha (add interval alpha))
        (++ i)
    );while end
    (append (list name) (reverse head) (rest tail))
)
```

Let us divide the fuzzy number *B* by the fuzzy number *A* at the Lisp prompt:

> (fl-fuzzy-div 'B/A B A 5)
: (B/A (2.133333333 0) (2.2 0.2) (2.269767442 0.4) (2.342857143 0.6)
(2.419512195 0.8) (2.5 1) (2.584615385 0.8) (2.673684211 0.6) (2.767567568 0.4)
(2.866666667 0.2) (2.971428571 0))

Again, the result of $A \times B$ and A/B is a fuzzy set with a discrete membership function, so, as the reader already knows, any intermediate membership degree can be obtained by means of the function *(fl-dset-membership?)*. For example, for knowing the membership degree for $x = 2.4$ of *B/A*, we only need to type:

> (fl-dset-membership? (fl-fuzzy-div 'B/A B A 5) 2.4)
: (B/A 0.7490909091)

6.7.2 More Numerical Operations on Fuzzy Sets

Aside the basic arithmetic operations described in the previous paragraphs, another interesting operation is the multiplication and division of fuzzy sets with triangular or trapezoidal membership functions by a real number *k*. In general, and remembering the definition of singular points given by (6-2), we have:

$$kA = k(x_1, x_2, x_3, x_4) = (kx_1, kx_2, kx_3, kx_4) \tag{6-16}$$

For dividing a fuzzy set with triangular or trapezoidal membership function by a real number *p*, we only need to do $k = 1/p$ and then use again the formula given by (6-16). Translating this into FuzzyLisp we obtain the function *(fl-fuzzy-factor)*. Shown in Code 6-16, it takes as parameters a fuzzy set and then a real number *k*:

```
;code 6-16
(define (fl-fuzzy-factor fset k, x1 x2 x3 x4 list-out)
    (setq x1 (mul k (nth 1 fset)))
    (setq x2 (mul k (nth 2 fset)))
    (setq x3 (mul k (nth 3 fset)))
    (setq x4 (mul k (nth 4 fset)))

    (if (>= x4 x1);normal case
        (setq list-out (list (nth 0 fset) x1 x2 x3 x4))
```

```
            ;else there is a negative number
            (setq list-out (list (nth 0 fset) x4 x2 x3 x1))
        ) ; if end
        list-out
)
```

It is easy to foresee the results produced by this function. When $k > 1$ it will perform a multiplication and conversely, when $k < 1$ it will perform a division. Having, for example a set $A : (A1 -2 3 3 8)$, then

> (fl-fuzzy-factor A 3)
: *(A1 -6 9 9 24)*

Dividing the same fuzzy set A by the real number 4.0 is as easy as multiplying it by 0.25:

> (fl-fuzzy-factor A 0.25)
: (A1 -0.5 0.75 0.75 2)

Another interesting FuzzyLisp function serves for shifting a fuzzy set towards left or right over the real axis X by an amount given by a real value x. Code 6-17 shows the function *(fl-fuzzy-shift)*:

```
;code 6-17
(define (fl-fuzzy-shift fset x)
    (list
            (nth 0 fset) ;the name of the set
            (add x (nth 1 fset))
            (add x (nth 2 fset))
            (add x (nth 3 fset))
            (add x (nth 4 fset))
    ) ; list end
)
```

Let us imagine we have a fuzzy set for defining a certain age representing those individuals that, although still young, are transiting to a mature age: *(young-plus 15.0 25.0 35.0 45.0)*. Then, if after a while we think it would be better to represent it with a five years old shift to the right we can now type:

> (fl-fuzzy-shift '(young-plus 15.0 25.0 35.0 45.0) 5.0)
: *(young-plus 20 30 40 50)*

This seems an innocent function at first, but its true strength appears when we use it dynamically from inside a program. That is, when some or many variable states representing a model change in such a way that we need to modify the very own definition of the fuzzy sets that take part of it while the program is running.

In a complementary way, these last functions inspire other one that allows a fuzzy set to grow or contract while remaining static on the real axis. This FuzzyLisp function is named *(fl-expand-contract-set)*. The function expands or contracts the support and nucleus of a triangular or trapezoidal fuzzy set by a factor *k*. As can be seen from Code 6-18, it first multiplies the fuzzy set by *k* and then moves it back to its original position:

```
;code 6-18
(define (fl-expand-contract-set fset k, center1 center2
result)
    ;calculate support's centre:
    (setq center1 (div (add (nth 2 fset) (nth 3 fset)) 2.0))
    (setq result (fl-fuzzy-factor fset k)) ;expand/contract
    ;calculate new support's centre:
    (setq center2 (div (add (nth 2 result) (nth 3 result))
    2.0))
    ;shift it to its original position
    (fl-fuzzy-shift result (mul (sub center2 center1) -1.0))
)
```

Let us play a bit with this function for demonstrating how does it work:

> **(fl-expand-contract-set '(a 0 1 1 2) 2.0)**
: *(a -1 1 1 3)*

> **(fl-expand-contract-set '(a -1 1 1 3) 0.5)**
: *(a 0 1 1 2)*

Figure 6.18 shows a graphical representation of these geometrical/numerical transformations on fuzzy sets. Identical behavior happens with fuzzy sets that have trapezoidal membership functions, as can be seen typing: *(fl-expand-contract-set '(a 0 1 2 3) 2.0)* → *(a -1.5 0.5 2.5 4.5)*, and then: *(fl-expand-contract-set '(a -1.5 0.5 2.5 4.5) 0.5)* → *(a 0 1 2 3)*.

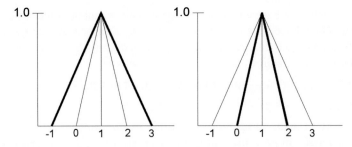

Fig. 6.18 Expansion/contraction of fuzzy sets with triangular membership functions. At *left* the obtained expansion is shown in *thick line*. At *right*, the obtained contraction

6.7.3 Fuzzy Averaging

After learning to calculate the sum of two fuzzy numbers A, B, we can extend these ideas to calculate the sum of n fuzzy numbers F_n, as expressed in Eq. (6-17):

$$S = F_1 + F_2 + \cdots + F_n = \sum F_i, \quad i = 1\ldots n \tag{6-17}$$

Then, for calculating the fuzzy average of n fuzzy numbers we shall have:

$$Fav = S/n = (F_1 + F_2 + \cdots + F_n)/n = \left(\sum F_i\right)/n \tag{6-18}$$

The reader should note that both S and Fav are both fuzzy numbers. This remark is important because for translating expressions (6-17) and (6-18) into FuzzyLisp functions we are going to use a data structure that is a set of fuzzy sets but it is not a fuzzy set in itself. An example will suffice for clarifying this concept: Let us take five fuzzy numbers from F_1 to F_5 by means of the following Lisp expressions: *(setq F1 '(set1 -2 0 0 2)), (setq F2 '(set2 3 5 5 7)), (setq F3 '(set3 6 7 7 8)), (setq F4 '(set4 7 9 11 12))* and *(setq F5 '(set5 8 10 10 12))*. All of them are fuzzy sets with triangular membership functions with the exception of F_4, which is a fuzzy interval. Now let us form the following data structure at the Lisp prompt:

> **(setq Fsets '(F1 F2 F3 F4 F5))**
: *(F1 F2 F3 F4 F5)*

The information for every F_i fuzzy set seems lost, but it is only a question of appearance. For checking, let us type, for example:

> **(eval (first Fsets))**
: *(set1 -2 0 0 2)*

With these ideas fixed on mind we can now introduce the FuzzyLisp function *(fl-fuzzy-add-sets)*. It takes two arguments: first the set of fuzzy sets to add and then a symbol for naming the resulting fuzzy set. Code 6-19 shows the function:

```
;code 6-19
(define (fl-fuzzy-add-sets fsets name, i n lst-out)
   (setq n (length fsets))
   (setq lst-out '())

   (setq lst-out
       (fl-fuzzy-add name (eval (nth 0 fsets)) (eval (nth 1
       fsets)))))
   (setq i 2); we have already added two fuzzy numbers
   ;now for the rest of fuzzy numbers:
   (while (< i n)
```

```
        (setq lst-out (fl-fuzzy-add name lst-out (eval (nth i
        fsets))))
        (++ i)
    ); while end
    lst-out
)
```

Let us call it from the Lisp prompt, assigning the result to the symbol SFs:

> (setq SFs (fl-fuzzy-add-sets Fsets 'Sum-of-Fs))
: (Sum-of-Fs 22 31 33 41)

Now, for calculating the fuzzy average of the fuzzy sets F_1 through F_5 we only would need to type:

> (fl-fuzzy-factor SFs (div 1 5))
: (Sum-of-Fs 4.4 6.2 6.6 8.2)

It is a nice result, but I think it would be even nicer if we could have a FuzzyLisp function for directly calculating fuzzy averages. Let us discover it in Code 6-20:

```
;code 6-20
(define (fl-fuzzy-average fsets name)
    (fl-fuzzy-factor
        (fl-fuzzy-add-sets  fsets  name)  (div  1.0  (length
        fsets))
    )
)
```

Testing it is straightforward:

> (fl-fuzzy-average Fsets 'Average)
: (Average 4.4 6.2 6.6 8.2)

Figure 6.19 shows graphically all the fuzzy sets involved into this last operation. Please note that the average is a fuzzy set with a trapezoidal membership function with a small nucleus as a result of the participation of the fuzzy set F_4.

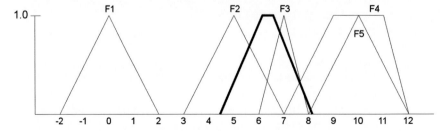

Fig. 6.19 Fuzzy sets F_1 through F_5. The fuzzy set representing their average is shown in *thick lines*

Sometimes it is useful to obtain a crisp, traditional real number from a fuzzy set. Let us imagine you have performed several arithmetic operations on fuzzy numbers and you wish to transform the obtained final fuzzy number into a crisp one. Maybe you need it in a commercial report, a scientific article or, in general, for using it for communicating a numerical value. The process for transiting from a fuzzy number to a real number is named defuzzification. Although we shall dedicate enough space to this important procedure of fuzzy sets theory on the next chapter, it seems now convenient to introduce simple defuzzification.

In general, for fuzzy numbers with symmetrical shaped membership functions, either a triangle or a trapezium, the resulting sharp number is simply the central position of their support, which coincides with the central point of their nuclei. For example, the defuzzification of the fuzzy number *(fnumber 1 3 3 5)* is 3, as it happens also with the fuzzy interval *(finterval 1 2 4 5)*. Things change a bit when we handle asymmetrical shaped fuzzy numbers or intervals such as *(fanumber 0 1 1 5)*, or *(fainterval 0 1 3 8)*. In these cases, the central position of their nuclei does not coincide with the central position of their support. Geometrically, and since we are dealing with normal fuzzy sets, it seems clear that the point x whose membership degree equals to one must have an important role to play in the defuzzified, resulting crisp number, but it is also true that the geometrical bias inherent to the asymmetrical shape of the characteristic function must be taken into account, too. Naming m as the point corresponding to the nucleus' average:

$$m = (x_2 + x_3)/2 \qquad\qquad (6\text{-}19)$$

then we have the following expressions for calculating simple defuzzification from fuzzy numbers and fuzzy intervals:

$$(x_1 + m + x_4)/3 \qquad\qquad (6\text{-}20a)$$

$$(x_1 + 2m + x_4)/4 \qquad\qquad (6\text{-}20b)$$

$$(x_1 + 4m + x_4)/6 \qquad\qquad (6\text{-}20c)$$

$$(x_1 + 6m + x_4)/8 \qquad\qquad (6\text{-}20d)$$

these expressions concede increasing weight to the nucleus' average of a fuzzy number or interval. Translating these formulas to Lisp language is easy, as shown in Code 6-21:

```
;code 6-21
(define (fl-simple-defuzzification fset mode, m)
    ;first, get nucleus average:
    (setq m (div (add (nth 2 fset) (nth 3 fset)) 2.0))

    ;for every mode, we give different weights to m:
    (case mode
```

```
      (1 (div (add (nth 1 fset) m (nth 4 fset)) 3.0))
      (2 (div (add (nth 1 fset) (mul m 2.0) (nth 4 fset)) 4.0))
      (3 (div (add (nth 1 fset) (mul m 4.0) (nth 4 fset)) 6.0))
      (4 (div (add (nth 1 fset) (mul m 6.0) (nth 4 fset)) 8.0))
  )
)
```

As parameters, the function *(fl-simple-defuzzification)* takes first a fuzzy number *fset* and then a number named *"mode"* between 1 and 4, corresponding to expressions [6-20x]. Let us see some examples with the asymmetrical shaped fuzzy number *(fanumber 0 1 1 5)*:

> **(fl-simple-defuzzification '(q 0 1 1 5) 1)**
: 2

> **(fl-simple-defuzzification '(q 0 1 1 5) 4)**
: 1.375

The second call to the function shows how using 3 as the *mode* for the function, the resulting crisp number is closer to the nucleus of the fuzzy number. Now the question is: what mode should be used in practical applications? The answer is simple: it all depends on the designer of the fuzzy model to develop. We shall find several advices for this and other similar questions later in this book.

6.8   Linguistic Variables

A linguistic variable is a finite collection of two or more fuzzy numbers or fuzzy intervals that, representing linguistic concepts, completely patches a given universe of discourse defined on a base variable. A base variable represents a range of values of real numbers from any measurable concept or magnitude such as age, temperature, voltage, currency exchange rates, acceleration, density…, etc. Linguistic variables contain fuzzy numbers that represent, in linguistic terms, value states from the base variable. The concept of "patching" is important in this definition and it implies that the ordered fuzzy numbers that form a linguistic variable intersect common spaces from the universe of discourse defined on the base variable. As a counterexample, the set of fuzzy numbers represented in Fig. 6.16 is not a linguistic variable because it does not patch any possible universe of discourse defined on the real interval [0,8], not to mention the fact that no linguistic meaning has been associated to every fuzzy set in the figure. Formally, the condition of "patching" in a linguistic variable LV formed by F_i fuzzy sets can be expressed as follows:

$$LV = \{F_i \mid F_i \cap F_{i+1} \neq \phi\} \qquad (6\text{-}20)$$

Lisp symbol	Lisp assignment
age1	*(setq age1 '(young 0 0 15 30))*
age2	*(setq age2 '(young-plus 15 30 30 45))*
age3	*(setq age3 '(mature 30 45 45 60))*
age4	*(setq age4 '(mature-plus 45 60 60 75))*
age5	*(setq age5 '(old 60 75 90 90))*

Table 6.5 Five fuzzy sets for building the linguistic variable lv-age

Let us consider again the concept of "age" in order to give a complete example of linguistic variable. Now we can say that "age" represents a base variable and we are going to define the universe of discourse U on it for a range from 0 to 90 years old. Table 6.5 shows some Lisp assignments for creating five fuzzy sets that completely patch U.

The linguistic labels associated to every fuzzy set are: "young", "young-plus", "mature", "mature-plus" and "old". For creating the linguistic variable *lv-age* in Lisp, we only need to include the following line of code in our program after the inclusion of the Lisp assignments shown in Table 6.7:

$$(setq \ lv\text{-}age \ '(age1 \ age2 \ age3 \ age4 \ age5))$$

Figure 6.20 represents graphically the linguistic variable *lv-age*.

From the simple examination of the figure, it can be seen that the five fuzzy sets that constitute the linguistic variable completely patch the universe of discourse where they are defined. Another interesting fact is also relevant in this arrangement of fuzzy sets: for any crisp value x belonging to the universe of discourse U, the sum of the all possible corresponding membership degrees $f_i(x)$ from the fuzzy sets in the linguistic variable equals exactly to one. That is:

$$\forall x \in U, \sum f_i(x) = 1.0 \tag{6-21}$$

When the fuzzy sets from a linguistic variable LV are arranged in such a way that expression (6-21) holds, we say this family of nonempty fuzzy sets are a *fuzzy partition* of the linguistic variable LV on U (Belohlavek and Klir 2011).

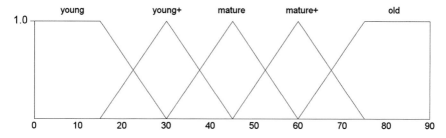

Fig. 6.20 Graphical representation of the linguistic variable lv-age

Since in the rest of this book we are going to use linguistic variables a lot, it seems convenient to have a FuzzyLisp function that at any time from the Lisp prompt we can list the fuzzy sets it contains. Such a function, named *(fl-list-sets)* is shown in Code 6-22:

```
;code 6-22
(define (fl-list-sets lv, i n fset)
    (setq i 0)
    (setq n (length lv))
    (while (< i n)
        (setq fset (eval (nth i lv)))
        (print fset)
        (++ i)
        (print "\n")
    )
    (println)
)
```

Its only argument is the name of a linguistic variable already stored in Lisp's memory. Let us test it:

> (fl-list-sets lv-age)
: (young 0 0 15 30)
: (young-plus 15 30 30 45)
: (mature 30 45 45 60)
: (mature-plus 45 60 60 75)
: (old 60 75 90 90)

The output from this function call is an equivalent to the material exposed in Table 6.5. As the reader already knows, the high degree of interaction between the language and the user in a Lisp session means that *(fl-list-sets)* is a FuzzyLisp function, that albeit simple, is frequently called when building a fuzzy logic based model since it allow us to have all the used linguistic variables under control in a highly readable format.

A function that helps to test the membership degrees $f_i(x)$ of a linguistic variable is the FuzzyLisp function *(fl-lv-membership?)*. It takes the name of an existing linguistic variable *lv* and a crisp value *x* belonging to the universe of discourse where *lv* is defined. The function, shown in Code 6-23, returns the membership degree for every fuzzy set included in the linguistic variable:

```
;code 6-23
(define (fl-lv-membership? lv x, i n fset answer)
    (setq i 0)
    (setq n (length lv))
    (while (< i n)
        ;first we obtain every fuzzy set
```

```
            (setq fset (eval (nth i lv)))
            ;now we call the function that performs the
            calculations
            (setq answer (fl-set-membership? fset x))
            (println answer)
            (++ i)
      );end while
      (println)
  )
```

Let us test it with a simple example:

> **(fl-lv-membership? lv-age 32)**
: *(young 0)*
: *(young-plus 0.8666666667)*
: *(mature 0.1333333333)*
: *(mature-plus 0)*
: *(old 0)*

As it happened with *(fl-list-sets)*, this function is well suited to interactively establish a dialogue with Lisp in order to test and extract information of the behavior of linguistic variables. However, a slight variation of it proves more flexible and powerful, especially for programming. Named *(fl-lv-membership2?)* it is shown in Code 6-24:

```
;code 6-24
(define (fl-lv-membership2? lv x, list_out i n fset)
    (setq list_out '())
    (setq i 0)
    (setq n (length lv))
    (while (< i n)
        ;here we obtain a fuzzy set:
        (setq fset (eval (nth i lv)))
        ;and now it's time to perform calculations:
        (setq list_out (append list_out
                (list (fl-set-membership? fset x)))))
        (++ i)
    );while end
    list_out
)
```

After testing it, the output is very similar to the one produced by *(fl-lv-membership?)*:

> **(fl-lv-membership2? lv-age 32)**
: ((young 0) (young-plus 0.8666666667) (mature 0.1333333333) (mature-plus 0) (old 0))

Table 6.6 Five fuzzy sets for building the linguistic variable lv-age2

Lisp symbol	Lisp assignment
agge1	*(setq agge1 '(young 0 0 0 90))*
agge2	*(setq agge2 '(young-plus 0 30 30 90))*
agge3	*(setq agge3 '(mature 0 45 45 90))*
agge4	*(setq agge4 '(mature-plus 0 60 60 90))*
agge5	*(setq agge5 '(old 0 90 90 90))*

However, the difference, while subtle, is important: *(fl-lv-membership2?)* returns the membership degrees as a list! This convenience will show by itself in the next chapter where we shall see how it helps to build an inference engine.

Let us see now an alternative way to describe the concept of age with a linguistic variable where the entire universe of discourse U is the support of every fuzzy set included in the linguistic variable. The respective Lisp assignments are shown in Table 6.6.

For constructing the linguistic variable named *lv-age2* in Lisp we only need to include the following line of code:

$$(setq \ lv\text{-}age2 \ '(agge1 \ agge2 \ agge3 \ agge4 \ agge5))$$

Figure 6.21 represents graphically this linguistic variable.

This example of linguistic variable for describing the concept of age seems bizarre at first. Interestingly, for any crisp value x belonging to its universe of discourse, the corresponding membership degree $f_i(x)$ from every fuzzy set in the linguistic variable is always bigger than zero in the open interval (0,90). In other words: no fuzzy set in the linguistic variable gets a null membership degree for any x crisp value in the open interval (0,90). We shall name *maximum fuzziness linguistic variable* to this type of linguistic variables. With the help of the function *(fl-lv-membership2?)*, let us compare the behavior of the linguistic variables *lv-age* and *lv-age2* for some values x. The comparison is shown in Tables 6.7a, b.

While every value x from the universe of discourse generates one or two corresponding membership degrees in the linguistic variable *lv-age*, it generates five membership degrees in *lv-age2*. If Aristotle had problems for starting to understand

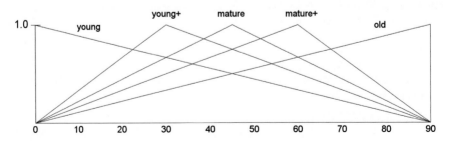

Fig. 6.21 Graphical representation of the linguistic variable lv-age2

Table 6.7a Some membership degrees for the fuzzy sets of the linguistic variable lv-age

x (age)	$\mu(young)$	$\mu(young+)$	$\mu(mature)$	$\mu(mature+)$	$\mu(old)$
1	1	0	0	0	0
18	0.8	0.2	0	0	0
35	0	0.66667	0.33333	0	0
55	0	0	0.33333	0.66667	0
84	0	0	0	0	1

Table 6.7b Some membership degrees for the fuzzy sets of the linguistic variable lv-age2

x (age)	$\mu(young)$	$\mu(young+)$	$\mu(mature)$	$\mu(mature+)$	$\mu(old)$
1	0.98888	0.03333	0.02222	0.01667	0.01111
18	0.8	0.6	0.4	0.3	0.2
35	0.61111	0.91667	0.77777	0.58333	0.38888
55	0.38888	0.58333	0.77777	0.91667	0.61111
84	0.0667	0.1	0.13333	0.2	0.93333

the fall of the 'law of non contradiction' as we saw in the previous chapter, the values represented in Table 6.7b would drive him crazy, probably. However, an interesting interpretation does exist: Nobody is completely young or old. A one year old children has an almost 1.0 membership degree to the fuzzy set of young people, but since his life has already been depleted by twelve months he is not an entirely new born being. And also, from the point of view of the fuzzy sets included in *lv-age2*, he has already started to be an old person, yet with an almost symbolic membership degree. The opposite is also true: An 84 years old person still keeps some youth in his mind. Which linguistic variable models better the concept of age? It depends. Normally, *lv-age* would be a well-formed fuzzy partition for representing age, but *lv-age2* models better a philosophical approach, showing an extraordinary smooth transition between linguistic terms.

If this were not enough, yet another way to describe the concept of age would be to use fuzzy sets with discrete characteristic functions, for example, by using bell shaped functions. Since FuzzyLisp supports discrete functions this is a good moment to show an example of it. Let us start writing some Lisp expression for representing five bell-shaped functions, *f1* to *f5* (remember the family of functions given by $\mu_4(x)$ in Table 6.1 and Fig. 6.2):

```
(setq f1 '(div (add 1.0 (cos (mul 0.0333 pi (sub x 0.0)))) 2.0))
(setq f2 '(div (add 1.0 (cos (mul 0.067 pi (sub x 30.0)))) 2.0))
(setq f3 '(div (add 1.0 (cos (mul 0.067 pi (sub x 45.0)))) 2.0))
(setq f4 '(div (add 1.0 (cos (mul 0.067 pi (sub x 60.0)))) 2.0))
(setq f5 '(div (add 1.0 (cos (mul 0.0333 pi (sub x 90.0)))) 2.0))
```

Now Table 6.8 shows the required expressions for creating every fuzzy set.

Table 6.8 Five bell-shaped fuzzy sets for building the linguistic variable lv-age-bells

Lisp symbol	Lisp assignment
dBell1	(setq dBell1 (fl-discretize-fx 'Young f1 20 0 30))
dBell2	(setq dBell2 (fl-discretize-fx 'Young + f2 20 15 45))
dBell3	(setq dBell3 (fl-discretize-fx 'Mature f3 20 30 60)
dBell4	(setq dBell4 (fl-discretize-fx 'Mature + f4 20 45 75))
dBell5	(setq dBell5 (fl-discretize-fx 'Old f5 20 60 90))

And finally, as usually, we only need to write the following expression for creating the linguistic variable:

$$(setq\ lv\text{-}age\text{-}bells\ `(dBell1\ dBell2\ dBell3\ dBell4\ dBell5))$$

This linguistic variable is graphically shown in Fig. 6.22.

Now we certainly need a function for obtaining the membership degree of a crisp value x to a linguistic variable composed by fuzzy sets with discrete characteristic functions. Code 6-25 shows the function *(fl-dlv-membership2?)*, a variation of *(fl-lv-membership2?)* designed to deal with discrete linguistic variables:

```
;code 6-25
(define (fl-dlv-membership2? lv x, list_out i n fset)
    (setq list_out '())
    (setq i 0)
    (setq n (length lv))
    (while (< i n)
        ;here we obtain a fuzzy set:
        (setq fset (eval (nth i lv)))
        ;and now it's time to perform calculations:
        (setq list_out (append list_out
                (list (fl-dset-membership? fset x))))
        (++ i)
    );while end
    list_out
)
```

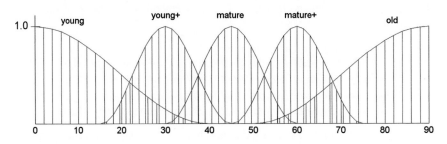

Fig. 6.22 Graphical representation of the linguistic variable lv-age-bells

Table 6.9 Some membership degrees for the fuzzy sets of the linguistic variable lv-age2

x (age)	μ(young)	μ(young+)	μ(mature)	μ(mature+)	μ(old)
1	0.99692	0	0	0	0
18	0.57821	0.09183	0	0	0
35	0.03948	0.74516	0.24812	0	0
55	0	0	0.24812	0.74516	0.03948
84	0	0	0	0	0.94550

The call to this function works in an absolutely similar way to *(fl-lv-membership2?)*:

> **(fl-dlv-membership2? lv-age-bells 23)**

: ((Young 0.1302642245) (Young + 0.5478879113) (Mature 0) (Mature + 0) (Old 0))

For comparison, Table 6.9 shows some membership values for some crisp values x.

After confronted with such a variety of possible linguistic variable types, I suspect the reader is now a bit lost when thinking about how to design a good linguistic model with fuzzy sets. As we have seen, there are many possible combinations of fuzzy numbers or fuzzy intervals to use. We can use triangles, trapeziums (both symmetrical or asymmetrical), discretized functions… etc. The variations seem almost infinite. Even so, we can give some simple hints for the overall design of linguistic variables in order to allow the reader to have a clear picture for some preliminary design ideas. Later, in Chap. 8, we shall find more sophisticated models.

One of the first things to decide is the number of fuzzy sets that a linguistic variable should have. This is known as the *granularity of a linguistic variable*, and deep theoretical studies have been conducted about this issue, see for example (Cordón et al. 2000; Chan and Wong 2006). However, and as a rule of thumb, we can say that the usual case is to use from three to five fuzzy sets in a linguistic variable, although there are some cases where only two fuzzy sets are enough. It is not usual to observe more than five fuzzy sets in a linguistic variable, not only because the increasing complexity of the resulting models, but from the fact that we humans tend to qualify things with no more than five adjectives. In fact many times is easier for us to use only three ones, such as "small", "medium" and "big", or "short", "medium" and "large". For example, if we speak about the speed of a car we can say it runs "slow", at "medium speed" or "fast". A richer description would be "very slow", "slow", "medium speed", "fast" and "very fast". It all depends from the nature of the model to design and the experience of the modeler, but for starting, from three to five fuzzy sets is a good recommendation.

Another complementary question is based on the distribution of the fuzzy sets over the universe of discourse. In the three previous examples of linguistic variables representing *age* the fuzzy sets have been distributed uniformly and in a symmetric mode with respect to the vertical axis of symmetry of the linguistic variable. While it is possible to design models where does exist some asymmetry with respect to the location of the fuzzy sets involved, for now we shall only consider regular spaced fuzzy sets.

About the question of what type of fuzzy sets are the best for building linguistic variables, we have already seen that creating bell-shaped or other fuzzy sets with discrete characteristic functions takes more time, it is more tedious and it also takes more computing time, although this last point, with the computing power of the current hardware systems is only a minor problem present only in very large fuzzy logic based models and/or when an almost instantaneous response from the built model is required, as is the case with critical fuzzy control systems. In any case, and for starting, we can recommend to use triangles and trapeziums as the desired membership functions for the fuzzy sets contained in a linguistic variable.

Finally, while maximum fuzziness linguistic variables present some advantages in modeling, as we shall see in the next chapter, using standard fuzzy partitions is a sure bet for starting to create well-behaved fuzzy logic based models. We should also mention the fact that for designing standard fuzzy partitions we can also use bell-shaped characteristic functions or other functions that satisfy the properties of fuzzy numbers.

6.9 Fuzzy Databases

In Chap. 4 we developed a small library of Lisp functions for managing flat databases in CSV (comma separated values) format. For testing the functions and evaluating its behavior we used a CSV database formed by the Messier objects, a collection of 110 deep-sky objects visible from the northern hemisphere with the aid of a small telescope. Since we have already seen how to load a CSV database into memory and perform simple queries on it, it seems now natural to enhance the initial set of Lisp functions in order to incorporate some "fuzzy" functionality.

Fuzzy databases are a branch of fuzzy logic that has grown up to the point to convert itself into a new discipline of Soft Computing (Pons et al. 2000; Galindo et al. 2006). Basically speaking, a Fuzzy database is a structured collection of data that contains fuzzy information and can be exploited using fuzzy queries. Needless to say, we have not enough space in this book for diving into the nowadays deep waters of fuzzy databases, but our exploration of the subject will cover at least the basics of the topic.

We shall concentrate in the simplest approach to fuzzy databases, that is, to choose a numerical field from a traditional, crisp database, apply a definition of one or more fuzzy sets on it, convert the crisp numerical values stored in the database's field into membership degrees and then store the obtained values into a new field. Albeit a simply strategy, it will show the power of adding fuzzy capabilities to crisp databases. The key FuzzyLisp function for bridging the gap between the world of

CSV databases and the world of fuzzy logic is the function *(fl-db-new-field)*, as shown in Code 6-26:

```
;code 6-26
(define (fl-db-new-field lst sharp-field fz-field fset mode,
                          i l position fz-value lst-out)
   (setq lst-out '())
   (setq l (length lst)) ;number of records in lst
   ;get the field's position reading the database header:
   (setq position (find sharp-field (nth 0 lst)))
   ;in the following line we append the new field's name:
   (setq lst-out (cons (append (nth 0 lst)
                       (list fz-field)) lst-out))

     ;then we copy the entire database,
     ;calculating every fuzzy value
   (setq i 1)
   (while (< i l)
       ;read the value from the sharp-field and fuzzify it:
       (case mode
          (1 (setq fz-value
              (last (fl-set-membership? fset
                     (float (nth position (nth i lst)))))))
              ))
          (2 (setq fz-value
              (last (fl-dset-membership? fset
                     (float (nth position (nth i lst)))))))
              ))
       ) ;end case

     ;then put it on the fz-field (last position):
       (setq lst-out
          (cons (append (nth i lst) (list (string fz-value)))
                     lst-out)
       )
       (++ i)
   ) ;while end

   (reverse lst-out)
) ;end function
```

Although the function seems a bit complex, it is in fact a variation of the function *(db-new-field)*, already seen in Sect. 4.3.3 with fuzzification features. As parameters it takes first the Lisp symbol that points to the entire database already loaded in the computer's memory, then the name of the database field whose values we wish to fuzzify, then the name of the field to be created where the membership degrees will be stored, then a FuzzyLisp expression representing a fuzzy set and finally an integer with two possible values: 1, for using triangular or trapezoidal membership functions or 2, for using a fuzzy set with discrete membership function.

For testing the function, we shall need to have both the fuzzylisp.lsp and csv.lsp files loaded into the NewLisp environment, both available from the accompanying web site of this book. After clicking on the green (run) button on the icons bar for both files all the functions will be available for use.

Let us suppose we wish to know which are the brightest objects in the Messier database because we are planning an observing session from the suburbs of a city where no dark skies are available. The first step is to load the CSV file containing the Messier database into memory by typing, as we already know, the following at the Lisp prompt:

> **(setq messier (db-load))**
: (("Name" "NGC" "Constellation" "Class" "Right ascension" "Declination" "Magnitude" "Angular size" "Burnham" "Remarks")
("M1" "1952" "Taurus" "Planetary nebula" "5 h 34.5 m" "22d 1 m" "8.2" "6x4" "!!" "SNR (1054) - Crab Nebula")

....
("M110" "205" "Andromeda" "Elliptical galaxy" "0 h 40.4 m" "41d 41 m" "9.4" "8.0x3.0" "" "E6 - companion of M31"))

Now, let us define a fuzzy set named, for example, *BM*, for representing "bright magnitude". In astronomy, brighter celestial objects have smaller numbers. In this moment we don't remember exactly the range of brightness for all the Messier objects, but we are sure that no one of them is brighter than Mv = -1, and on the other hand we know that Mv = 5 is not anymore a bright magnitude for celestial objects, so we decide to create an asymmetrical trapezium by-left as the desired characteristic function for the fuzzy set *BM* by typing:

> **(setq BM '(bright-magnitude -1 -1 3 5))**
: (bright-magnitude -1 -1 3 5).

The name of the field where magnitude values stored on the database is *"Magnitude"*, and let us name *"fz-magnitude"* the name of the field to create where all the fuzzy values will be stored after being fuzzified by the fuzzy set *BM*. It is now time to call the function *(fl-db-new-field)* as follows:

> **(setq messier (fl-db-new-field messier "Magnitude" "fz-magnitude" BM 1))**
: (("Name" "NGC" "Constellation" "Class" "Right ascension" "Declination" "Magnitude" "Angular size" "Burnham" "Remarks" "fz-magnitude")
("M1" "1952" "Taurus" "Planetary nebula" "5 h 34.5 m" "22d 1 m" "8.2" "6x4" "!!" "SNR (1054) - Crab Nebula" "0")

...

("M110" "205" "Andromeda" "Elliptical galaxy" "0 h 40.4 m" "41d 41 m"
"9.4" "8.0x3.0" "" "E6 - companion of M31" "0"))

For every record, that is, for every celestial object in the database, *(fl-db-new-field)* has added the corresponding membership degree in the new field *"fz-magnitude"*. If we type, for example: *(nth 6 messier)* → *("M6" "6405" "Scorpius" "Open cluster" "17 h 40.1 m" "-32d 13 m" "4.2" "15" "!!" "n: 80 - Butterfly Cluster" "0.4")*, that is, the Open Cluster M6 in Scorpius has received a membership degree $\mu(x) = 0.4$ to the fuzzy set *BM* representing Bright Magnitude. For obtaining the brightest celestial objects in the Catalogue now we can use a threshold value *t* for querying the database. Here, $t = 0.75$ is a good value, so let us call the function *(db-filter)*:

> **(db-filter messier '(>= "fz-magnitude" 0.75))**
: (("Name" "NGC" "Constellation" "Class" "Right ascension" "Declination"
"Magnitude" "Angular size" "Burnham" "Remarks" "fz-magnitude")
("M7" "6475" "Scorpius" "Open cluster" "17 h 53.9 m" "-34d 49 m" "3.3"
"80" "!!" "nmbr:80-fine naked-eye OC" "0.85")
("M42" "1976" "Orion" "Diffuse nebula" "5 h 35.4 m" "-5d 27 m" "2.9"
"66x60" "!!!"
"Great Orion Nebula" "1")
("M44" "2632" "Cancer" "Open cluster" "8 h 40.1 m" "19d 59 m" "3.1"
"95" "!!" "Praesepe - Beehive Clustr" "0.95")
("M45" "" "Taurus" "Open cluster" "3 h 47 m" "24d 7 m" "1.2" "110" ""*
"nmber: 130 - The Pleiades" "1"))

That is great. M7, M42, M44 and M45 have been filtered as the brightest celestial objects in the Messier Catalogue. From experience, these are deep-sky objects that can be observed with a small refractor or binoculars from light-polluted suburban skies. By the way, from the Lisp prompt you could save this query typing, for example:

> **(setq query1 '(db-filter messier '(>= "fz-magnitude" 0.75)))**
: (db-filter messier '(>= "fz-magnitude" 0.75))

And later, when you need to use the query, you only would need to type *(eval query1)* at the Lisp prompt, getting the same result from above. Even more: you can store several queries in a new document in the NewLisp environment (File->New tab) and then save it to disk with a specific name, such as, for example, "my-queries.qry". Soon you could have a practical fuzzy database management system at your disposal.

Now, let us compare the query we have just made against a "crisp" query. From the definition of the fuzzy set *BM*, we can observe that its support is defined in the interval [-1,5]. Because this, let us take, for example, Mv = 5 as the inferior limit for obtaining the brighter objects from the Messier Catalogue in a traditional way typing the following at the Lisp prompt: *(db-filter messier '(<= "Magnitude" 5))*

Table 6.10 A comparison of the fuzzy query *(>= "fz-magnitude" 0.75)* versus the crisp query *(<= "Magnitude" 5)* for the brightest celestial objects in the Messier Catalogue

(>= "fz-magnitude" 0.75)	(<= "Magnitude" 5)	fz-magnitude	Mv
M7	M7	0.85	3.3
M42	M42	1.0	2.9
M44	M44	0.95	3.1
M45	M45	1.0	1.2
–	M6	0.4	4.2
–	M24	0.2	4.6
–	M25	0.2	4.6
–	M31	0.1	4.8
–	M39	0.2	4.6
–	M41	0.2	4.6
–	M47	0.25	4.5

For saving space we have not written here the output of this query, but it returns eleven celestial objects. Table 6.10 compares the output from the crisp and fuzzy queries.

Let us observe, for example, M31, the Andromeda galaxy. With a visual magnitude Mv = 4.8 it has a membership degree $\mu(x) = 0.1$ to the fuzzy set defined by *BM*, so it has not been included in the output of the fuzzy query with a threshold value $t = 0.75$ but it has been included in the output of the crisp query. We must say at this point that M31 can be really hard to observe from an observing place with light-polluted skies, and it is not, definitely, one of the brightest Messier objects. What is the main and conceptual difference between the two outputs? The difference dwells from the fact that the very definition of the fuzzy set *BM* encompasses expert knowledge. As an astronomer I know beforehand the features of celestial objects and can model an appropriate fuzzy set for describing bright deep-sky objects. On the other hand, the crisp, traditional query *(<= "Magnitude" 5)* has less intrinsic information.

The attentive reader can argue that I have arbitrarily selected a crisp threshold value $t = 0.75$ to the resulting membership degrees, and it is a good observation, but first, since all the records in the database receive a membership degree from 0.0 to 1.0, we must select those records that have the highest membership degrees in order to finish to perform our query and second, a value $t = 0.75$ is usually a good one as a threshold when dealing with filtering membership degrees.

If you are not still convinced, we can push even further our exposition. Let us define a fuzzy set *FT* for representing a fuzzy-threshold value by typing the following at the Lisp prompt: *(setq FT '(fuzzy-threshold 0.7 0.9 1.0 1.0))*. Now we shall create another field in the database that will filter the records by applying the fuzzy set *FT* on it. Let us type the following expression at the keyboard:

(setq messier (fl-db-new-field messier "fz-magnitude" "after-threshold" FT 1))

Lisp will update the *messier* database with the new field named "*after-treshold*", filling it with the membership degree calculations performed by using *FT*. Now, let us finally type:

> **(db-filter messier '(> "after-threshold" 0.0))**

: (("Name" "NGC" "Constellation" "Class" "Right ascension" "Declination" "Magnitude" "Angular size" "Burnham" "Remarks" "fz-magnitude" "after-threshold")

("M7" "6475" "Scorpius" "Open cluster" "17 h 53.9 m" "-34d 49 m" "3.3" "80" "!!" "nmbr:80-fine naked-eye OC" "0.85" "0.75")

("M42" "1976" "Orion" "Diffuse nebula" "5 h 35.4 m" "-5d 27 m" "2.9" "66x60" "!!!"

"Great Orion Nebula" "1" "1")

("M44" "2632" "Cancer" "Open cluster" "8 h 40.1 m" "19d 59 m" "3.1" "95" "!!" "Praesepe - Beehive Clustr" "0.95" "1")

("M45" "" "Taurus" "Open cluster" "3 h 47 m" "24d 7 m" "1.2" "110" "*" "nmber: 130 - The Pleiades" "1" "1"))

As you can see, the query returns the same celestial objects than in Table 6.10: M7, M42, M44 and M45. In the next chapter we shall revisit again fuzzy databases, getting even "smarter" results.

6.10   As a Summary

Together with the previous chapter, the reader should have acquired by now a good perspective on the theory of fuzzy sets. All the FuzzyLisp functions developed in these last sections will have probably helped to consolidate this new knowledge. As usually we are going now to summarize these ideas and concepts in the following paragraphs.

- Continuous membership functions of the type $x \rightarrow f(x)$, where x is a point on the real axis and its image $f(x)$ is a real number defined in the closed interval [0,1] are of the maximum interest for defining fuzzy sets. Among this family of functions, those whose shape is a triangle or a trapezium are the most used for fuzzy modeling.
- Both triangular and trapezoidal shaped membership functions can be defined with only four singular points on the real axis, x_1, x_2, x_3, and x_4. By adequately combining these values we can obtain a good variety of characteristic functions, up to the point that in many practical uses no substantial difference does exist between models developed with triangular and trapezoidal shaped functions and other more sophisticated geometrical shapes.
- We call FuzzyLisp Standard Set Representation (FLSSR) to those Lisp expressions that, representing triangular or trapezoidal functions, are of the type *(fuzzy-set-name x1 x2 x3 x4)*. All the fuzzy sets with a FLSSR representation

have at least one point x_i on the real axis such as its image $f(x_i) = 1$. When fuzzy sets satisfy this property we call them *Normal* fuzzy sets. Those fuzzy sets that do not satisfy it are called *Subnormal* fuzzy sets. Additionally, we call *height* of a fuzzy set to its maximum membership degree. The height of a normal fuzzy set is, hence, 1.0.

- We call *Support* of a fuzzy set to the interval resulting from the geometrical projection of its membership function over the real axis. Likewise we call *Nucleus* of a fuzzy set to the interval resulting from the geometrical projection of the section of its characteristic function whose membership degree equals one. The nucleus of a triangular shaped, normal fuzzy set, is exactly one and only one point over the real axis. An alpha-cut is an interval defined on the real axis after sectioning a characteristic function by a horizontal line $y = \alpha$, where $\alpha \in [0,1]$, in such a way that every crisp value x belonging to the interval satisfies the expression $f(x) \geq \alpha$. The following relationship between nucleus (k), an alpha-cut (a) and the support of a fuzzy set (s) always holds: $k \leq a \leq s$. Every fuzzy set with a FLSSR representation can be defined by means of only two alpha-cuts.

- Aside triangular and trapezoidal shaped membership functions, FuzzyLisp has an alternative way of representing fuzzy sets named FuzzyLisp Discrete Set Representation (FLDSR). This form takes a finite list composed by sublists of elements x_i belonging to the universe of discourse where they are defined and their images $\mu(x_i)$ with the following structure: *(fuzzy-set-name $(x_1 \mu(x_1))$ $(x_2 \mu (x_2))$ ... $(x_n \mu(x_n)))$*. The key FuzzyLisp functions in a FLDSR are *(fl-discretize)*, *(fl-dset-membership?)* and *(fl-discretize-fx)*. These functions open an entire universe of possibilities for fuzzy set modeling.

- In a similar way as in the previous chapter, we define Complement, Union and Intersection of fuzzy sets within a FLSSR by means of the following expressions, respectively: $A' = 1 - \mu_A(x)/x \in [x_1, x_4]$, $A \cup B = max[\mu_A(x), \mu_B(x)]/ x \in [x_{1A}, x_{4B}]$ and $A \cap B = min[\mu_A(x), \mu_B(x)]/x \in [x_{1A}, x_{4B}]$. In general, the complement, union and intersection of fuzzy sets have not a FLSSR, but they all indeed have a discrete representation, FLDSR. Three FuzzyLisp functions returns the membership degree of any crisp value x belonging to A', $A \cup B$, $A \cap B$: *(fl-set-complement-membership?)*, *(fl-set-union-membership?)* and *(fl-set-intersect-membership?)*.

- Fuzzy numbers are probably the most important type of fuzzy sets for building practical applications. Every fuzzy number must satisfy the following three properties: (a) A fuzzy number must be a normal fuzzy set, that is, there exists at least a crisp value x belonging to the support of the fuzzy set whose image $\mu(x)$ equals one. (b) A fuzzy number must be a convex fuzzy set and (c) The support of a fuzzy number must be bounded.

- The strategy to perform arithmetic calculations with fuzzy numbers is based on the following procedure: Take some alpha-cuts from the fuzzy numbers involved in the calculations, perform the adequate arithmetic operations on the resulting intervals and then build the resulting fuzzy number from the obtained intervals. In this book way pay attention to sum, subtraction, multiplication and division of fuzzy numbers. Additionally, we focus also on multiplying a fuzzy number by a crisp, real value k and then on fuzzy averaging. Finally, while the fuzzy addition and subtraction of two fuzzy numbers always produce a fuzzy number with a FLSSR, the multiplication and division of two fuzzy numbers does not. On the other hand, any arithmetic operation on fuzzy sets can be defined in the frame of a FLDSR.

- A linguistic variable LV is a finite collection of two or more fuzzy numbers or fuzzy intervals that, representing linguistic concepts, completely patches a given universe of discourse defined on a base variable. The concept of "patching" in a linguistic variable composed by F_i fuzzy sets is important in this definition and it comes defined by the following expression: $LV = \{F_i \,|\, F_i \cap F_{i+1} \neq \phi\}$.

- In FuzzyLisp we can define linguistic variables by using either a FLSSR or a FLDSR, but we cannot mix both types of representation at the same time in the same linguistic variable. A workaround to this is to "translate" first a triangular or trapezoidal membership function into a FLDSR by means of the function *(fl-discretize)* and then build the entire linguistic variable with fuzzy sets with discrete characteristic functions.

- Fuzzy databases are one of the most exciting practical applications of fuzzy-logic nowadays. A Fuzzy database is a structured collection of data that contains fuzzy information and can be exploited by using fuzzy queries. Our exploration of fuzzy databases in this chapter covers the simplest approach to the topic, that is: To choose a numerical field from a traditional, crisp database, apply a definition of one or more fuzzy sets on it, convert the crisp numerical values stored in the database's field into membership degrees and then store the obtained values into a new field. Even so, the power of this technique has been shown when applied, as a practical example, to the Messier database of Deep Sky objects.

In the following chapter we shall apply all the material that we have learnt in this and in the previous chapter towards the understanding of the key concepts of fuzzy logic. As we shall see, if we have understood well the theory of fuzzy sets, the concepts of fuzzy logic are a natural step forwards.

References

Belohlavek, R., Klir, G.: Concepts and Fuzzy Logic. MIT Press, Cambridge (2011)
Chan, A., Wong, A.: A fuzzy approach to partitioning continuous attributes for classification. IEEE Trans. Knowl. Data Eng. **18**(5), 715–719 (2006)
Cordón, O., Herrera, F., Villar, P.: Analysis and guidelines to obtain a good uniform fuzzy partition granularity for fuzzy rule-based systems using simulated annealing. Int. J. Approximate Reasoning **25**(3), 187–215 (2000)
Galindo, J., Urrutia, A., Piattini, M.: Fuzzy Databases: Modeling, Design and Implementation. Idea Group Publishing, Hershey (2006)

Gray, A.: Lord Kelvin: An Account of His Scientific Life and Work. Hard Press Publishing (2014)
Guyton, A., Hall, J.: Textbook of Medical Physiology. Saunders (2005)
Klir, G., Yuan, B.: Fuzzy Sets and Fuzzy Logic. Theory and Applications. Prentice-Hall, New York (1995)
Pons, O., Vila, M., Kacprzyk, J. (eds.): Knowledge Management in Fuzzy Databases. Physica-Verlag, Heidelberg (2000)

Chapter 7
Fuzzy Logic

7.1 Introduction

This chapter is probably the densest one in this book and at the same time the more demanding from the reader. It deals with logic and then with a superset of it, fuzzy logic, so it covers a lot of material that will need to be digested with calm. Even so, and for reasons of space, it only scratches the surface of this branch of mathematics. It must be taken into account that a deep revision or a complete introduction to logic usually needs an entire book, so we have opted to show a concise and clear set of concepts that we hope will satisfy the audience. For example, for fuzzy implication we expose the Lukasiewicz operator as the most classic one in fuzzy logic (Kundu and Chen 1994), but the reader should be aware that there is much more fuzzy logic theory out there that can be easily found in theoretical books.

The material already covered in Chaps. 5 and 6 allow us to start the chapter with a brief introduction to propositional logic that quickly leads us to fuzzy logic, paying attention to the logical operators in fuzzy compound propositions for conjunction, disjunction, negation and implication. Later, fuzzy hedges are exposed, exploring at the same time some of their most interesting properties. Then, the fuzzy compound expression of the type "if x is *A* and y is *B* then z is *C*" paves the way to Fuzzy Rule Based Systems, FRBS.

Aside describing the concept of knowledge database and the classic architecture that makes up a FRBS, we discuss some defuzzification methods, discovering Singletons at the same time as a useful tool for creating practical fuzzy logic based applications. Along the chapter new and powerful FuzzyLisp functions will allow us to develop some real fuzzy logic models at the end of the chapter.

© Springer International Publishing Switzerland 2016
L. Argüelles Méndez, *A Practical Introduction to Fuzzy Logic using LISP*,
Studies in Fuzziness and Soft Computing 327,
DOI 10.1007/978-3-319-23186-0_7

7.2   The Beginning of Logic

More than 2000 years ago, in a European region bathed by the east Mediterranean waters a country, or better said, a civilization, was flourishing. Greece not only consolidated the study of mathematics and astronomy, but brought to history the development of democracy and then other astonishing consequences of human reasoning such as philosophy and within it, the study of logic. Although the controversy about whether logic is a part of philosophy or merely an instrument for progressing in philosophy continued for centuries, the truth is that even today logic is taught in the universities as a curriculum for philosophy studies. Maybe surprisingly, the study of philosophy and then logic is the reason behind the interest of many actual philosophers on fuzzy logic.

As it is well known, the most important philosopher from antique Greece was Aristotle (384BC–322BC). Maybe the most influential philosopher ever, Aristotle was the founder of logic, a structured pattern of thinking that makes it possible to arrive to new knowledge from previous established information. Central to Aristotle's Logic is his theory of syllogisms, from the Greek term *sullogismos* (deduction). In his work titled "Prior Analytics", Aristotle describes what a syllogism is: "A syllogism is speech (*logos*) in which, certain things having been supposed, something different from those supposed results of necessity because of their being so". In this sentence, the terms "things supposed" mean what we call a premise, while the terms "results of necessity" are what we understand as a conclusion. Translating it into English, we can say that a syllogism is a logic construction where after some premises are given we can reach a conclusion, or a logic construction that transform some given premises into a conclusion. As an example, the following syllogism is one of the most famous ones:

- all men are mortal
- all Athenians are men
  → all Athenians are mortal

In this syllogism we have two premises: "all men are mortal" and "all Athenians are men". The syllogism produces a conclusion: "all Athenians are mortal". As homage to syllogisms, we can express these premises in Lisp simply typing the following at the Lisp prompt: *(setq c1 '(men mortal)), (setq c2 '(athenians men))*. For obtaining the conclusion we can write a simple Lisp function for dealing with this naïve type of syllogisms. The function, named *(syllogism)* is shown in Code 7-1:

```
;code 7-1
(define (syllogism c1 c2)
   (if (= (first c1) (last c2))
     (list (first c2) (last c1))
     nil
   )
)
```

For trying this function we only need to type the following at the Lisp prompt:

> **(syllogism '(men mortal) '(athenians men))**
: *(athenians mortal)*

This type of syllogism is only one from the 24 valid types of syllogisms. The reader can find all of them in a good book of Logic and it would not be a hard task to translate them all to Lisp functions. In fact, such a Lisp project should be appealing to students of logic, simply because it would be a good thing for the study of reasoning itself (McFerran 2014; Priest 2001).

7.3 Modern Bivalent Logic

After the brilliant period of philosophy in Greece, no true revolution happened in the history of logic until the beginning of the nineteenth century, when the method of proof used in mathematics was applied to the concepts of logic. Moreover, Aristotelian logic was unable to satisfy mathematical creations at all because many arguments used for developing logical mathematical constructions are based on formulations of the type "if… then…", and these formulations were absent in Aristotelian logic. This section of the chapter deals with propositional logic, an important branch of modern formal logic that is a solid foundation in classical texts of artificial intelligence (e.g.: Russell and Norvig 2009).

Propositional logic is interested in logic relationships that are based on the study of propositions, and then in the creation of other propositions by means of logical operators. The main question for propositional logic is thus the following one: What can we do with propositions and with combining them by using logical operators? In this question there are two key words: propositions and logical operators.

A proposition is natural language (Aristotle would name it again *logos*), expressing an assertion that can be either true or false, hence the term "bivalent", but not true and false at the same time". Simple propositions take the following form:

x is P

here, x is called the subject of the proposition, and P is called the predicate of the proposition, usually expressing a property or feature of x. The following are valid propositions:

Tigers are mammals
The Montblanc is the highest mountain in the Alps
France is a European country
The "Jupiter" Symphony was written by Beethoven

The three first propositions are true, while the last one is false (Mozart would hate us if stated otherwise!).

When x is any subject belonging to an universe of discourse X, then a predicate P converts itself automatically into a function defined on X, forming a different proposition for every subject x. We shall represent by P(x) to this type of functions, named propositional functions. From different values x_1, x_2, ..., x_n we can obtain different propositions $p(x_1)$, $p(x_2)$, ..., $p(x_n)$. Any of them can be also expressed by:

$$x_i \text{ is } P$$

and, as already stated, anyone of them can be true or false until the variable itself is instanced. As an example, let X be the universe of discourse of some European countries X = {France, Spain, Germany} and P a predicate meaning "x is a Country bathed by the Mediterranean Sea". Then, the propositional function P(x) produces the following propositions:

$p(x_1)$ = "France is a Country bathed by the Mediterranean Sea"
$p(x_2)$ = "Spain is a Country bathed by the Mediterranean Sea"
$p(x_3)$ = "Germany is a Country bathed by the Mediterranean Sea"

Propositions $p(x_1)$ and $p(x_2)$ have a truth value equal to true, while proposition $p(x_3)$ has a truth value equal to false. Note again that the propositional function P(x) = "x is a Country bathed by the Mediterranean Sea" on its own is neither false nor true, and only when we substitute x by a member from the Universe of discourse the resulting proposition becomes true or false.

Things become even more interesting when two or more simple propositions are combined, forming what is known as a compound proposition. In this case, the truth-value of a compound proposition results from the truth-values of every simple proposition and the way they are connected, that is, from the type of logical connective, or logical operator, used for establishing the link between simple propositions. The main logical operators are the following ones:

Conjunction: Given two propositions p, q, we call their conjunction to the compound proposition "p and q", denoting them by the logical expression "p ∧ q". This compound proposition is true if and only if both p and q propositions are true. As an example, the following compound proposition is true: "Japan is a country and Kyoto is a Japanese city". The following one is false: "The sun is the brighter sky object in the solar system and it revolves around Earth", as for example, Ptolemy affirmed.

Disjunction: Given two prepositions p, q, we call their disjunction to the compound proposition "p or q", denoting them by the logical expression "p ∨ q". This compound proposition is true when at least one of both propositions, p or q, is true, or, obviously, when both are true. For being false it needs that both propositions are false. An example of true compound proposition linked by a disjunction is: "four is bigger than two or eight is a prime number" (it does not matter than eight is not a prime number, it suffices with the true proposition "four is bigger than two". On the other hand: "Pi is an integer or a circle has two centres" is a false compound proposition.

Negation: From any proposition p we shall call "negation of p" to the proposition "not p", denoting it by the expression "¬p". The proposition ¬p will be true when p is false and will be false when p is true. As an example, after stating the proposition p = "the Sun is a star", then ¬p, that is, "the Sun is not a star", is false.

Implication: Also known as "conditional proposition", it is expressed by the sentence "p implies q", and it is usually denoted by the expression "p → q". When p → q, it means that the conditional proposition is true except when the proposition p is true and q is false. In other words, **if p (also named the "antecedent" in implication) is true, then q (the consequent) must be also true for satisfying p → q being true**. As an example, the implication "if I breath then I produce CO_2" is true. At the other extreme, please note that we have lots of expressions in common language where both p and q are false, for example: "if you are the Queen of Saba then I am Solomon". These types of implications, where both the antecedent and consequent are false, are always true, because the logical structure of the conditional is satisfied.

If we define the truth value of a proposition by means of using the number 1 for expressing true and using 0 for expressing false, we can express formally these logical operators in the following way:

- Conjunction: $p \wedge q = \min(p,q)$ (7-1)

- Disjunction: $p \vee q = \max(p,q)$ (7-2)

- Negation: $\neg p = 1-p$ (7-3)

- Implication: $p \rightarrow q = \min(1, 1 + q - p)$ (7-4)

And, hereafter, we can construct a truth table, an arranged table of values that enumerates all the possible truth-values combinations of p_i propositions. Table 7.1 shows the truth table for conjunction, disjunction, negation and implication from two propositions p, q. In this table we represent false by the number 0 and true by the number 1.

Translating Table 7.1 into Lisp is easy because Lisp (and all traditional programming languages) use standard bivalent logic, including the functionality of "and", "or" and "not", as we have learnt in Part I of this book. For completing the architecture of Table 7.1 we only need to create a simple function named *(implication)*, as shown in Code 7-2:

Table 7.1 Truth table for conjunction, disjunction, negation and implication

p	q	p ∧ q	p ∨ q	¬p	¬q	p → q
1	1	1	1	0	0	1
1	0	0	1	0	1	0
0	1	0	1	1	0	1
0	0	0	0	1	1	1

False = 0, True = 1

Table 7.2 A translation of Table 7.1 into Lisp

p	q	(and p q)	(or p q)	(not p)	(not q)	(implication p q)
(setq p true)	(setq q true)	True	True	Nil	Nil	True
(setq p true)	(setq q nil)	Nil	True	Nil	True	Nil
(setq p nil)	(setq q true)	Nil	True	True	Nil	True
(setq p nil)	(setq q nil)	Nil	Nil	True	True	True

```
;code 7-2
 (define (implication p q)
    (if (and p (not q))
            nil
            true
    )
 )
```

Typing the following Lisp expressions row by row and from left to right at the Lisp prompt we obtain the complete contents of Table 7.2.

Now it is time to make an important assertion. Maybe one of the most important ones in this chapter. A sentence that will help the reader to give a quantum leap towards the understanding of fuzzy logic:

> **There is an isomorphism between the logical operators "and", "or", "not", "implication" and the set operations "intersection", "union", "complement" and "inclusion".**

In practical terms this means that all the material exposed in Sect. 5.2 about the classic theory of sets is of entire application in the field of logic after the simple substitution of the concept of membership or not membership of an element x to a set A and the concept of true or false when referring to a proposition p.

7.4 Fuzzy Logic

Let us now consider the proposition p = "John is old" when John is 55 years old. Under a traditional point of view in logic, in order to know if the proposition p is true or false, we need to define the predicate of p, that is, a definition of an old person. Arbitrarily, let us assume that an old person is a person older than 50 years old. Taking into account this definition, then proposition p is true. If we reflect about this, we immediately realize the isomorphism enunciated some lines above: When we say "John is old", we implicitly declare that John belongs to the set of old persons, defined for example as the set of all persons older than 50 years old. That is, since John belongs to the set of old persons, then proposition p is true.

However, and as we discussed extensively in Chap. 5, in fuzzy sets theory the membership of an element x to a fuzzy set S is a question of degree, expressed in the closed interval [0,1]. In fact, the isomorphism continues between the membership degree of an element x to a fuzzy set *A* and the truth degree of a predicate and then of a proposition *P*, thus conceptually filling the gap between fuzzy sets and fuzzy logic (Klir and Yuan 1995). This line of reasoning gives birth to the concept of fuzzy proposition whose definition results from introducing some important touches in the definition of a traditional proposition:

A fuzzy proposition is natural language declaring an assertion that has implicitly associated a truth-value expressed by a real number in the closed interval [0,1]. Simple fuzzy propositions have the following structure:

$$x \text{ is } P$$

The reader should note that the subject *x* of a fuzzy proposition *P* is usually not fuzzy. What distinguishes a classic proposition from a fuzzy proposition is the characterization of the predicate *P*. The fuzziness of *P* is what generates fuzzy propositions. Hereafter we shall refer to fuzzy propositions writing them in italics: e.g.: *p, q, r*.

These are sound paragraphs, but these ideas are what really allow us to transit from classical logic to fuzzy logic. Fortunately we already have learnt lots of concepts, definitions, and Lisp constructions on fuzzy sets, and all this material is of immediate application in fuzzy logic. We can even say that we started, without saying it so, to speak about fuzzy logic from Sect. 5.3 of this book.

In order to evaluate a simple fuzzy proposition *p*, we need to know the definition of its predicate, which will be generally given by the definition of a fuzzy set. Let, for example *P*, be a fuzzy set representing "old persons" expressed in FuzzyLisp by the construction *(setq P '(old 50 90 90 90))*. For calculating the truth-value of the proposition *p* = "John is old" when John is 55 years old, we can simply type at the Lisp prompt:

> **(last (fl-set-membership? P 55))**
: *0.125*

So the resulting truth-value of proposition *p* is 0.125.

In this moment, and more from an aesthetic or even also semantic point of view than a real disadvantage, you can argue that the name of the function *(fl-set-membership?)* does not seem appropriate for evaluating fuzzy propositions. Thanks to Lisp this is no problem. We can type the following expression at the Lisp prompt:

> **(setq fl-truth-value? fl-set-membership?)**
: *(lambda ....) (output omitted)*

and now we can write:

> **(last (fl-truth-value? P 55))**
: *0.125*

what we have made is to create an alias for the FuzzyLisp function *(fl-set-mem-bership?)*. If you wish so, you can create new aliases from all the existing FuzzyLisp functions for your own practical applications. The rest of this book will not use aliases. By the way, I would like to seize the opportunity to say that if English is not your mother tongue you can translate all the FuzzyLisp functions to your language. If you are, for example, a German reader, you could rename *(fl-set-membership?)* to:

> **(setq fl-Wahrheitswert? fl-set-membership?)**
: (lambda ....) (output omitted)

and then again:

> **(last (fl-Wahrheitswert? P 55))**
: 0.125

After translating all the FuzzyLisp functions, you could put all these definitions into a single file named, for example, fl-deutsch.lsp and then, for creating your fuzzy models you would only need to include the following two lines of code at the beginning of your application:

```
(load "fuzzylisp.lsp")
(load "fl-deutsch.lsp")
```

And hereafter you can call all the functions in your own language. By the way, all the original names of the FuzzyLisp functions as described in Appendix II will be always available at your fingertips despite the translation.

7.5 Logical Connectives in Fuzzy Propositions

In the same way we have defined logical connectives in order to form compound propositions in classical logic, we can now extend these ideas to compound fuzzy propositions. Let us review them:

The <u>Conjunction</u> of two fuzzy propositions p, q, represented by $p \wedge q$, is the result of the minimum truth-value, Tv, of both p and q. Expressed formally:

$$Tv(p \wedge q) = min(Tv(p), Tv(q)) = min(\mu_A(x), \mu_B(y)) \tag{7-5}$$

where A and B are fuzzy sets representing the predicates associated to the fuzzy propositions p and q, respectively, while x and y are feature values associated to their respective subjects. As an example, we can have the following two fuzzy propositions: p = "John is old" and q = "Eva is young". As in the previous example, let us define a fuzzy set P for representing the concept of old persons by means of the Lisp expression (setq P '(old 50 90 90 90)) and then a fuzzy set Q representing the concept of a young person by means of the Lisp expression (setq Q '(young 0 0

15 30)). If John is 55 years old, that is, x = 55 and Eva is 18 years old, that is, y = 18, we shall have:

$$(last\ (fl\text{-}set\text{-}membership?\ P\ 55)) \rightarrow 0.125$$
$$(last\ (fl\text{-}set\text{-}membership?\ Q\ 18)) \rightarrow 0.8$$

That is, the fuzzy proposition *p* has a truth-value 0.125 and the fuzzy proposition *q* has a truth-value 0.8. Now using expression (7-5) we arrive to the conclusion that the compound fuzzy proposition "John is old and Eva is young" has a truth-value equal to 0.125.

Writing a NewLisp function for obtaining the truth value of a compound fuzzy proposition with a logical connective "and" is straightforward, as shown in Code 7-3:

```
;code 7-3
(define (fl-truth-value-p-and-q? P Q x y, a b)
   (setq a (last (fl-set-membership? P x)))
   (setq b (last (fl-set-membership? Q y)))
   (min a b)
)
```

Trying the function confirms our previous result:

> **(fl-truth-value-p-and-q? P Q 55 18)**
: 0.125

In a similar way, we define the Disjunction of two fuzzy propositions *p*, *q*, represented by *p* ∨ *q*, as the result of the maximum truth-value of both *p* and *q*. Expressed formally:

$$Tv(p \lor q) = max(Tv(p), Tv(q)) = max(\mu_A(x), \mu_B(y)) \qquad (7\text{-}6)$$

where again *A* and *B* are fuzzy sets representing the predicates associated to the fuzzy propositions *p* and *q*, respectively, while *x* and *y* are feature values associated to their respective subjects. Using the same example from some lines above, the fuzzy proposition "John is old or Eva is young" will have a truth value *Tv(p ∨ q) = 0.8*. Let us prove it by creating the function *(fl-truth-value-p-or-q)*, as shown in Code 7-4:

```
; Code 7-4
(define (fl-truth-value-p-or-q? P Q x y, a b)
(setq a (last (fl-set-membership? P x)))
(setq b (last (fl-set-membership? Q y)))
(max a b)
)
```

Testing the function with the same supplied data P: *(old 50 90 90 90)*, Q: *(young 0 0 15 30))*, x = 55 years old and y = 18 years old, we obtain:

> **(fl-truth-value-p-or-q? P Q 55 18)**
> : *0.8*

The Negation of a fuzzy proposition *p*, denoted by ¬*p*, is the result of subtracting the truth-value of *p* from one. Formally:

$$Tv(\neg p) = 1 - Tv(p) = 1 - (\mu_A(x)) \qquad (7\text{-}7)$$

where *A* is the fuzzy set representing the predicate associated to the fuzzy proposition *p* and *x* is the feature value associated to its subject. From all the logical connectives used on fuzzy propositions, this is the easier one to implement in Lisp, as shown in Code 7-5:

```
;Code 7-5
;returns the truth value of the negation of fuzzy
proposition P
(define (fl-truth-value-negation-p? P x)
    (sub 1.0 (last (fl-set-membership? P x))))
)
```

For testing the function, we shall calculate the truth-value of the following propositions: "John is not old", "Eva is not young", using the same definitions of fuzzy sets *P* and *Q* as above. For John we shall have x = 55 and for Eva we shall have x = 18 years old. Then

> **(fl-truth-value-negation-p? P 55)**
> : *0.875*

> **(fl-truth-value-negation-p? Q 18)**
> : *0.2*

As expected, the expression that returns the truth-value of a fuzzy conditional proposition or fuzzy implication *p* → *q* is the most complex one from the four logical connectives from this section:

$$Tv(p \rightarrow q) = min(1, 1 + Tv(q) - Tv(p)) = min(1, 1 + \mu_B(y) - \mu_A(x)) \qquad (7\text{-}8)$$

As usually, *A* and *B* are fuzzy sets representing the predicates associated to fuzzy propositions *p* and *q*, respectively, while *x* and *y* are the feature values associated to their respective subjects. Code 7-6 shows the code for calculating fuzzy conditional propositions:

```
;Code 7-6
(define (fl-truth-value-fuzzy-implication-p-q? P Q x y, a b)
    (setq a (last (fl-set-membership? P x)))
    (setq b (last (fl-set-membership? Q y)))
    (min 1.0 (sub (add 1.0 b) a))
)
```

Now, let us suppose Laura (10) is a daughter of John (55). For calculating the truth value of the fuzzy conditional proposition "if John is old then Laura is young", we only need to type:

> **(fl-truth-value-fuzzy-implication-p-q? P Q 55 10)**
: *1*

The only way to get a zero truth value from a fuzzy implication is when the truth value of the fuzzy proposition p is exactly 1.0 and the truth value of the fuzzy proposition q is exactly 0.0, as it happens with classical implication. In our example, given the definitions of fuzzy sets P and Q, then for any pair x, y included into the supports of their respective fuzzy sets this only happens if x = 90 years old (John) and y = 30 years old (Laura). Expressed into words: "if John is old then Laura is young" means, for these values x, y that the proposition "John is old" has a truth value equal to 1.0, but Laura, from the definition of "being young" in the fuzzy set Q grows and grows until she is not anymore a young person, hence the proposition "Laura is young" ultimately reaches a truth value equal to 0.0 resulting in a zero truth value for the fuzzy implication $p \to q$. Testing it:

> **(fl-truth-value-fuzzy-implication-p-q? P Q 90 30)**
: *0*

Table 7.3 shows several truth values as a result of $p \to q$ for the fuzzy conditional implication "if John is old then Laura is young" when different values x, y are used as the feature values associated to their respective subjects, John and Laura. In the first row several values x_i are given, while the first column represents several y_i values. The shaded cells in the table tell us that when Laura grows and approaches the right limit of the support of fuzzy set Q, then the truth value of the fuzzy implication "if John is old then Laura is young" decreases until becoming null. Once again we must remark that John being 90 and Laura being 30 years old is entirely possible (in this instance, Laura would have been conceived when John was 60), but the fuzzy implication has a null truth value because the definitions of fuzzy sets P and specially Q, where Laura at 30 is not anymore a young person:

It is interesting to note that the conjunction/disjunction of two fuzzy propositions, and also the fuzzy implication, are fuzzy relations. For example, the aforementioned conjunction "John is old and Eva is young" is an instance of the general fuzzy compound proposition "x is A and y is B" that ultimately produces a truth value (isomorphic to membership degree) in the closed interval [0,1]. Remembering the expression [5-24] for defining a fuzzy relation:

Table 7.3 A truth table for different combinations of values x, y in a fuzzy implication $p \to q$

$P \to q$	50	60	70	80	90
0	1	1	1	1	1
10	1	1	1	1	1
20	1	1	1	0.91667	0.66667
30	1	0.75	0.5	0.25	0

$$R = \{((x, y), \mu_R(x, y)) \mid (x, y) \in AxB, \mu_R(x, y) \in [0, 1]\}$$

We can see that the general compound fuzzy proposition "x is A and y is B" is a mapping from the pair x,y to the closed interval [0,1] where A and B are the fuzzy sets representing the predicates associated to the individual fuzzy propositions "x is A", "y is B", respectively. In the case of the logical connective conjunction, the mapping is given by the expression:

$$\mu_R(x, y) = \min(\mu_A(x), \mu_B(y))$$

The same line of reasoning is valid for disjunction and fuzzy implication in fuzzy compound propositions.

Another important consideration in fuzzy propositions arises when we ask ourselves about the nature of the fuzzy sets associated to them. Do fuzzy propositions need associated fuzzy sets with a continuous characteristic function? Absolutely not. In fact we can use also fuzzy sets with discrete characteristic functions, so we can use both FuzzyLisp Standard or Discrete Set Representations, FLSSR, FLDSR for building fuzzy propositions. As an example, let us again observe the fuzzy proposition p = "John is old", this time defining a FLDSR for P. Let us type at the Lisp prompt:

> **(setq P (fl-discretize '(old 50 90 90 90) 4))**
: (old (50 0) (60 0.25) (70 0.5) (80 0.75) (90 1) (90 1) (90 1) (90 1) (90 1))

Now, for calculating the truth vale of p we only need to type:

> **(last (fl-dset-membership? P 55))**
: 0.125

The same is also valid for compound fuzzy propositions. In this case we need slight modifications to the functions *(fl-truth-value-p-and-q?)*, *(fl-truth-value-p-or-q?)*, *(fl-truth-value-negation-p?)* and *(fl-truth-value-fuzzy-implication-p-q?)* as shown in Code 7-3b, Code 7-4b, Code 7-5b and Code 7-6b, respectively:

```
; Code 7-3b
(define (fl-dtruth-value-p-and-q? dP dQ x y, a b)
    (setq a (last (fl-dset-membership? dP x)))
    (setq b (last (fl-dset-membership? dQ y)))
    (min a b)
)

; Code 7-4b
(define (fl-dtruth-value-p-or-q? dP dQ x y, a b)
    (setq a (last (fl-dset-membership? dP x)))
    (setq b (last (fl-dset-membership? dQ y)))
    (max a b)
)
```

```
;Code 7-5b
(define (fl-dtruth-value-negation-p? dP x)
   (sub 1.0 (last (fl-dset-membership? dP x)))
)

;Code 7-6b
(define (fl-dtruth-value-fuzzy-implication-p-q? dP dQ x y,
a b)
   (setq a (last (fl-dset-membership? dP x)))
   (setq b (last (fl-dset-membership? dQ y)))
   (min 1.0 (sub (add 1.0 b) a))
)
```

Let us test them quickly after defining a fuzzy set Q with a discrete characteristic function for representing "young people":

> **(setq Q (fl-discretize '(young 0 0 15 30) 4))**
:(young (0 1) (0 1) (0 1) (0 1) (0 1) (3.75 1) (7.5 1) (11.25 1) (15 1) (18.75 0.75) (22.5 0.5) (26.25 0.25) (30 0))

Then, considering again the fuzzy propositions p = "John is old" and q = "Eva is young" when John is 55 years old and Eva is 18 we have:

- "John is old and Eva is young":
> **(fl-dtruth-value-p-and-q? P Q 55 18)**
: 0.125

- "John is old or Eva is young":
> **(fl-dtruth-value-p-or-q? P Q 55 18)**
: 0.8

- "Eva is not young"
> **(fl-dtruth-value-negation-p? Q 18)**
: 0.2

Finally, for the proposition q = "Laura is young", when Laura (ten years old), being a daughter of John, we have:

"if John is old then Laura is young"
> **(fl-dtruth-value-fuzzy-implication-p-q? P Q 55 10)**
: 1

Another example (Bojadziev and Bojadziev 1999) will help to understand even better the fuzzy implication $p \rightarrow q$, in this case using fuzzy sets with discrete characteristic functions. Let P and Q be fuzzy sets formed respectively by the following Lisp expressions:

```
(setq P '(high-score (0 0) (20 0.2) (40 0.5) (60 0.8) (80 0.9) (100 1)))
(setq Q '(good-credit (0 0) (20 0.2) (40 0.4) (60 0.7) (80 1) (100 1)))
```

Where P represents the concept of "high score" applied to a bank customer and Q represents a certain type of measure about the quality of a hypothetical loan. We can form the fuzzy propositions p = "x is high score" and q = "y is good credit" and then the fuzzy propositional function "if x is high score then y is good credit". When, for example, x = 85 and y = 75 we can type at the Lisp prompt:

> **(fl-dtruth-value-fuzzy-implication-p-q? P Q 85 75)**
: *1*

As the reader can appreciate, there is no practical difference between using a FLSSR or a FLDSR for evaluating truth-values from fuzzy propositions. When the related fuzzy sets have a triangular or trapezoidal membership function results are the same. When other continuous functions are discretized for representing membership functions the differences are minimal.

7.6   Fuzzy Hedges

Fuzzy hedges are linguistic modifiers applied to fuzzy predicates. This short definition soon reveals its consequences because if we say they are applied to fuzzy predicates it immediately means they are also applied to fuzzy sets related to fuzzy predicates. Then it is easy to follow that they will affect fuzzy propositions and then their truth-values, too. In general, a fuzzy hedge H can be represented by an unary operator on the closed interval [0,1] in such a way that a fuzzy proposition of the type "x is P" converts in:

$$x \text{ is } HP$$

If A is a fuzzy set associated to a fuzzy predicate P, the following expression shows how A is modified by H for every element x (subject) in A:

$$HA = H(\mu_A(x)) \qquad (7\text{-}9)$$

The most used fuzzy hedges are the linguistic modifiers H_1, "very" and H_2, "fairly", defined respectively by the following expressions:

$$H_1 : H_1(\mu_A(x)) = (\mu_A(x))^2 \qquad (7\text{-}10)$$

$$H_2 : H_2(\mu_A(x)) = (\mu_A(x))^{1/2} \qquad (7\text{-}11)$$

As an example, if we take the fuzzy proposition p: "Joe is old", with a truth value $Tv(p) = 0.8$ that is, if x = Joe has a 0.8 membership degree to the fuzzy set A representing old People, then the fuzzy expression q: "Joe is very old" will have a truth value:

$$\mathrm{Tv}(q) = H_1(\mu_A(x)) = (\mu_A(x))^2 = 0.64$$

On the other hand, the fuzzy proposition r: "John is fairly old" will have a truth value: $\mathrm{Tv}(r) = H_2(\mu_A(x)) = (\mu_A(x))^{1/2} = 0.89$. From the intrinsic nature of the functions $y = x^2$ and $y = x^{1/2}$ it is easy to follow that the linguistic hedge H_1, "very", decreases the membership degree of an element x to a fuzzy set A when $\mu_A(x) \neq 1.0$ and ultimately, the truth value of an associated fuzzy proposition p. Conversely, the linguistic hedge H_2, "fairly", increases the membership degree in fuzzy sets and then in any associated fuzzy proposition when $\mu_A(x) \neq 1.0$. In fuzzy logic theory, we say a linguistic hedge is a Strong modifier when $H(\mu_A(x)) < \mu_A(x)$ and a Weak modifier when $H(\mu_A(x)) > \mu_A(x)$. Hence, the fuzzy hedge "very" is a strong modifier, while the fuzzy hedge "fairly" is a weak modifier.

FuzzyLisp has a function named *(fl-dset-hedge)* that applies a fuzzy hedge to a discrete fuzzy set. Again from the very nature of the functions $y = x^2$ and $y = x^{1/2}$ the resulting fuzzy set after applying a linguistic modifier on it can not be generally represented by a triangular or trapezoidal characteristic function, so the function always returns the transformed fuzzy set with a FLDSR. This feature invites to design the function in such a way that the input fuzzy set has also a FuzzyLisp Discrete Set Representation. Code 7-7 shows this function:

```
;Code 7-7
(define (fl-dset-hedge dset hedge, i n list-out sublist)
   (setq list-out (first dset));we conserve the set's name
   (setq i 1 n (length dset))
   (while (< i n)
      (setq sublist (nth i dset))
      (case hedge
        (FAIRLY
           (setq list-out (cons
              (list (first sublist) (sqrt (last sublist)))
                       list-out))
        );end FAIRLY
        (VERY
           (setq list-out (cons
              (list (first sublist) (pow (last sublist)))
                       list-out))
        );end VERY
      );end case
    (++ i)
  );while end
  (reverse list-out)
)
```

(fl-dset-hedge) takes two arguments: The first one is a fuzzy set with a FLDSR and the second one is a Lisp symbol representing the desired fuzzy hedge to apply

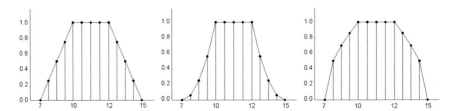

Fig. 7.1 From *left* to *right* a fuzzy set *A* and the fuzzy sets "*VeryA*" and "*FairlyA*"

on it, either FAIRLY or VERY. If our initial fuzzy set has a FLSSR representation we can first make a call to the function *(fl-discretize)* and then use the resulting discretized fuzzy set. Let us see an example of the linguistic modifiers "very" and "fairly" applied to the fuzzy set *(A1 7 10 12 15)*:

> **(fl-dset-hedge (fl-discretize '(A1 7 10 12 15) 4) 'VERY)**
: (A1 (7 0) (7.75 0.0625) (8.5 0.25) (9.25 0.5625) (10 1) (10.5 1) (11 1) (11.5 1)
(12 1) (12.75 0.5625) (13.5 0.25) (14.25 0.0625) (15 0))

> **(fl-dset-hedge (fl-discretize '(A1 7 10 12 15) 4) 'FAIRLY)**
: (A1 (7 0) (7.75 0.5) (8.5 0.7071067812) (9.25 0.8660254038) (10 1) (10.5 1)
(11 1)
(11.5 1) (12 1) (12.75 0.8660254038) (13.5 0.7071067812) (14.25 0.5) (15 0))

it suffices to numerically compare the results from these two functions calls to *(fl-dset-hedge)* to realize the impact of both fuzzy hedges on the fuzzy set defined by the Lisp expression *(A1 7 10 12 15)*, but a graphical representation shows it even better, as can be easily seen in Fig. 7.1.

Needless to say, we can experiment also with fuzzy sets that require from scratch of a discrete characteristic function, as it is the case with a bell shaped function. As we already know, we start defining the function f in Lisp:

> **(setq f '(div (add 1.0 (cos (mul 2.0 pi (sub x 2.0)))) 2.0))**
: (div (add 1 (cos (mul 2 pi (sub x 2)))) 2)

now we discretize it:

> **(setq dBell (fl-discretize-fx 'Bell f 10 1.5 2.5))**
: (Bell (1.5 0) (1.6 0.09549150283) (1.7 0.3454915028) (1.8 0.6545084972) (1.9
0.9045084972) (2 1) (2.1 0.9045084972) (2.2 0.6545084972) (2.3 0.3454915028)
(2.4 0.09549150283) (2.5 0))

and then we can apply a linguistic modifier on it. In this case, "very":

> **(fl-dset-hedge dBell 'VERY)**
: (Bell (1.5 0) (1.6 0.009118627113) (1.7 0.1193643785) (1.8 0.4283813729)
(1.9   0.8181356215)   (2   1)   (2.1   0.8181356215)   (2.2   0.4283813729)   (2.3
0.1193643785) (2.4 0.009118627113) (2.5 0))

Something very interesting happens when we reiteratively apply a fuzzy hedge on a fuzzy set A. Let us define, for example, a fuzzy number "close to seven" expressed in Lisp as: *(about-seven 6.5 7.0 7.0 7.5)*:

> **(setq A '(about-seven 6.5 7.0 7.0 7.5))**
: (about-seven 6.5 7 7 7.5)

First we discretize this fuzzy number defined by a triangular membership function. We use an adequate resolution in order to better appreciate the expected effects of reiteration:

> **(setq dA (fl-discretize A 10))**
: (about-seven (6.5 0) (6.55 0.1) (6.6 0.2) (6.65 0.3) (6.7 0.4) (6.75 0.5) (6.8 0.6) (6.85 0.7) (6.9 0.8) (6.95 0.9) (7 1) (7.05 0.9) (7.1 0.8) (7.15 0.7) (7.2 0.6) (7.25 0.5) (7.3 0.4) (7.35 0.3) (7.4 0.2) (7.45 0.1) (7.5 0))

applying the fuzzy hedge "Very" we obtain $H_1(\mu_{dA}(x))$:

> **(setq VERY-7 (fl-dset-hedge dA 'VERY))**
: (about-seven (6.5 0) (6.55 0.01) (6.6 0.04) (6.65 0.09) (6.7 0.16) (6.75 0.25) (6.8 0.36) (6.85 0.49) (6.9 0.64) (6.95 0.81) (7 1) (7.05 0.81) (7.1 0.64) (7.15 0.49) (7.2 0.36) (7.25 0.25) (7.3 0.16) (7.35 0.09) (7.4 0.04) (7.45 0.01) (7.5 0))

now we apply it again, obtaining $H_1(H_1(\mu_{dA}(x)))$:

> **(setq VERY-VERY-7 (fl-dset-hedge VERY-7 'VERY))**
: (about-seven (6.5 0) (6.55 0.0001) (6.6 0.0016) (6.65 0.0081) (6.7 0.0256) (6.75 0.0625) (6.8 0.1296) (6.85 0.2401) (6.9 0.4096) (6.95 0.6561) (7 1) (7.05 0.6561) (7.1 0.4096) (7.15 0.2401) (7.2 0.1296) (7.25 0.0625) (7.3 0.0256) (7.35 0.0081) (7.4 0.0016) (7.45 0.0001) (7.5 0))

the fuzzy set "*very-very-very-dA*", $H_1(H_1(H_1(\mu_{dA}(x))))$ is again:

> **(setq VERY-VERY-VERY-7 (fl-dset-hedge VERY-VERY-7 'VERY))**
: (about-seven (6.5 0) (6.55 1e-08) (6.6 2.56e-06) (6.65 6.561e-05) (6.7 0.00065536) (6.75 0.00390625) (6.8 0.01679616) (6.85 0.05764801) (6.9 0.16777216) (6.95 0.43046721) (7 1) (7.05 0.43046721) (7.1 0.16777216) (7.15 0.05764801) (7.2 0.01679616) (7.25 0.00390625) (7.3 0.00065536) (7.35 6.561e-05) (7.4 2.56e-06) (7.45 1e-08) (7.5 0))

and after only four iterations, we obtain $H_1(H_1(H_1(H_1(\mu_{dA}(x)))))$, that is, the fuzzy set "*very-very-very-very-dA*":

> **(setq VERY-VERY-VERY-VERY-7 (fl-dset-hedge VERY-VERY-VERY-7 'VERY))**
: (about-seven (6.5 0) (6.55 1e-16) (6.6 6.5536e-12) (6.65 4.3046721e-09) (6.7 4.294967296e-07) (6.75 1.525878906e-05) (6.8 0.0002821109907) (6.85 0.003323293057) (6.9 0.02814749767) (6.95 0.1853020189) (7 1)

(7.05   0.1853020189)   (7.1   0.02814749767)   (7.15   0.003323293057)   (7.2 0.0002821109907) (7.25 1.525878906e-05) (7.3 4.294967296e-07) (7.35 4.3046721e-09) (7.4 6.5536e-12) (7.45 1e-16) (7.5 0))

Figure 7.2 shows the comparison of the original fuzzy set *dA* and the result of applying the fuzzy hedge "very" reiteratively at four iterations.

The simple observation of Fig. 7.2 explains the following corollary: When n tends to infinity, the n-iteration of the hedge VERY on a fuzzy number *A* produces another special fuzzy set *A'* whose nucleus and support is established exactly at x_0, where its membership degree equals one:

$$VERY^n(A) \rightarrow (x_0, \mu(x_0) = 1.0)$$

In a similar way, when n tends to infinity, the n-iteration of the hedge VERY on a fuzzy interval *I*, expressed by a trapezoidal membership function, produces another fuzzy interval [a, b], where a = x_2 and b = x_3, as suggested in Fig. 7.3.

Interestingly, the concept of fuzzy hedges is not entirely new for us. In the previous chapter, Sect. 6.7.2, we created the function *(fl-expand-contract-set)* that allows us to expand or to contract the support and nucleus of a fuzzy set by a real number k. The resulting fuzzy set remains centered on its original position. As an example, for the fuzzy interval represented by the Lisp expression *(a 2 3 5 6)*, we would have, for k = 2 and k = 0.5, the following results, respectively:

> **(fl-expand-contract-set '(a 2 3 5 6) 2.0)**
: (a 0 2 6 8)

> **(fl-expand-contract-set '(a 2 3 5 6) 0.5)**
: (a 3 3.5 4.5 5)

The comparison between the resulting fuzzy sets can be appreciated in Fig. 7.4. Let us see what happens when k = 0, shown at right in the same figure:

> **(fl-expand-contract-set '(a 2 3 5 6) 0)**
: (a 4 4 4 4)

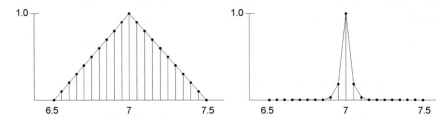

Fig. 7.2 A fuzzy set *dA* and the resulting fuzzy set after applying four times the fuzzy hedge "very" on it

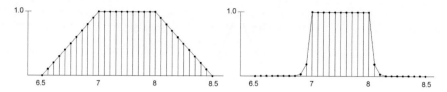

Fig. 7.3 A fuzzy interval *dI* and the resulting fuzzy set after applying four times the fuzzy hedge "very" on it

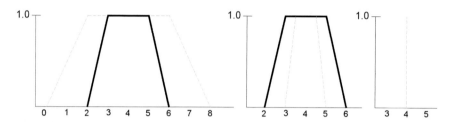

Fig. 7.4 A comparison between a fuzzy interval and some transformations after applying the FuzzyLisp function *(fl-expand-contract-set)* on it. Transformed fuzzy sets are shown in *dashed lines*

As can be seen, the result of using a value k = 0 as the second parameter in the FuzzyLisp function *(fl-expand-contract-set)* is the equivalent to apply the n-iteration of the hedge VERY on a fuzzy set. These peculiar fuzzy sets where their nucleus and support are established at a unique point x_0 on the real axis such as $(x_0, \mu(x_0) = 1.0)$ have an important and special role to play in fuzzy logic, and we shall dedicate enough space to them later in this chapter.

Speaking rigorously, the function *(fl-expand-contract-set)* is not the same thing as using a pure fuzzy hedge, but it also concentrates or spreads-out a fuzzy set, and in some fuzzy modeling circumstances can be a valid alternative to the standard "very" and "fairly". Another valid strategy to represent the meaning of these fuzzy hedges is to conveniently use the FuzzyLisp function *(fl-fuzzy-shift)*. For example, if we represent the fuzzy set *Old* by the triangular membership function *(setq Old '(age-old 75 80 80 85))*, we can define "very old" by means of the following function call:

> **(setq Very-old (fl-fuzzy-shift Old 5))**
: (age-old 80 85 85 90)

And now, for keeping the semantic FuzzyLisp structure:

> **(setq (nth 0 Very-old) 'age-very-old)**
: age-very-old

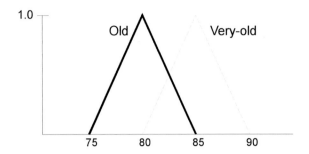

Fig. 7.5 The emerging of a linguistic variable from the use of the fuzzy set Old and a shift of it towards *right*, obtaining the fuzzy set Very-old

and then, evaluating the resulting symbol: *Very-old* → *(age-very-old 80 85 85 90)*

As can be seen in Fig. 7.5 this strategy ultimately invites ourselves to understand a linguistic variable as the result of starting with a seminal fuzzy set and then add other fuzzy sets to the linguistic variable by means of applying several shifts towards left and/or right on that set. The reader should note the huge existing flexibility in designing linguistic variables as we started to appreciate in the previous chapter.

7.7   Fuzzy Systems from Fuzzy Propositions

Exactly at 20:17:40 UTC on July 20th, 1969, Neil Armstrong landed the Eagle on the Moon. He had to land the Lunar Module semi-automatically because the resulting landing target stored on the onboard computer was not well suited for a good descent after visual inspection. Buzz Aldrin was providing him a continuous readout of altitude and velocity data from the instruments on the console and finally, the commandant of the Apollo XI mission got a successfully touchdown when less than 30 s of fuel remained available (Kranz 2009). We can guess that for getting a smooth descent he used his enormous experience as a test pilot and also the fruits of an extensive training at NASA's simulators. Even without being aware, lots of fuzzy compound propositions were inside his brain. The following one was probably one of them:

"If altitude is low and descent speed is high then thrust must be strong"

This type of fuzzy propositions can be generalized with the following expression:

$$(p \wedge q) \rightarrow r \qquad\qquad (7\text{-}12)$$

Or, expressed more clearly:

$$\text{If } x \text{ is } A \text{ and } y \text{ is } B \text{ then } z \text{ is } C \qquad\qquad (7\text{-}13)$$

Until now, all the fuzzy propositions analyzed so far in this chapter, both the fuzzy sets representing predicates and the feature values associated to their respective subjects are perfectly determined, that is, they have known parameters and values that allow to calculate the truth value of a given proposition. Now we introduce a new situation where the feature value z in the consequent of expression (7-13) is unknown. This is new with respect to all the previous fuzzy compound propositions exposed so far in this chapter, but opens a door to new techniques in fuzzy logic that deserve to be explored since expression (7-13) is perhaps one of the most useful and practical ones in the theory. As an example, and again in Armstrong's mind: Given an altitude x and a descent speed y from the readouts, what must be the value z for thrust engine? Let us do a formal analysis of expressions (7-12) and (7-13) first.

Since all the parameters in the antecedent x, y, A, B from expression (7-13) are known, having into account (7-5) we can write:

$$
\begin{aligned}
Tv(p \wedge q) &= min(\mu_A(x), \mu_B(y)) \\
Tv((p \wedge q) \rightarrow r) &= min(1, 1 + \mu_C(z) - min(\mu_A(x), \mu_B(y)))
\end{aligned}
\qquad (7\text{-}14)
$$

Since z is unknown, then $\mu_C(z)$ is unknown, too, so Eq. (7-14) is impossible to solve. However, if we assume that the truth-value of the antecedent is the same as the truth-value of the consequent in (7-13), we can write:

$$\mu_C(z) = min(\mu_A(x), \mu_B(y)) \qquad\qquad (7\text{-}15)$$

and substituting in (7-14):

$$Tv((p \wedge q) \rightarrow r) = min\,(1, 1 + \mu_C(z) - \mu_C(z)) \qquad\qquad (7\text{-}16)$$

and this simplifies to:

$$Tv((p \wedge q) \rightarrow r) = 1 \qquad\qquad (7\text{-}17)$$

This is a pivotal result for practical uses of fuzzy logic: When we assume that the truth value of the antecedent is the same as the truth value of the consequent in compound fuzzy propositions of the type expressed in (7-13) then the truth-value of $(p \wedge q) \rightarrow r$ is always one, that is, the fuzzy implication $(p \wedge q) \rightarrow r$ always holds. This should be not new for us, since we already got a hint about this point in Sect. 7.3 when we stated that if the antecedent is true in an implication then the consequent must be also true for satisfying the implication.

Expression (7-15) guarantees that we can obtain the membership degree of z to the fuzzy set C, but we still do not know the crisp value z in (7-13). Following our previous example, Armstrong still wouldn't know the thrust engine to apply for landing the Eagle. Let us see what can be done with the material learnt so far.

For representing "low altitude" in landing the Eagle, we can create, for example, a triangular fuzzy set H by means of the Lisp expression *(setq H '(low-altitude 30 100 100 170))*, where the magnitude is expressed in meters. In the same way, "high descent speed", in m/s, can be described with a triangular fuzzy set dH with the expression *(setq dH '(high-descent-speed 100 200 200 300))*. The whole thrust available in the Eagle propulsion system can be expressed by means of a range from 0.0 to 100.0 percent, so we can formulate "strong thrust" with the fuzzy set sT in the following way: *(setq sT '(strong-thrust 50 75 75 100))*. Now, giving some readouts h and s representing altitude and descending speed, respectively, expression (7-13) converts to:

$$\text{If h is } H \text{ and s is } dH \text{ then z is } sT$$

When readouts from the instruments on the Eagle are, for example, h = 79 m, and s = 190 m/s, then we can directly evaluate the antecedent in the rule just typing the following expression at the Lisp console:

> **(fl-truth-value-p-and-q? H dH 79 190)**
: *0.7*

Now, using (7-15), we have that the membership degree of z to sT is $\mu_{sT}(z) = 0.7$. In this moment, the question is: Given a 0.7 membership degree to the fuzzy set "strong thrust", what is the resulting crisp value z? A first answer comes from the FuzzyLisp function *(fl-alpha-cut)* as follows:

> **(fl-alpha-cut sT 0.7)**
: *(strong-thrust 67.5 82.5)*

Both 67.5 and 82.5 are crisp values of thrust that have a 0.7 membership degree to the fuzzy set "strong thrust", but Armstrong cannot spend his time doubting between applying a 67.5 % or a 82.5 % of thrust to the Eagle spacecraft. How can we solve this situation? The answer is not far from us. In fact is hidden inside the text written in the first paragraph of this section: "Even without being aware, lots of fuzzy compound propositions were inside the brain of Armstrong". That is: in general, for obtaining an adequate and useful crisp value from expression (7-13) we need several fuzzy compound propositions working together. In other words, we need several values and fuzzy sets z_1, C_1, z_2, C_2, ... z_n, C_n for creating a defuzzification procedure that finally will result into a crisp value. With a unique fuzzy proposition of the type "If x is A and y is B then z is C" the more we can say is that z has a $\mu_C(z)$ membership degree to C.

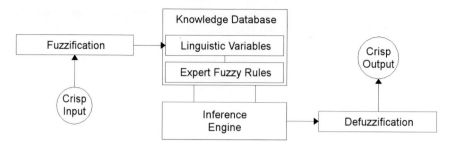

Fig. 7.6 Architecture of a Fuzzy Rule-Based System, FRBS

7.7.1 Fuzzy Rule-Based Systems

Fuzzy Rule-Based Systems, FRBS, are logical constructions that bring together several fuzzy-logic based processes and structures. The architecture of a FRBS is shown in Fig. 7.6.

The simple observation of Fig. 7.6 shows that a FRBS is a system that receives information (crisp input), processes it and then produces a crisp result. Usually the crisp input is based on available numerical information, resulting from measuring features of a physical system (e.g., density, speed, temperature, etc.) or numerical values representing any measurable entity (e.g., population, life expectancy at birth, GPD, etc.). The crisp output represents a numerical magnitude that is initially unknown but can be calculated by the inner workings of the fuzzy system. It is important to note that both the crisp input and crisp output can be formed by one or several numerical values, although in this chapter we shall be specially focused on FRBS with two numerical values as the crisp input and only one numerical output.

The first process in a fuzzy system is to fuzzify the crisp input (numerical data) and then translate this information into membership degrees. This process is known as "Fuzzification" and it is not new for us because we have already seen it in previous sections of this book. In fact, the key FuzzyLisp functions that allows to fuzzify any crisp numerical value are the functions *(fl-set-membership?)*, *(fl-dset-membership?)* and then their derived functions *(fl-lv-membership2?)* and *(fl-dlv-membership2?)*.

A Knowledge Database consists of several linguistic variables and a set of fuzzy propositions. Hereafter we shall call "rules" or "expert fuzzy rules" to the set of fuzzy propositions inside a FRBS. The term "expert" is important in this definition because every rule encompasses human knowledge, usually elicited from experts in a definite field of knowledge. Human knowledge is also embedded in the very definition of the linguistic variables at play, as we saw in the previous chapter when we spoke about fuzzy databases, so linguistic variables are also part of the knowledge database in a fuzzy system. Traditionally in fuzzy logic theory only the collection of expert fuzzy rules makes up a knowledge database, but this definition is incomplete because, as already mentioned, the expert knowledge does not come only from the fuzzy rules in the system, but from the meaning of the fuzzy sets that form the linguistic variables of the system.

The inference engine in a fuzzy system has a simple mission, as its name suggests: To make inferences. Until the inference engine starts to work, the only things we have in a fuzzy system are membership degrees, lots of fuzzy sets and expert fuzzy rules "floating up in the air". The inference engine is the glue that interacts with all these elements. It reads a rule, identifies its antecedent and consequent, finds the definition of a fuzzy set, gets the required membership degree, notes down this temporal data, obtains more membership degrees until the antecedent is computationally complete and then performs the logical inference in the rule, passing the obtained numerical values to the consequent. Then it processes the following rule and repeats the same procedure until finishing the complete set of rules. Using the same notation as in (7-13), the output of the inference engine in a FRBS has the following structure:

From Rule$_1$: "z is C_1" has an associated truth value $\mu_{C1}(z)$
From Rule$_2$: "z is C_2" has an associated truth value $\mu_{C2}(z)$

...

From Rule$_n$: "z is C_n" has an associated truth value $\mu_{Cn}(z)$

Or, written it formally, we can say that the inference engine ultimately produces the following set of numerical values (membership degrees):

$$Fuzzy\ Output = \{\mu_{C1}(z), \mu_{C2}(z), \ldots, \mu_{Cn}(z)\} \tag{7-18}$$

The procedure for obtaining a crisp output from the set of membership degrees in (7-18) is called "Defuzzification" and we shall discuss it immediately after stating an important remark: In a FRBS there is not a nucleus, that is, we can not say that the inference engine, or the knowledge database is the nucleus of the system. A Fuzzy Rule Based System has a holistic nature where all the subsystems work in a status of synergy. This is one of the strongest strengths of fuzzy systems.

7.7.2   Defuzzification

Let us do an analogy between the set of fuzzy rules of a knowledge database and the board of directors of a Club. In this analogy, every rule is represented by a member of the board where every member has a definite opinion on a certain subject matter that is represented in the analogy by a membership degree $\mu_{Ci}(z)$. The members of the board have a complete analysis and integral representation of the subject matter, but a final agreement must be reached, a final decision that democratically must take into account every member's opinion. This situation describes rather well the expression (7-18): We have a complete set of membership degrees as the result of the inferences made in the fuzzy system, but we need to come back to the realm of crisp values in order to get a final, definite result. Just imagine the Club is trying to

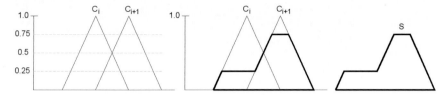

Fig. 7.7 Creating a geometrical shape from membership degrees in the fuzzy sets belonging to consequents in fuzzy rules

approve a final budget for the next year: they need a crisp quantity. The partners must know how much money the Club will have at its disposal for the next year. We need to obtain the z, crisp output from the system.

There are several defuzzification methods in fuzzy logic theory. For convenience in our exposition, we shall assume initially that only some expert fuzzy rules from the FRBS haven been fired in such a way that only two $\mu_{Ci}(z)$, $\mu_{Ci+1}(z)$ membership degrees from the fuzzy output expressed by (7-18) have a value bigger than zero. Then, if C_i, C_{i+1} are fuzzy sets belonging to the consequents of the knowledge database of fuzzy rules, we can create a geometrical shape as shown in Fig. 7.7.

As can be seen at left in the figure, we have chosen $\mu_{Ci}(z) = 0.25$, $\mu_{Ci+1}(z) = 0.75$ on the fuzzy sets C_i, C_{i+1}, producing a sort of mountain-shaped form S than can be easily observed at right.

An additional step in the process of defuzzification is shown in Fig. 7.8, where the shape S can be imagined, for a better exposition, as a materialized solid form made of metal, wood or any solid and rigid material you can name in this moment. This materialized shape S has, obviously, a center of gravity, G. Projecting the two-dimensional point G over the horizontal axis we obtain a real value x that represents the crisp, defuzzified numerical value z.

This method of defuzzification, known as the "Centroid Method" has a strong advantage that dwells from its intuitive formulation: it is very easy to understand. In fact, this is the reason why we have chosen it as the first one of the several defuzzification methods available. However it has two important disadvantages. First, it can become tedious to calculate, especially if the shape of the characteristic functions of the fuzzy sets C_i are Gaussian, bell-shaped, etc. The second disadvantage is that the Centroid Method is unable to provide a crisp output value z at the

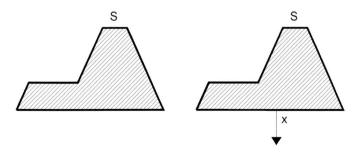

Fig. 7.8 Materializing the shape S and getting a crisp output x

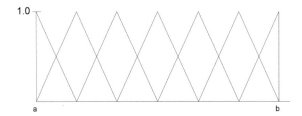

Fig. 7.9a A linguistic variable composed by seven fuzzy sets defined in the range [a, b]

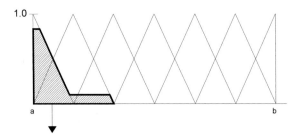

Fig. 7.9b Extreme a at *left* is never reached using the Centroid Method of defuzzification. The same happens with the extreme b at *right*

extremes of the linguistic variable where the fuzzy sets C_i are defined. Figures 7.9a and 7.9b illustrate this problem.

In Fig. 7.9a we see a linguistic variable composed by seven fuzzy sets. All of them have an isosceles triangular characteristic function except the two ones located at the extremes of the universe of discourse of the linguistic variable. These ones have a right triangle as characteristic functions in order to adapt the linguistic variable to its extremes a, b.

Very interestingly, Fig. 7.9b shows a situation where despite any possible value of the membership degrees in the consequents of a FRBS, a crisp value obtained by means of the Centroid Method of defuzzification never reaches the extremes of the linguistic variable. The figure shows a shape whose center of gravity is not far from the extreme a, but it never reaches it, even if only the leftmost fuzzy set had an associated membership degree equal to one and the rest of fuzzy set in the linguistic

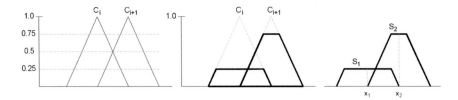

Fig. 7.10 Creating two geometrical shapes S_1, S_2 from membership degrees

variable had a null associated membership degree. The crisp output never reaches the extreme. FuzzyLisp does not support the Centroid Method of defuzzification because these two reasons.

Another well-known defuzzification method is the so named "Weighted average method". This defuzzification strategy is easily understood if we observe first Fig. 7.10. Again using only two $\mu_{Ci}(z)$, $\mu_{Ci+1}(z)$ membership degrees from the fuzzy output expressed by (7-18), then, if C_i, C_{i+1} are fuzzy sets belonging to the consequents of the knowledge database of fuzzy rules, we can create two geometrical shapes S1, S2 as shown in the figure.

Now, taking the vertical axis of symmetry of every shape we obtain the crisp values x_1, x_2 just on the real axis. If the associated membership degrees are, respectively, $\mu_1(z)$, $\mu_2(z)$, then the resulting defuzzified crisp value z is given by the formula:

$$z = (x_1\mu_1(z) + x_2\mu_2(z))/(\mu_1(z) + \mu_2(z)) \tag{7-19}$$

In general, for n membership degrees from the fuzzy output expressed by (7-18), the Weighted average method uses the following generalized expression:

$$z = \sum(x_i m_i(z))/\sum \mu_i(z) \tag{7-20}$$

The main advantage of the weighted average method is its easy implementation, although this method is only valid for cases where the output membership functions are symmetrical in shape. A variation of the weighted average method results from substituting membership degrees to "weights" associated to every shape obtained from fuzzy sets belonging to the consequents of the knowledge database of fuzzy rules.

As can be seen in Fig. 7.11, to every resulting shape S_i we obtain a weight w that is proportional to the area of the shape. In this example, the area of S_2 is 2.14 times the area of S_1. Applying these weights in the center of the basis of their respective

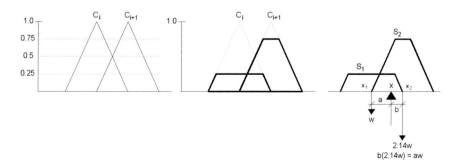

Fig. 7.11 Applying weights to shapes

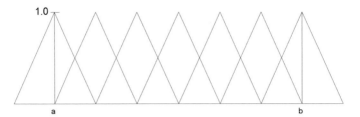

Fig. 7.12 Defuzzification: reaching the extremes

shapes we obtain a set of weights w_i. The defuzzificated crisp value z corresponds to a value x (shown in the figure by an arrow pointing upwards) on the real axis that equilibrates all the weights at play. Then, expression (7-20) converts to:

$$z = \sum x_i w_i / \sum w_i \qquad (7\text{-}21)$$

The two exposed weighted average methods are easy to calculate, and the second one adds the concept of "weighting area" already implicit in the Centroid Method of defuzzification. However, we still cannot reach the extremes of the output linguistic variables, unless we strategically design its fuzzy sets as shown in Fig. 7.12.

This arrangement of fuzzy sets in a linguistic variable can seen bizarre at first, but does not violate any theoretical matter on fuzzy sets and solves the biggest inconvenience of the three defuzzification methods exposed so far.

7.7.2.1   Singletons

When fuzzy sets in a linguistic variable defining the output of a FRBS are formed by isosceles triangles and are arranged strategically as in Fig. 7.12, the two exposed versions of the weighted average method for defuzzification always apply the resulting membership degrees/weights on the vertical axis of symmetry of their characteristic functions. This suggests the use of another type of fuzzy set named "Singleton" that we are going to introduce just now with the following definition:

In fuzzy sets theory a singleton S is a special fuzzy set whose membership function is equal to zero except at a point x, where its membership degree is exactly equal to one. A singleton is always associated to a crisp real number x_0 in such a way that its characteristic function can be defined formally as:

$$\mu S(x) = 1 \quad \text{if } x = x_0$$
$$\mu S(x) = 0 \quad \text{if } x \neq x_0$$

The FuzzyLisp function *(fl-expand-contract-set)* already heralded Singletons in Sect. 7.6 of this chapter (source code number 6-18) when it takes a fuzzy set in

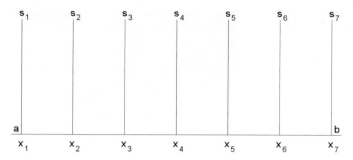

Fig. 7.13 Substituting isosceles shaped characteristic functions by singletons, S_i

FLSSR representation as its first argument and applies zero as its second argument. Again at the Lisp prompt:

> **> (fl-expand-contract-set '(a 2 3 5 6) 0)**
> *: (a 4 4 4 4)*

As can be seen from the list returned by Lisp, both the nucleus and support of a singleton is a single point on the real axis. The graphical representation of a singleton S can be seen at the right of Fig. 7.4. Under a practical point of view the list *(a 4 4 4 4)* is not computationally efficient for representing singletons, so in FuzzyLisp they are represented by expressions of the following type:

$$(singleton\text{-}label\, x_0)z$$

After several assignments of the type *(setq s_i (singleton-label$_i$ x_i))* a linguistic variable composed by singletons can be built by means of the following expression: *(setq singleton-LV '(s_1 s_2 ... s_n))*. For our practical interests, we can now represent the previous linguistic variable composed by seven triangular shaped fuzzy sets shown in Fig. 7.12 by means of seven singletons S_1 ... S_7 as shown in Fig. 7.13.

Since every singleton can be seen as a geometrical spike resting fixed on a real point x_i, we can apply the weighted average method of defuzzification for singletons in order to obtain the crisp output of a FRBS. After some reflection you will able to appreciate that, for defuzzification purposes, Fig. 7.13 is essentially the same as 7.12 in disguise. This is the reason behind formula (7-20) serves perfectly well for the defuzzification procedure if singletons are used as fuzzy sets in the consequents of every fuzzy rule in a FRBS. In fact, the output of an inference engine in a FRBS will have now the following structure:

From Rule$_1$: "z is S_1" has an associated truth value $\mu_1(z)$ *in point* x_1
From Rule$_2$: "z is S_2" has an associated truth value $\mu_2(z)$ *in point* x_2
...
From Rule$_n$: "z is S_n" has an associated truth value $\mu_n(z)$ *in point* x_n

and finally, for obtaining the crisp output z of the system, we can write again:

$$z = \sum (x_i \mu_i(z)) / \sum \mu_i(z) \qquad (7\text{-}22)$$

Michio Sugeno from Japan, one of the most important names in Fuzzy Logic, has been one of the researchers that first explored the elegance and computational power of singletons in fuzzy systems. As we shall see immediately, singleton defuzzification is the standard defuzzification method used by FuzzyLisp (Fig. 7.14).

7.8 Modeling FRBS with FuzzyLisp

Section 7.7 has been strongly theoretical, and it is now the time to introduce a few important FuzzyLisp functions in order to model Fuzzy Rule Based Systems. Since this is a practical book following a "learning by doing" approach, the next paragraphs will show the reader how to model the workings of an air-conditioner controller by means of fuzzy logic techniques. In this moment it is not important for us to model the best, finest air-conditioner controller available, but to show how to organize and process all the required information for the controller, or, expressed with other words, introduce a practical application example where the FRBS architecture exposed in Fig. 7.6 is translated into real FuzzyLisp code.

As it is well known, an air-conditioner system's goal is to maintain the temperature of a given enclosure at a constant value, where the enclosure can be a room, a building, an aircraft cockpit or the interior of a car, to name only some examples. The basic information needed to accomplish the goal is usually based on two values: actual temperature t and temperature variation *delta-t* obtained from two readings from a thermometer obtained every s seconds. Other interesting input values could be the quantity of solar radiation received by the enclosure and the time of the day and even the day of the year, but t and *delta-t* will be enough for our simple model. After the readings are made, our fuzzy controller will calculate the needed output airflow and temperature, AFT, to stabilize the enclosure's

Fig. 7.15 Temperature and Delta-temperature are the inputs to the air-conditioner controller. The output is temperature and air flow, AFT

temperature at a temperature goal, let us say, T = 21 °C. A general formulation of the problem is given by the following expression:

$$AFT = f(t, delta\text{-}t) \qquad (7\text{-}23)$$

Figure 7.15 shows graphically a representation of the intended fuzzy controller. If the reader is versed on differential equations then is highly probable that a ring has sounded in his head after the simple observation of expression (7-23). However no one differential equation will be needed to obtain a good working air-conditioner controller.

Since we have two input values, we shall create two specific linguistic variables to represent them. The first step is to define the ranges of these two linguistic variables: For representing temperature we shall use a range of 0–50 °C, while for temperature variation we shall use a range from −2.0 to 2.0 °C per minute. Now we need to create a partition on the linguistic variables by means of defining suitable fuzzy sets on them. Tables 7.4a and b show the fuzzy sets that make up the input linguistic variables lv-temperature and lv-delta-t, respectively.

In a similar way, Table 7.4c shows the fuzzy singletons that make up the output linguistic variable AFT (Air Flow Temperature). The final, crisp output of the controller will be positive if hot air is needed to stabilize the system. Conversely, it

Table 7.4a Fuzzy sets for the linguistic variable lv-temperature

Fuzzy set label	Membership function	Lisp representation
Very-cold	0, 0, 6, 11	*(setq T1 '(very-cold 0 0 6 11))*
Cold	6, 11, 16, 21	*(setq T2 '(cold 6 11 16 21))*
Optimal	16, 21, 21, 26	*(setq T3 '(optimal 16 21 21 26))*
Hot	21, 26, 31, 36	*(setq T4 '(hot 21 26 31 36))*
Very-hot	31, 36, 50, 50	*(setq T5 '(very-hot 31 36 50 50))*

Table 7.4b Fuzzy sets for the linguistic variable lv-delta-t

Fuzzy set label	Membership function	Lisp representation
Decreasing	−2, −2, −0.2, 0	*(setq dT1 '(decreasing -2-2 -0.2 0))*
No-change	−0.2, 0, 0, 0.2	*(setq dT2 '(no-change -0.2 0 0 0.2))*
Increasing	0, 0.2, 2, 2	*(setq dT3 '(increasing 0 0.2 2 2))*

Fuzzy set label	Singleton at x_0	Lisp representation
Cold-strong	−100	*(cold-strong -100)*
Cold-medium	−50	*(cold-medium -50)*
Stop	0	*(stop 0)*
Heat-medium	50	*(heat-medium 50)*
Heat-strong	100	*(heat-strong 100)*

Table 7.4c Fuzzy sets for the output linguistic variable AFT (Air Flow Temperature)

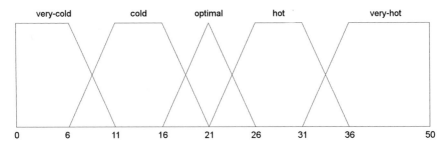

Fig. 7.16a Input linguistic variable lv-temperature

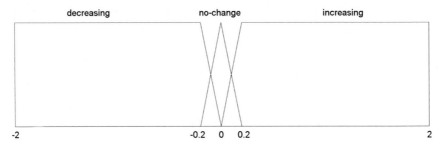

Fig. 7.16b Input linguistic variable lv-delta-t

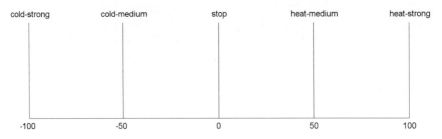

Fig. 7.16c Output linguistic variable AFT

will be negative if cold air is required. The numerical magnitude expresses the required flow of air in a range from 0 to 100 %, that is, from null to maximum output of the system.

Figures 7.16a, 7.16b and 7.16c show a graphical representation of the input and output variables of the air-conditioner controller.

After all the required input and output linguistic variables have been established we need a set of expert rules of the type given by expression (7-13) that links fuzzy sets from the input linguistic variables to the fuzzy singletons in the output. One example of rule for our fuzzy controller is the following one:

> If t is very-cold and delta-t is decreasing then AFT is heat-strong

In plain English this rule means that if the temperature of the room is very cold and the temperature is decreasing, then the controller must react supplying hot air at a strong high rate flow. In Code 7-8 we can see the complete FuzzyLisp application for our air-conditioner fuzzy controller:

```
;code 7-8
(load "my-path/fuzzy-lisp.lsp")
;fuzzy sets and linguistic variables definitions:
(setq T1 '(very-cold 0 0 6 11))
(setq T2 '(cold 6 11 16 21))
(setq T3 '(optimal 16 21 21 26))
(setq T4 '(hot 21 26 31 36))
(setq T5 '(very-hot 31 36 50 50))
(setq lv-temperature '(T1 T2 T3 T4 T5))

(setq dT1 '(decreasing -2 -2 -0.2 0))
(setq dT2 '(no-change -0.2 0 0 0.2))
(setq dT3 '(increasing 0 0.2 2 2))
(setq lv-delta-t '(dT1 dT2 dT3))

(setq AFT '(
     (cold-strong -100)
     (cold-medium -50)
     (stop 0)
     (heat-medium 50)
     (heat-strong 100))
)

;fuzzy rules section:

(setq rules-controller '((lv-temperature lv-delta-t AFT)
     (very-cold decreasing heat-strong AND-product)
     (very-cold no-change heat-strong AND-product)
     (very-cold increasing heat-strong AND-product)

     (cold decreasing heat-strong AND-product)
     (cold no-change heat-medium AND-product)
```

```
(cold increasing heat-medium AND-product)

(optimal decreasing heat-medium AND-product)
(optimal no-change stop AND-product)
(optimal increasing cold-medium AND-product)

(hot decreasing cold-medium AND-product)
(hot no-change cold-medium AND-product)
(hot increasing cold-strong AND-product)

(very-hot decreasing cold-strong AND-product)
(very-hot no-change cold-strong AND-product)
(very-hot increasing cold-strong AND-product))
)
```

After the first lines of comments, the expression *(load "my-path/fuzzy-lisp.lsp")* tells NewLisp where to find FuzzyLisp. Here you will have to replace "my-path" for the path where the file fuzzy-lisp.lsp is located in your computer.

The following lines in the code create the fuzzy sets and linguistic variables already exposed in Tables 7.4a, b and c for the controller. Please note the used structure for the output linguistic variable AFT. Since singletons can be formed by very simple Lisp expressions, the output linguistic variable can be expressed in a more compact way when compared to the input linguistic variables.

After all the fuzzy sets and linguistic variables are created, then it comes the fuzzy rules section with a very definite ordering. Several important details must be remarked in this structure of rules. First, the complete set of fuzzy rules need a Lisp symbol for a proper identification. In this case we have chosen "rules-controller". In the same line of code, we have declared the names of the two input variables and the output linguistic variable, as can be seen in the very first sublist: *(lv-temperature lv-delta-t AFT)*. FuzzyLisp needs to know beforehand the ordering of the linguistic variables in the body of rules. Expressed with other words: The first rule in the body of rules in FuzzyLisp is in itself not a rule but an enumeration of the used linguistic variables.

Another important point is the use of the symbol AND-product in the rules. FuzzyLisp allows the designer to use several logical connectives in fuzzy rules. If the symbol AND-min is used, then expression (7-5) will be used for inferences, that is:

$$Tv(p \wedge q) = min(Tv(p), Tv(q)) = min(\mu_A(x), \mu_B(y))$$

However, when using AND-product, then the following inference is applied:

$$Tv(p \wedge q) = Tv(p) \cdot Tv(q) = \mu_A(x) \cdot \mu_B(y) \qquad (7\text{-}24)$$

That is, FuzzyLisp multiplies the membership degrees $\mu_A(x)$, $\mu_B(y)$. This usually produces a very smooth, continuous crisp output in the system. For expressing an "or" logical connective FuzzyLisp uses the symbol OR-max as defined by expression (7-6). Other symbols could be added in future versions of FuzzyLisp representing different ways of conjunction and disjunction (the interested reader will find more information in fuzzy-logic theoretical books under the terms "t-norms" and "t-conorms". The study of t-norms and t-conorms is outside the scope of this book). Another important remark emerges when observing the last rule:

(very-hot increasing cold-strong AND-product))

Just note the double closing parenthesis. As you know, in every Lisp program the number of left parenthesis equals the number of right parenthesis. Since the construction of a set of expert fuzzy rules in FuzzyLisp usually results into a relatively complex expression, special care should be put on matching left and right parenthesis. If a fuzzy model developed with FuzzyLisp crashes or does not behaves well, the first place to look is in the declaration part of rules and linguistic variables.

Code 7-8 represents the complete knowledge database for the fuzzy controller. Now, for fulfilling the architecture of a FRBS as shown in Fig. 7.6, we need a fuzzification-defuzzification mechanism and an inference engine. All these software components are included as functions in FuzzyLisp and now it is time to uncover them.

Maybe the most important function in FuzzyLisp is *(fl-translate-rule)*. This function is the nucleus of its inference engine. It takes a rule at a time from the fuzzy rules body, performs the adequate inferences and translates it into membership degrees. As arguments, it needs the first sublist from the body of rules where the enumeration of the used linguistic variables is expressed. In our controller example, the needed sublist is *(lv-temperature lv-delta-t AFT)*. The second argument is the rule to "translate", as for example, *(optimal increasing cold-medium AND-product)*. Finally, the third and fourth arguments are the crisp input numerical values to the FRBS. In our example these values correspond to a given crisp values of temperature *t* and temperature variation *delta-t*. For example, $t = 22C$ and *delta-t* $= 0.25$ C/min. Code 7-9 shows the function:

```
;code 7-9
(define (fl-translate-rule header rule x y,
    lv1 lv2 lv3 fset1 fset2 fset3 mu1 mu2 s
    operator m1 m2 wi wixi)

    ;extract linguistic variables from the header:
    (setq lv1 (nth 0 header) lv2 (nth 1 header) lv3 (nth 2
        header))
    ;extract fuzzy sets from the rule:
    (setq fset1 (nth 0 rule) fset2 (nth 1 rule) fset3 (nth 2
        rule))
    ;extract the operator in the rule:
    (setq operator (nth 3 rule))

    ;get first lv, calculate memberships given x and then
    ;select the membership degree corresponding to the first
```

```
;set in the rule:
(setqmu1 (assoc fset1 (fl-lv-membership2? (eval lv1) x)))

;now do the same for second lv, y and second set in the
    rule:
(setqmu2 (assoc fset2 (fl-lv-membership2? (eval lv2) y)))

;now get the appropiate singleton, s:
(setq s (assoc fset3 (eval lv3)))

;extract the membership degrees:
(setq m1 (last mu1) m2 (last mu2))
(case operator
    (AND-product
        ;apply AND-product by making wi=m1xm2:
        (setq wi (mul m1 m2)) (setq wisi (mul wi (last s)))
    )
    (AND-min
        ;apply AND-min by making wi=min(m1,m2):
        (setq wi (min m1 m2)) (setq wisi (mul wi (last s)))
    )
    (OR-max
        ;apply OR-max by making wi=max(m1,m2):
        (setq wi (max m1 m2)) (setq wisi (mul wi (last s)))
    )
) ;case end
;finally return the numerical translation of the rule
(list m1 m2 wi wisi)
)
```

Function *(fl-translate-rule)* begins reading the header of the body rules *(first sublist)* and extracting the linguistic variables from the header and then the fuzzy sets at play. Later, several calls to the function *(fl-lv-membership2?)* are responsible for the fuzzification of the crisp input, storing the obtained membership degrees in the internal variables *mu1* and *mu2*. The singleton from the consequent in the rule is identified as it happens also with the operator located at the end of the analyzed rule. The second part of the function works by means of a *case* Lisp statement, selecting the appropriate inference to be performed on the rule. Finally, the function returns a list containing the membership degrees to the first and second fuzzy sets in the antecedent of the rule, then the result of the performed numerical inference, and finally the numerical value after multiplying it by the corresponding singleton in the consequent. Let us take as example rule number nine:

(optimal increasing cold-medium AND-product)

Calling the function we obtain:

> **(fl-translate-rule (first rules-controller) (nth 9 rules-controller) 22 0.25)**
: (0.8 1 0.8 -40)

The value 0.8 is the membership degree of 22 °C to the fuzzy set *optimal*, while 1 is the membership degree of 0.25 °C per minute to the fuzzy set *increasing*. The third value is the result of multiplying both membership degrees (inference dictated by the symbol *AND-product*), and the last value is obtained after multiplying 0.8 by the value associated to the singleton *cold-medium* (-50). Just note the appropriate use of the Lisp functions *(first)* and *(nth)* in the previous expression. This gives us a hint to create a function that will evaluate all the fuzzy rules contained in the knowledge database. This function, named *(fl-translate-all-rules)* is shown in Code 7-10:

```
;code 7-10
(define (fl-translate-all-rules set-of-rules x y, n i
    list-out)
  (setq list-out '())
  (setq n (length set-of-rules))
  (setq i 1);indexed on the first rule

  (while (< i n)
    (setq list-out (cons
            (fl-translate-rule (first set-of-rules)
            (nth i set-of-rules) x y) list-out))
    (++ i)
  );while end
  (reverse list-out)
)
```

as can be easily noted observing the code, this function scans all the rules from the body of rules and returns a list containing all the calculations made for each of them. As parameters, the function takes the name of the complete body of rules and two crisp input values to feed the system. Let us try the function at the Lisp prompt:

> **(fl-translate-all-rules rules-controller 22 0.25)**
: ((0 0 0 0) (0 0 0 0) (0 1 0 0) (0 0 0 0) (0 0 0 0) (0 1 0 0) (0.8 0 0 0) (0.8 0 0 0) (0.8 1 0.8 -40) (0.2 0 0 0) (0.2 0 0 0) (0.2 1 0.2 -20) (0 0 0 -0) (0 0 0 0) (0 1 0 0))

From the output we can see that rules #9 and #12 have been fired in the example when *t* = 22 °C and *delta-t* = 0.25 °C/min are used as crisp inputs. This is easy to observe because the last element in their respective sub-lists is not null. In fact, calling the functions *(fl-translate-rule)* and *(fl-translate-all-rules)* from the Lisp prompt is a good strategy when we whish to debug and/or in depth analyze a fuzzy application made with FuzzyLisp. Now it is time to take the output of the function *(fl-translate-all-rules)* and apply a defuzzification procedure to it. A new FuzzyLisp function, named *(fl-defuzzify-rules)* is dedicated specifically to this task, as shown in Code 7-11:

```
; Code 7-11
 (define (fl-defuzzify-rules translated-rules,
          i n sum-wi sum-wixi)
   (setq i 0)
   (setq sum-wi 0) (setq sum-wixi 0)
   (setq n (length translated-rules))
   (while (< i n)
       (setq sum-wi (add sum-wi (nth 2 (nth i translated-
           rules))))
       (setq sum-wixi (add sum-wixi
                       (nth 3 (nth i translated-rules))))
       (++ i)
   )
   (div sum-wixi sum-wi)
 )
```

The function takes as arguments the output list of the function *(fl-translate-all-rules)* and then traverses it in order to apply the defuzzification mathematical expression (7-22), providing a crisp numerical result to the fuzzy system. Let us try this function at the Lisp prompt:

> **(fl-defuzzify-rules (fl-translate-all-rules rules-controller 22 0.25))**
: *-60*

So, our model tells that when temperature is $t = 22$ °C and variation of temperature is *delta-t* = 0.25 °C/min, the airflow temperature AFT must be AFT = −60. This is a great result, but once again, the call to the function *(fl-defuzzify-rules)* suggests the development a more comfortable to use one, named *(fl-inference)*, as shown in Code 7-12:

```
;code 7-12
(define (fl-inference rules x y)
    (fl-defuzzify-rules (fl-translate-all-rules rules x y))
)
```

This function puts it all together, allowing the NewLisp user to obtain the crisp output of a FRBS using only three parameters: The name of the complete body of

Table 7.5 Airflow temperatures AFT obtained as the crisp result of the fuzzy air-conditioner controller

	0	3	8.5	13.5	18.5	23.5	28.5	33.5	36.5	50
−1	100	100	100	100	75	0	−50	−75	−100	−100
−0.15	100	100	93.75	87.5	62.5	−6.25	−50	−75	−100	−100
−0.10	100	100	87.5	75	50	−12.5	−50	−75	−100	−100
−0.05	100	100	81.25	62.5	37.5	−18.75	−50	−75	−100	−100
0	100	100	75	50	25	−25	−50	−75	−100	−100
0.05	100	100	75	50	18.75	−37.5	−62.5	−81.25	−100	−100
0.10	100	100	75	50	12.5	−50	−75	−87.5	−100	−100
0.15	100	100	75	50	6.25	−62.5	−87.5	−93.75	−100	−100
1	100	100	75	50	0	−75	−100	−100	−100	−100

Input temperatures are shown in the first row. Temperature variation is shown in the first column

rules and then the crisp input to the system, that is, two numerical values. The function call for our air-conditioner controller example cannot be simpler:

> **(fl-inference rules-controller 22 0.25)**
: *-60*

In practical terms, this means that for developing fuzzy-models based on the architecture shown in Fig. 7.6 where two crisp values are used as input, a FuzzyLisp user only needs to declare fuzzy sets, linguistic variables and a set of expert fuzzy rules with the structure exposed in Code 7-8 and then, by simple callings to the function *(fl-inference)* all the internal processes of fuzzification, knowledge data-base management and defuzzification are automatically performed. In Chap. 1 of this book we stated that Lisp is a powerful language. I hope that once we have reached this milestone the reader will agree.

For extensively testing our air-conditioner fuzzy controller we can make successive calls to the function *(fl-inference)*, obtaining, for example, the results shown in Table 7.5. In the first row you can see some values for temperatures, *t*, while the first column at left represents some values for temperature variation, *delta-t*. Needless to say, the rest of values in the table represent output airflow temperatures, AFT.

Albeit impressive, this table requires 90 function calls to *(fl-inference)*, and this is a tedious task to perform manually. Fortunately, Lisp, besides allowing us to express our ingenuity and mathematical creativity, gives us also, as it happens with other computer languages, the opportunity to automatize repetitive work. This is the philosophy behind the function *(fl-3d-mesh)*. This FuzzyLisp function is in fact a procedure for creating tables as the one shown above these lines. Let us observe it in Code 7-13:

```
;code 7-13
(define (fl-3d-mesh namefile rules nx ny,
                    x1 x2 y1 y2 header lv1 lv2 stepx stepy x y)
```

```
(setq header (nth 0 rules))
;get the extremes x1 x2 of the first lv
(setq lv1 (nth 0 header) lv2 (nth 1 header))
(setq x1 (nth 1 (eval (first (eval lv1)))))
(setq x2 (last (eval (last (eval lv1)))))
;now get the extremes y1 y2 of the second lv
(setq y1 (nth 1 (eval (first (eval lv2)))))
(setq y2 (last (eval (last (eval lv2)))))

(setq stepx (div (sub x2 x1) nx))
(setq stepy (div (sub y2 y1) ny))
(setq x x1 y y1)
(println "Writing 3Dmesh ...")
(device (open namefile "write"))
(while (<= x x2)
    (while (<= y y2)
        (print x "," y "," (fl-inference rules x y) "\n")
        (setq y (add y stepy))
    ); end while y
    (setq y y1); reset y value
    (setq x (add x stepx))
);end while x
(close (device))
(println "3Dmesh written to file")
)
```

Basically speaking, this function scans the range of two linguistic variables used as antecedents in a FRBS by making successive calls to the function *(fl-inference)* with changing input crisp values x,y inside the range of the linguistic variables at play. Let us try a call example in order to better understand how does it work:

>(fl-3d-mesh "air-conditioner-controller.csv" rules-controller 20 20)
: Writing 3Dmesh ...
3D mesh Written to File

As the output of the function call suggests, *(fl-3d-mesh)* writes a file to the computer's hard disk where all the calculations resulting from adequate calls to *(fl-inference)* are stored. As arguments, this function requires a string representing a filename for the desired output file, the name of the complete body of rules of the fuzzy model and then two numerical values nx, ny that inform the function how many discretization steps are needed for building the "table" of output data. As a rule of thumb, values of nx, ny between 20 and 50 are enough for any practical application. Let us take a view to the contents of the output file air-conditioner-controller.csv:

```
0,-2,100
0,-1.8,100
0,-1.6,100
...

22.5,-0.4,20
22.5,-0.2,20
22.5,0,-15
22.5,0.2,-65
22.5,0.4,-65
...

50,1.6,-100
50,1.8,-100
50,2,-100
```

As can be seen, the file has a comma separated value (csv) format, so you can open it not only in any text editor program such as Notepad or TextEdit, but also in Excel and many other programs that accept csv-type files for data importation. Since every line can be interpreted as the coordinates of a 3D point in space, many technical programs allows to create a 3D visualization of the interpolated three dimensional surface, as we can appreciate in Fig. 7.17.

While the base of the figure is represented by a bi-dimensional plane formed by *t* and *delta-t* (corresponding to the linguistic variables *lv-temperature* and *lv-delta-t*,

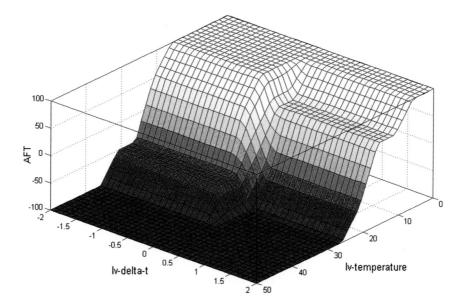

Fig. 7.17 Output surface of the fuzzy air-conditioner controller

respectively), the height of every point belonging to the 3D surface represents the output airflow temperature calculated by the fuzzy controller, AFT. As can be seen, dark areas in the surface represent low, negative values of AFT, while lighter ones represent positive AFT values. Several remarks must be made at this point: The first one is the steep region located about the middle of the surface. This is caused mainly because the relatively small support of the fuzzy set *no-change* in the input linguistic variable *lv-delta-t* and the large support of the fuzzy sets *decreasing* and *increasing* in the same linguistic variable. The second remark points to the presence of some flat regions in the output surface that we shall name "terraces", including the big decks corresponding to AFT = −100 and AFT = 100. While there are lots of creative work and parameters to choose when creating a FRBS, one thing is certain: If trapezoidal fuzzy sets are used on the linguistic variables that make up a fuzzy application then terraces will always appear on the resulting output surface. The existence of terraces is not, by itself, an inconvenient, and can be just the desired design of the fuzzy application, but if you don't desire terraces in your output surfaces you will need to use only triangular membership functions in your models. The reader is invited to modify and play with the fuzzy sets that make up the linguistic variables *lv-temperature* and *lv-delta-t*. As an example, the fuzzy set *no-change* can be expanded by means of using the FuzzyLisp function (*fl-expand-contract-set*) from the Lisp prompt, the fuzzy sets *decreasing* and *increasing* can be modified, shortening their support and nuclei, etc. The best advice in order to acquire experience in designing fuzzy systems is to play with them and observe their behavior. In the rest of this chapter some extra advice will be given for this conceptual challenge and also in the next chapter, but nothing replaces experience and experimentation. FuzzyLisp will help you in every step you make, but, as they say, there is no such a thing as a free lunch.

7.9 Fuzzy Logic in Motor Racing: Scoring in Regularity Rallies

Regularity rallies are a form of motor sport with strong tradition both in Europe and North America. The goal in this type of competition is not to complete an itinerary or a circuit in the shortest possible time but to complete a predefined route (usually with open traffic) on a precise time specified by the organization of the race. Let us imagine we are the organizers of a regularity rally where sponsors have asked us for some innovation in the regulations. Figure 7.18 represents a sort of ring, 18 km. in length, composed by three route sections. The first one starts in point A and follows a regional, low traffic road, until reaching point B. In this point the race goes along a national, heavy traffic road until reaching point C, where again another regional road leads to complete the ring at Finish (point A). The usual main rule for a race of this type is to estipulate an ideal time t for completing the route and then to time every participant car. Cars start the route not at the same time but with a time difference between them, e.g., one minute apart. After the race is over, every car C_i

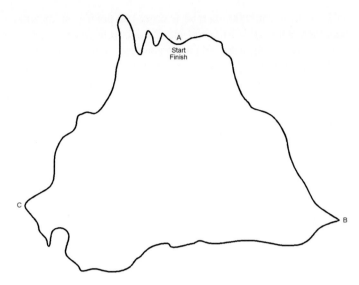

Fig. 7.18 Regularity rally track sections

will get a register time $t \pm dt_i$ and the car with the lower dt_i time will be the winner. For example, if we think about an average goal speed of 60 km/h, then the ring should be completed in exactly 18 min. Increasing deviations from this goal time will get decreasing ranking positions in the final classification.

A more complex rule can be derived from establishing two stages for the race in such a way that every car will complete the route in two lap times t_1, t_2. Then, the winner will be the car with the smaller difference *delta-time = abs(t_1 − t_2)*. As an example, if a car has registered lap times t_1 = *17:57* and t_2 = *18:13*, (format is minutes: seconds) then its overall race time will be *delta-time = 16 s*.

We are confident this is an improvement with respect to ordinary race regulations in regularity rallies, but after some reflection, we decide to go further. This time using fuzzy logic. First, we create two sharp rules in order to bind the model:

- Cars with lap time t1 or t2 < 16:00 will be eliminated
- Cars with lap time t1 or t2 > 20:00 will be eliminated

This is so because we do not want pilots achieving extremely high speeds (remember, roads are open to traffic in this type of competition). On the other hand slow cars are not the essence of motor sport, so an inferior time limit seems appropriate to apply. The design goal for our scoring model is to offer the pilots a variety of strategies to choose from, always having regularity in mind, but also offering them the possibility to opt for an aggressive strategy (higher speeds), a defensive, conservative strategy (lower speeds), or something in between. In any case, extreme aggressive strategies, and at a lesser extent, extreme defensive strategies, even with lap times well between 16 and 20 min, will have an implicit penalty. For example, if the pilot of a given car decides to go for lap times *t1* = 16:00:00 and *t2* = 16:00:00, then he will get 100 points, the maximum score. But if he gets, for example *t1 = 16:00:00* and *t2 = 16:00:05*, then he will get exactly zero points in the scoring. In some way the organizers are happy to have aggressive strategies in the race, but they give the teams the following warning: "if you want to run fast you must be extremely regular". The regulation completes with the following overall fuzzy rules:

- If you run really fast then you will need to be extremely regular to score the highest in the race.
- If you run relatively slow then you will need to be certainly regular to score high in the race.
- If you run at a moderate speed, then you will have the bigger tolerance with respect to delta-time, allowing great scoring in the race.

In order to create a fuzzy model for scoring the regularity rally, we shall create two linguistic variables as input (lap-time and delta-time) and an output linguistic variable named "points" for representing score. The membership functions and Lisp representation of the fuzzy set that make up these linguistic variables are shown in Tables 7.6a, b and c. Time units are expressed in seconds:

Table 7.6a   Fuzzy sets for the linguistic variable "lap-time"

Fuzzy set label	Lisp representation of membership function
Very-quick	*(setq t1 '(very-quick 960 960 980 1000))*
Quick	*(setq t2 '(quick 980 1000 1040 1060))*
Slow	*(setq t3 '(slow 1040 1060 1100 1120))*
Rather-slow	*(setq t4 '(rather-slow 1100 1120 1160 1180))*
Very-slow	*(setq t5 '(very-slow 1160 1180 1200 1200))*

Table 7.6b   Fuzzy sets for the linguistic variable "delta-time"

Fuzzy set label	Lisp representation of membership function
Delta-very-small	*(setq d1 '(delta-very-small 0 0 2 5))*
Delta-small	*(setq d2 '(delta-small 2 5 5 10))*
Delta-medium	*(setq d3 '(delta-medium 5 10 10 20))*
Delta-big	*(setq d4 '(delta-big 10 20 20 40))*
Delta-very-big	*(setq d5 '(delta-very-big 20 40 60 60))*

Table 7.6c   Fuzzy sets for the output linguistic variable "points"

Fuzzy set label	Singleton at x_0	Lisp representation
Very-low	0	*(very-low 0)*
Low	25	*(low 25)*
Medium	50	*(medium 50)*
High	75	*(high 50)*
Very-high	100	*(very-high 100)*

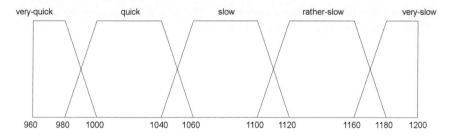

Fig. 7.19a Input linguistic variable lap-time, in seconds

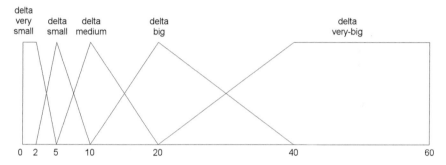

Fig. 7.19b Input linguistic variable delta-time, in seconds

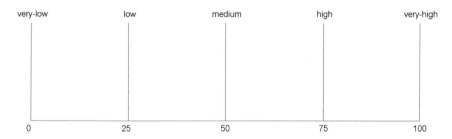

Fig. 7.19c Output linguistic variable representing score from 0 to 100

These fuzzy sets and linguistic variables are graphically shown in Figs. 7.19a, 7.19b and 7.19c.

The input variable "lap-time" has nothing special to comment, and, as can be seen, is composed by trapezoidal shaped fuzzy sets. The second input variable, "delta-time" is composed by fuzzy sets with both triangular and trapezoidal membership functions. The most important feature in the design of this linguistic variable is the non-uniform distribution of its fuzzy sets. These appear "compressed" towards the left because an important reason: As designers we wish the scoring model to be sensitive to differences of delta-time, and extremely sensitive when delta-time values are small and very small, hence the concentration of fuzzy sets towards the left of the linguistic variable. The rules for organizing the linguistic variables in the model are shown in Code 7-14:

```
;code 7-14
(setq rules-cars '((lap-time delta-time points)
    (very-quick delta-very-small very-high AND-product)
    (very-quick delta-small very-low AND-product)
    (very-quick delta-medium very-low AND-product)
    (very-quick delta-big very-low AND-product)
    (very-quick delta-very-big very-low AND-product)
    (quick delta-very-small high AND-product)
    (quick delta-small high AND-product)
    (quick delta-medium high AND-product)
    (quick delta-big medium AND-product)
    (quick delta-very-big very-low AND-product)

    (slow delta-very-small very-high AND-product)
    (slow delta-small very-high AND-product)
    (slow delta-medium very-high AND-product)
    (slow delta-big very-high AND-product)
    (slow delta-very-big medium AND-product)

    (rather-slow delta-very-small very-high AND-product)
    (rather-slow delta-small high AND-product)
    (rather-slow delta-medium high AND-product)
    (rather-slow delta-big high AND-product)
    (rather-slow delta-very-big very-low AND-product)

    (very-slow delta-very-small high AND-product)
    (very-slow delta-small high AND-product)
    (very-slow delta-medium medium AND-product)
    (very-slow delta-big low AND-product)
    (very-slow delta-very-big very-low AND-product))
)
```

Running the model by means of the FuzzyLisp functions *(fl-inference)* and *(fl-3d-mesh)* we obtain, after exporting the obtained data to a surface graphing application, the output surface from Fig. 7.20.

We can observe that small values of delta-time lead to the highest scores, and also an abrupt collapse in scoring when lap-time is very small and delta-time starts to increase (rightmost region of the Figure), as we desired on the initial design of the model. However, it is also easy to observe the presence of several terraces on the output surface. For obtaining a good classification in a sport competition this is not a desirable result because it may lead to many possible *ex-aequo* instances among the participants. As we commented in the previous section, the presence of

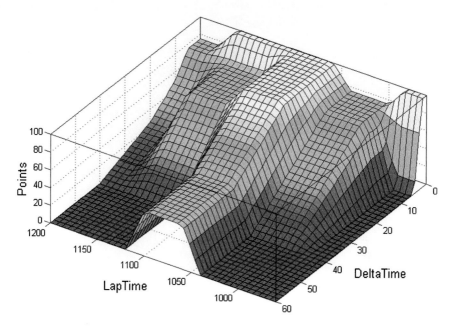

Fig. 7.20 Output surface of the scoring model

Table 7.7 Changing fuzzy sets for the linguistic variable "lap-time"

Fuzzy set label	Lisp representation of membership function
Very-quick	*(setq t1 '(very-quick 960 960 960 1200))*
Quick	*(setq t2 '(quick 960 1020 1020 1200))*
Slow	*(setq t3 '(slow 960 1080 1080 1200))*
Rather-slow	*(setq t4 '(rather-slow 960 1140 1140 1200))*
Very-slow	*(setq t5 '(very-slow 960 1200 1200 1200))*

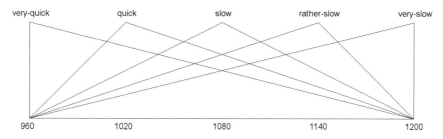

Fig. 7.21 Input linguistic variable lap-time, modified

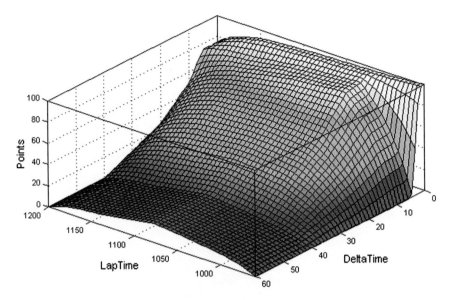

Fig. 7.22 Improved output surface of the scoring model

these terraces is due to the existence of trapezoidal membership functions in the
linguistic variable lap-time. A workaround to this problem is to use triangles for the
membership functions in the linguistic variable. Let us experiment what happens
when using a maximum fuzziness linguistic variable where all the fuzzy sets at play
share a common support. Table 7.7 shows the new linguistic variable lap-time.

The new sets are shown in Fig. 7.21.

Running again the model by means of the FuzzyLisp functions *(fl-inference)* and
(fl-3d-mesh) we obtain a new output surface as shown in Fig. 7.22.

Now we can realize the improvement in smoothness we have got on the output
surface. The overall design of the model is still there, but now we have got a more
adequate result. In any case, the use of maximum fuzziness linguistic variables in
fuzzy modeling sometimes generates too much "smoothness" and many times the
use of standard triangular membership functions for the fuzzy sets in the model is
desirable. The reader is again invited to play with the model, transforming it and
observing the results. The complete Lisp code for this application is shown in Code
7-15:

```
;code 7-15
(load "my-path/fuzzy-lisp.lsp")
;fuzzy sets and linguistic variables definitions:
;lap time, all times in seconds
;For qualifying, time must be between 16 and 20 min
;that is, from 960 to 1200 s
(setq t1 '(very-quick 960.0 960.0 960.0 1020.0))
```

```
(setq t2 '(quick 960.0 1020.0 1020.0 1200.0))
(setq t3 '(slow 960.0 1080.0 1080.0 1200.0))
(setq t4 '(rather-slow 960.0 1140.0 1140.0 1200.0))
(setq t5 '(very-slow 960.0 1200.0 1200.0 1200.0))
(setq lap-time '(t1 t2 t3 t4 t5))

;delta-time, all times in seconds, from 0 s to 60 s
(setq d1 '(delta-very-small 0 0 2 5))
(setq d2 '(delta-small 2 5 5 10))
(setq d3 '(delta-medium 5 10 10 20))
(setq d4 '(delta-big 10 20 20 40))
(setq d5 '(delta-very-big 20 40 60 60))
(setq delta-time '(d1 d2 d3 d4 d5))

(setq points '(
    (very-high 100.0)
    (high 75.0)
    (medium 50.0)
    (low 25.0)
    (very-low 0.0))
)
;fuzzy rules section:

(setq rules-cars '((lap-time delta-time points)
    (very-quick delta-very-small very-high AND-product)
    (very-quick delta-small very-low AND-product)
    (very-quick delta-medium very-low AND-product)
    (very-quick delta-big very-low AND-product)
    (very-quick delta-very-big very-low AND-product)

    (quick delta-very-small high AND-product)
    (quick delta-small high AND-product)
    (quick delta-medium high AND-product)
    (quick delta-big medium AND-product)
    (quick delta-very-big very-low AND-product)

    (slow delta-very-small very-high AND-product)
    (slow delta-small very-high AND-product)
    (slow delta-medium very-high AND-product)
    (slow delta-big very-high AND-product)
    (slow delta-very-big medium AND-product)

    (rather-slow delta-very-small very-high AND-product)
    (rather-slow delta-small high AND-product)
    (rather-slow delta-medium high AND-product)
```

```
(rather-slow delta-big high AND-product)
(rather-slow delta-very-big very-low AND-product)

(very-slow delta-very-small high AND-product)
(very-slow delta-small high AND-product)
(very-slow delta-medium medium AND-product)
(very-slow delta-big low AND-product)
(very-slow delta-very-big very-low AND-product))
)
```

Before finishing this section, let's imagine that it's raining the day of the motor race. Needless to say, lap times should be changed accordingly in order to allow the pilots to race safely. For obtaining a new definition of the linguistic variable "lap-time" we can write a function that internally calls the FuzzyLisp function *(fl-fuzzy-shift)*. Such a function is shown in Code 7-16:

```
;code 7-16
(define (shift-lap-time x)
    (setq t1 (fl-fuzzy-shift t1 x))
    (setq t2 (fl-fuzzy-shift t2 x))
    (setq t3 (fl-fuzzy-shift t3 x))
    (setq t4 (fl-fuzzy-shift t4 x))
    (setq t5 (fl-fuzzy-shift t5 x))
    true
)
```

Now, for shifting two minutes (120 s) the linguistic variable lap-time, we only need to type at the Lisp prompt:

> (shift-lap-time 120)
: true

and for testing the modified fuzzy-sets we can type:

> (fl-list-sets lap-time)
: (very-quick 1080 1080 1080 1140)
(quick 1080 1140 1140 1320)
(slow 1080 1200 1200 1320)
(rather-slow 1080 1260 1260 1320)
(very-slow 1080 1320 1320 1320)

As the reader can imagine, it is the responsibility of the organization of the race to give every racing team a copy of the output surfaces in the form of scoring tables (both in wet and dry conditions) to allow them to choose their race strategy well ahead of time. Table 7.5 is a simplified example of how to share a fuzzy model.

Usually more resolution is required, but this is easy to achieve by means of using the function *(fl-3d-mesh)*.

Incidentally, the use of function *(fl-3d-mesh)* allows the reader to use FuzzyLisp as a general purpose fuzzy modeling toolbox that produces neutral text files representing the knowledge embedded into the model. Later, if you use Java, C#, or any other programming language you only need to load the produced text files into your applications and then, by using bilinear interpolation, you automatically incorporate fuzzy-logic features in them.

7.10 FRBS Using Fuzzy Sets with Discrete Membership Functions

In Sects. 7.8 and 7.9 we have introduced two example fuzzy models based on a FuzzyLisp Standard Set Representation (FLSSR), that is, models whose fuzzy sets have either a triangular or trapezoidal characteristic function. However, in Chap. 6 we emphasized too on fuzzy sets with a discrete characteristic function (FLDSR). It would be a pity that after the effort of writing dedicated FuzzyLisp functions to this type of fuzzy sets we would not have a way to develop fuzzy rule based systems with them.

For such an undertaking we need only a slight variation of the functions *(fl-translate-rule)*, *(fl-translate-all-rules)*, *(fl-inference)* and *(fl-3d-mesh)* and then some changes in the Lisp structure of a fuzzy model. The names of the new FuzzyLisp functions for handling systems composed by discrete fuzzy sets are shown in the right column from Table 7.8.

As can be seen in the table, the simple addition of the letter "*d*" inside the name of the function signals its proper use. The Lisp source code of these functions is practically identical to their "continuous" counterparts, so we shall not show it here in order to save space. You can freely examine it after downloading the file fuzzylisp.lsp from the book's Web site. The important thing to remember is that both variations of the functions have exactly the same arguments as inputs and they produce the same structure of data in their output. Essentially their use is identical. There is only one restriction: You cannot mix continuous and discrete fuzzy sets for a given set of fuzzy rules. That is, if you decide to use an input linguistic variable

Table 7.8 FuzzyLisp functions for developing Fuzzy Rule Based Systems

Functions for FLSSR	Functions for FLDSR
(fl-translate-rule)	*(fl-dtranslate-rule)*
(fl-translate-all-rules)	*(fl-dtranslate-all-rules)*
(fl-inference)	*(fl-dinference)*
(fl-3d-mesh)	*(fl-3d-dmesh)*

Functions for dealing with continuous fuzzy sets are shown on the left column. Functions for discrete fuzzy sets are shown on the right column composed by fuzzy sets with discrete membership functions, then the other input variable must be composed also by discrete fuzzy sets. The use of singletons for the output share exactly the same structure in both cases.

Code 7-17 shows a variation of the model of an air-conditioner controller where bell-shaped fuzzy sets are used for the input linguistic variables lv-temperature and lv-delta-t. Please note the needed definitions f$_i$ of every bell function and the now mandatory use of the function *(fl-discretize)* in order to follow an adequate FuzzyLisp sets definition. For example, the fuzzy set "optimal" needs first an adequate function definition, expressed by:

$$(setq\ f3\ '(div\ (add\ 1.0\ (cos\ (mul\ 0.2\ pi\ (sub\ x\ 21.0)))))\ 2.0))$$

and then we only need to create the corresponding fuzzy set by means of the following expression:

$$(setq\ T3\ (fl\text{-}discretize\text{-}fx\ 'optimal\ f3\ 20\ 16\ 26))$$

the membership function of the fuzzy set "optimal" *(T3)* is thus centered on x = 21, and its support is defined in the interval [16,26]. By means of the function *(fl-discretize)* we establish also the used resolution on the resulting fuzzy set, in this case n = 20. Note also that the fuzzy rules remain unchanged, as it happens also with the singletons for the consequents in the rules, being identical as the ones shown in Table 7.4c.

```
;code 7-17
(load "my-path/fuzzy-lisp.lsp")

;functions and definitions for first linguistic variable:
(setq f1'(div (add 1.0 (cos (mul 0.0909 pi (sub x 0.0))))) 2.0))
(setq f2 '(div (add 1.0 (cos (mul 0.13333 pi (sub x 13.5))))
    2.0))
(setq f3 '(div (add 1.0 (cos (mul 0.2 pi (sub x 21.0))))) 2.0))
(setq f4 '(div (add 1.0 (cos (mul 0.13333 pi (sub x 28.5))))
    2.0))
(setq f5 '(div (add 1.0 (cos (mul 0.05263 pi (sub x 50.0))))
    2.0))

(setq T1 (fl-discretize-fx 'very-cold f1 20 0 11))
(setq T2 (fl-discretize-fx 'cold f2 20 6 21))
(setq T3 (fl-discretize-fx 'optimal f3 20 16 26))
(setq T4 (fl-discretize-fx 'hot f4 20 21 36))
(setq T5 (fl-discretize-fx 'very-hot f5 20 31 50))
(setq lv-temperature '(T1 T2 T3 T4 T5))

;functions and definitions for second linguistic variable:
```

```
(setq f6 '(div (add 1.0 (cos (mul 0.5 pi (sub x -2.0))))) 2.0))
(setq f7 '(div (add 1.0 (cos (mul 1.0 pi (sub x 0.0))))) 2.0))
(setq f8 '(div (add 1.0 (cos (mul 0.5 pi (sub x 2.0))))) 2.0))

(setq dT1 (fl-discretize-fx 'decreasing f6 20 -2.0 0.0))
(setq dT2 (fl-discretize-fx 'no-change f7 20 -1.0 1.0))
(setq dT3 (fl-discretize-fx 'increasing f8 20 0.0 2.0))
(setq lv-delta-t '(dT1 dT2 dT3))

(setq AFT '(
     (cold-strong -100)
     (cold-medium -50)
     (stop 0)
     (heat-medium 50)
     (heat-strong 100))
)

(setq rules-controller '((lv-temperature lv-delta-t AFT)
    (very-cold decreasing heat-strong AND-product)
    (very-cold no-change heat-strong AND-product)
    (very-cold increasing heat-strong AND-product)

    (cold decreasing heat-strong AND-product)
    (cold no-change heat-medium AND-product)
    (cold increasing heat-medium AND-product)

    (optimal decreasing heat-medium AND-product)
    (optimal no-change stop AND-product)
    (optimal increasing cold-medium AND-product)

    (hot decreasing cold-medium AND-product)
    (hot no-change cold-medium AND-product)
    (hot increasing cold-strong AND-product)

    (very-hot decreasing cold-strong AND-product)
    (very-hot no-change cold-strong AND-product)
    (very-hot increasing cold-strong AND-product))
)
```

For running the model when, for example, temperature is 21 °C and temperature variation is 0.5 °C/min, we only need to type:

> (fl-dinference rules-controller 21 0.5)
: -11.32704597

Table 7.9 Airflow temperatures AFT obtained as the crisp result of a variation of the fuzzy air-conditioner controller using *discrete* membership functions

	0	7.5	13.5	17.5	21	22.5	28.5	32.5	38	50
−2	100	100	100	84.26	49.99	39.26	−50	−51.80	−100	−100
−1.5	100	100	100	84.26	49.99	39.26	−50	−51.80	−100	−100
−1.0	100	100	100	84.26	49.99	39.26	−50	−51.80	−100	−100
−0.5	100	88.67	61.33	45.58	11.33	4.74	−50	−51.80	−100	−100
0	100	85.35	50	34.26	0	−5.37	−50	−51.80	−100	−100
0.5	100	85.35	50	30.69	−11.33	−16.69	−61.33	−62.72	−100	−100
1.0	100	85.35	50	18.51	−49.99	−55.37	−100	−100	−100	−100
1.5	100	85.35	50	18.51	−49.99	−55.37	−100	−100	−100	−100
2.0	100	85.35	50	18.51	−49.99	−55.37	−100	−100	−100	−100

Input temperatures are shown in the first row. Temperature variation is shown in the first column

Looking into the details at the model is easy with the function *(fl-dtranslate-all-rules)*:

> **(fl-dtranslate-all-rules rules-controller 21 0.5)**
 : ((0 0 0 0) (0 0.5 0 0) (0 0.1464466094 0 0) (0 0 0 0) (0 0.5 0) (0 0.1464466094 0 0) (1 0 0 0) (1 0.5 0.5 0) (1 0.1464466094 0.1464466094 -7.322330472) (0 0 0 0) (0 0.5 0 0) (0 0.1464466094 0 0) (0 0 0 0) (0 0.5 0 0) (0 0.1464466094 0 0))

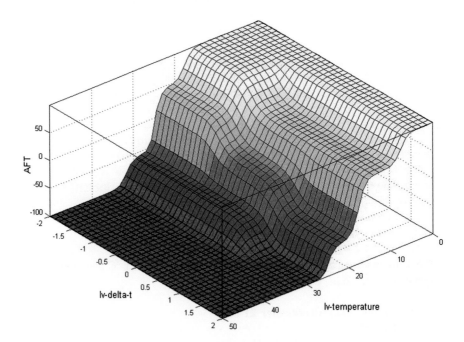

Fig. 7.23 Output surface of the modified fuzzy air-conditioner controller

In Table 7.9 we can have a first look at the behavior of the model. It offers different results when comparing it with the previous model whose linguistic variables are composed by fuzzy sets with triangular or trapezoidal membership functions. This is so because an important reason: while it is easy to replace triangles by bell-shaped functions, trapeziums are more difficult, especially when their support is large with respect to the whole range of the linguistic variable they belong. Due to this, the supports of the fuzzy sets belonging to the linguistic variable lv-delta-t have been more equally spaced.

Figure 7.23 shows the surface output of the model. If we compare it to the surface output shown in Fig. 7.17 we shall note immediately the effect caused by the new design of the output linguistic variable lv-delta-t, but at the same time it is easy to realize the resemblance between the two outputs, showing a familiar appearance. Besides having, as stated, identical singletons for the consequents and similar fuzzy sets in the input linguistic variable lv-temperature, the rules belonging to the knowledge database are exactly the same. Models based on bell shaped fuzzy sets exhibit a smoother output surface when compared to triangular and trapezoidal shaped fuzzy sets, but not a big difference of character is observed between them. On the other hand, creating FRBS with fuzzy sets based on discrete membership functions require more code in FuzzyLisp, and more internal calculations in general. In the rest of the book we shall use triangular and trapezoidal membership functions in the application models.

7.11   As a Summary

This chapter has allowed us to develop a conceptual transition from fuzzy sets theory to fuzzy logic. As already stated, this transition has been really smooth due to the existent isomorphism between membership degree to a (fuzzy) set and the truth-value of a (fuzzy) proposition. The following paragraphs summarize the main ideas exposed on this chapter:

- In general terms, logic is a structured pattern of thinking that makes it possible to arrive to new knowledge from previous established information. In traditional logic, a proposition is natural language expressing an assertion that can be either true or false, but not true and false at the same time.
- Simple propositions take the form "x is P" where x is called the subject of the proposition and P is called the predicate of the proposition, usually expressing a property or feature of x. After a proposition is evaluated, a truth-value results from the evaluation. In classic logic there are only two possible results: either "true" or "false".
- In compound propositions their truth value results from the truth values of every simple proposition taking part in it and the way they are connected, that is, from the type of logical connective, or logical operator, used for establishing the link

between simple propositions. The main logical operators are "conjunction, "disjunction", "negation" and "implication".

- Conjunction: Given two propositions p, q, we call their conjunction to the compound proposition "p and q", denoting them by the logical expression "p ∧ q". This compound proposition is true if and only if both p and q propositions are true.

- Disjunction: Given two prepositions p, q, we call their disjunction to the compound proposition "p or q", denoting them by the logical expression "p ∨ q". This compound proposition is true when at least one of both propositions, p or q, is true, or, obviously, when both are true. For being false it needs that both propositions are false.

- Negation: From any proposition p we shall call "negation of p" to the proposition "not p", denoting it by the expression "¬p". The proposition ¬p will be true when p is false and will be false when p is true.

- Implication: Also known as "conditional proposition", it is expressed by the sentence "p implies q", and it is usually denoted by the expression "p → q". When p → q, it means that the conditional proposition is true except when the proposition p is true and q is false. In other words, if p (also named the "antecedent" in implication) is true, then q (the consequent) must be also true to fulfill a conditional proposition.

- There is an important relationship between set theory and logic in such a way that an isomorphism do exist between the logical operators "and", "or", "not", "implication" and the set operations "intersection", "union", "complement" and "inclusion", respectively. In practical terms this means that all the material exposed in Sect. 5.2 of this book about the classic theory of sets is of entire application in the field of logic after the simple substitution of the concept of membership or not membership of an element x to a set A and the concept of true or false when referring to a proposition p. Naturally, this isomorphism extends also to fuzzy sets and fuzzy logic.

- A fuzzy proposition is natural language declaring an assertion that has implicitly associated a truth-value expressed by a real number in the closed interval [0,1]. Simple fuzzy propositions have the following structure: x is P. It should be noted that the subject x of a fuzzy proposition P is usually not fuzzy. What distinguishes a classic proposition from a fuzzy proposition is the characterization of the predicate P. The fuzziness of P is what generates fuzzy propositions. In fact, fuzzy propositions result from the existing isomorphism between fuzzy sets and fuzzy predicates. In order to evaluate a simple fuzzy proposition p, we need to know the definition of its predicate, which is generally given by the definition of a fuzzy set.

- The conjunction of two fuzzy propositions p, q, represented by $p \wedge q$, is the result of the minimum truth value of both p and q, that is: $Tv(p \wedge q) = min(\mu_A(x), \mu_B(y))$, where A and B are the fuzzy sets representing the predicates associated to the fuzzy propositions p and q, respectively, while x and y are the feature values associated to their respective subjects.

- The disjunction of two fuzzy propositions p, q, represented by $p \vee q$, is the result of the maximum truth value of both p and q, that is: $Tv(p \vee q) = max(\mu_A(x), \mu_B(y))$, where again A and B are the fuzzy sets representing the predicates associated to the fuzzy propositions p and q, respectively, while x and y are the feature values associated to their respective subjects.
- The negation of a fuzzy proposition p, denoted by $\neg p$, is the result of subtracting the truth-value of p from one, that is: $Tv(\neg p) = 1 - (\mu_A(x))$.
- The fuzzy conditional proposition, denoted by $p \rightarrow q$ is the most complex one from the four logical exposed connectives. Its truth-value is given by the expression $Tv(p \rightarrow q) = min(1, 1 + \mu_B(y) - \mu_A(x))$. As it happens with the previous fuzzy connectives, A and B are the fuzzy sets representing the predicates associated to the fuzzy propositions p and q, respectively, while x and y are the feature values associated to their respective subjects.
- Fuzzy hedges are linguistic modifiers applied to fuzzy predicates, so they also modify fuzzy sets related to fuzzy predicates. As a consequence, fuzzy hedges affect fuzzy propositions and then their truth-values, too. In general, a fuzzy hedge H can be represented by an unary operator on the closed interval [0,1] in such a way that a fuzzy proposition of the type "x is P" converts into: x is HP. If A is a fuzzy set associated to a fuzzy predicate P, then we have: $HA = H(\mu_A(x))$. The most used fuzzy hedges are the linguistic modifiers H_1, "very" and H_2, "fairly", defined respectively by the following expressions: H_1 : $H_1(\mu_A(x)) = (\mu_A(x))^2$ and H_2 : $H_2(\mu_A(x)) = (\mu_A(x))^{1/2}$.
- When n tends to infinity, the n-iteration of the hedge VERY on a fuzzy number A produces another special fuzzy set A' whose nucleus and support is established exactly at x_0, where its membership degree equals one, that is: $VERY^n(A) \rightarrow (x_0, \mu(x_0) = 1.0)$. This happens also using the FuzzyLisp function (fl-expand-contract-set) when its second argument k equals zero. These peculiar fuzzy sets, named singletons, play an important role in fuzzy logic.
- Fuzzy compound expressions of the type "if x is A and y is B then z is C" where A, B, C are fuzzy sets and x, y and z are crisp numerical values (z being the only data unknown in the expression) are certainly one of the most useful and practical ones in fuzzy logic theory.
- Fuzzy Rule-Based Systems, FRBS, are logical constructions that bring together several fuzzy-logic based processes and structures. The most usual structure in a FRBS is a set of fuzzy compound propositions of the type "if x is A and y is B then z is C", named "rules" or "expert rules". A FRBS is a system that receives information (crisp input), processes it and then produces a crisp result. The crisp output represents a numerical magnitude that is initially unknown but can be calculated by the inner workings of the fuzzy system.
- A Knowledge Database consists of several linguistic variables and a set of fuzzy propositions (expert rules). Traditionally in fuzzy logic theory only the collection of expert fuzzy rules makes up a knowledge database, but this definition is incomplete because the expert knowledge does not come only from the fuzzy rules in the system, but also from the meaning of the fuzzy sets that make up the linguistic variables of the system.

- The inference engine's task in a fuzzy system is, simply put, to make inferences. Until the inference engine starts to work, the only things we have in a fuzzy system are membership degrees, several fuzzy sets and expert fuzzy rules "floating up in the air". The inference engine is the glue that interacts with all these elements.
- The procedure for obtaining a crisp output from an output set of membership degrees is called "Defuzzification". There are several defuzzification methods in fuzzy logic theory. The "Centroid Method" has a strong advantage that dwells from its intuitive formulation: it is very easy to understand. However it has two important disadvantages. First, it can become tedious to calculate, and second, the Centroid Method is unable to provide a crisp output value z at the extremes of the linguistic variable where the output fuzzy sets are defined. Another well-known defuzzification method is the so named "Weighted average method". In general, for n membership degrees from a fuzzy output the weighted average method uses the expression $z = \Sigma(x_i \mu_i(z))/\Sigma\mu_i(z)$ for calculating the crisp output result.
- As previously suggested, a singleton S is a special fuzzy set whose membership function is equal to zero except at a point x_0, where its membership degree is exactly equal to one. In FuzzyLisp they are represented by expressions of the type *(singleton-label x_0)*. Since every singleton can be seen as a geometrical spike resting fixed on a real point x_i, we can apply the weighted average method of defuzzification for singletons in order to obtain the crisp output of a FRBS. Singleton defuzzification is the standard defuzzification method used by FuzzyLisp.
- Along the chapter new FuzzyLisp functions have been introduced for developing practical fuzzy logic models. The functions *(fl-translate-rule)*, *(fl-translate-all-rules)*, *(fl-inference)* and *(fl-3d-mesh)* provide powerful management of FRBS when the fuzzy sets contained in their knowledge database are formed by triangular or trapezoidal membership functions. An alternate set of functions named *(fl-dtranslate-rule)*, *(fl-dtranslate-all-rules)*, *(fl-dinference)* and *(fl-3d-dmesh)* are offered for dealing with FRBS in those cases when the fuzzy sets contained in their knowledge database have discrete characteristic functions.
- Functions *(fl-3d-mesh)* and *(fl-3d-dmesh)* can be understood as a bridge between FuzzyLisp and any other programming language. The text files produced by these functions can be loaded into any software project and then, by using bilinear interpolation, all the expert knowledge from the previously developed fuzzy models can be incorporated in those software projects. As an example, this opens the possibility for using FuzzyLisp as a software tool for developing intelligent apps for smartphones and other mobile devices.

After all the theory and fundamental FuzzyLisp functions for dealing with fuzzy logic inferences and FRBS management have been exposed, we have dedicated a good number of pages to develop two practical examples of fuzzy models: A fuzzy air-conditioner controller and a fuzzy model for scoring a motor sport regularity rally. These models are simple but have served us well for discussing their construction and for exploring some approaches of fuzzy model design. We shall continue to explore the development of more sophisticated fuzzy models in the last chapter of this book.

References

Bojadziev, G., Bojadziev, M.: Fuzzy Logic for Business, Finance and Management. World Scientific (1999)

Klir, G., Yuan, B.: Fuzzy Sets and Fuzzy Logic: Theory and Applications. Prentice Hall, Englewood Cliffs, NJ (1995)

Kranz, G.: Failure is Not an Option: Mission Control from Mercury to Apollo 13 and Beyond. Simon & Schuster, New York (2009)

Kundu, S., Chen, J.: Fuzzy logic or Lukasiewicz Logic: A clarification. Lecture Notes in Computer Science, vol 869 (1994)

McFerran, D.: A Syllogism Playbook: Using PLN for Deductive Logic. McFerran (2014)

Priest, G.: Logic: A Very Short Introduction. Oxford University Press, Oxford (2001)

Russell, S., Norvig, P.: Artificial Intelligence: A Modern Approach, 3rd edn. Prentice-Hall, Englewood Cliffs, NJ (2009)

Chapter 8
Practical Projects Using FuzzyLisp

8.1 Introduction

In the last sections of the previous chapter we introduced some simple examples of practical fuzzy logic based applications where a FRBS always had two crisp inputs and a crisp output. In this chapter we shall discover models of a bigger complexity that, more than examples, are intended to be a guideline for more ambitious goals. These models, aside practical, pretend to be inspiring for the reader. Engineers and scientists will soon find some ideas for using FuzzyLisp in their respective fields of knowledge, but any attentive reader will find new routes of exploration in order to tackle complex problems from the real world. There are three practical projects in this chapter chosen from astronautics, astronomy and medicine. All of them are exposed in a clear and friendly way, and it is expected that any reader of this book will find them interesting.

Section 8.2 introduces a simplified version of the Moon landing problem where a physical model (that is, a set of traditional kinematics equations) describe the descent movement of the NASA Lunar Module and then a FRBS takes control in order to get a smooth touchdown on Earth's satellite, being thus a combined exercise of simulation systems and fuzzy control. In Sect. 8.3 a sophisticated FRBS architecture is introduced for speech synthesis where double stars are the excuse for showing how to automatically generate expert reports in natural language from a set of numerical data. This is the most complex project in the chapter, but suggests many real applications in practically every science and branch of technology, from biology to robotics. Finally, in Sect. 8.4, the reader is invited to an introduction to adaptive fuzzy models using floating singletons while developing a model to interpret spirometric results for detecting Chronic Obstruction Pulmonary Disease (COPD), having into account not only numerical data from a spirometry but also smoking. Floating singletons theory is introduced adequately before the model's development, being the last piece of fuzzy theory in this book.

© Springer International Publishing Switzerland 2016
L. Argüelles Méndez, *A Practical Introduction to Fuzzy Logic using LISP*,
Studies in Fuzziness and Soft Computing 327,
DOI 10.1007/978-3-319-23186-0_8

Since the complexity of these models, as already stated, has grown with respect to the models developed in the previous chapter, the same has taken place with the corresponding Lisp code for every project. All the code is shown in the chapter with the appropriate comments here and there, but now the reader does not need to type everything at the keyboard. The "Lisp vision" is already on the reader's mind and you will find the complete programs in the book's web site. Do enjoy these projects. Maybe they will be the ultimate key for future developments in your life.

8.2 Landing the Eagle: Simulation and Fuzzy Control in Engineering

Back in 1984 I bought a rather sophisticated (by those days) programmable electronic calculator. Its programming language, embedded in its ROM circuitry, was a tiny dialect of BASIC, one of the more simple computer languages in the 80s. Despite its limitations as a computer language, every manufacturer of small personal computers offered a version of BASIC in their systems, and the small calculator was not an exception. In order to gain some experience with the system and its implementation of BASIC I decided to write a program to simulate the landing of the Eagle module on the Moon. After more than thirty years the program still runs on the calculator, as can be seen in Fig. 8.1.

Since the RAM memory of the device was less than 2K (yes, the reader is reading well, that's about 0.002 MB of RAM) the model was rather simple and some assumptions were needed to develop it. First a vertical descending trajectory

Fig. 8.1 A simple Moon landing simulator running on a 30+ years old programmable electronic calculator. Photograph by the author

was assumed: The simulated Lunar Module (LM) was initially placed at a random height between 2500 and 6000 m over the Moon surface and no control attitude was implemented. Also, the LM started its landing trip at an initial descending velocity of 300 m/s (that would be near Mach 1 speed in terrestrial atmosphere). Under these conditions, two kinematic equations describe the movement of the LM:

$$v = v_0 + g_m t$$
$$s = s_0 + v_0 t + 0.5\left(g_m t^2\right)$$

where v_0 is the initial velocity, s_0 the initial traversed space and then v and s are the resulting velocity and space after some elapsed time t. The Moon gravity acceleration, approximately equal to 1/6 of that of the Earth gravity is a constant $g_m = 1.625$ m/s^2. The variable mass of the LM due to fuel consumption is not considered in the model.

In this simulation, the user (the command pilot at the LM) must decide an upward thrust p in order to compensate both the descent speed of the LM and the pull of the Moon gravity for getting a smooth touchdown. It is assumed that a thrust p produces a positive and counteracting acceleration a on the LM. Under such conditions, and observing Fig. 8.2, the above equations transform into these ones:

$$v = v_0 + (g_m - a)t$$
$$h = h_0 - v_0 t - 0.5(g_m - a)t^2$$

The original simulation discretized descending time in 2 s intervals, that is, after a reading of height and velocity is made, the user decides how much thrust must be

Fig. 8.2 A graphical representation of the LM descent simulation

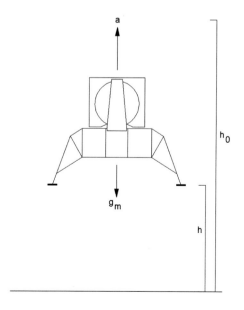

Table 8.1 Kinematic expressions describing the LM's movement

Kinematic equations	Equivalent kinematic expressions in Lisp
$v = v_0 + 2(g_m - a)$	*(setq v (sub v (mul 2.0 (sub a 1.625))))*
$h = h_0 - 2v_0 - 2(g_m - a)$	*(setq h (add (sub h (mul 2.0 v)) (mul 2.0 (sub a 1.625))))*

applied, and then, that thrust is applied along 2 s in such a way that the above equations convert into the expressions shown in Table 8.1.

It is important to note that the thrust system (a rocket) actuates continuously in the simulation along 2 s and then waits for another new thrust command. In other words: after a thrust command is launched, nothing can stop the system actuation in the next 2 s. Fuel consumption is considered too: It is assumed that fuel consumption is directly proportional to thrust magnitude. The simulation has several possible outcomes:

- The LM gets a smooth touchdown. In this simulation, a smooth touchdown is considered when the LM final speed is less or equal than 5 m/s.
- The LM reaches the Moon at an excessive speed, that is, bigger than 5 m/s. A fatal crash is the ultimate result of this outcome.
- The LM runs out of fuel. In this case, the most probable scenario is to enter into the previous outcome after some free fall time. Neil Armstrong almost got into this outcome in the Apollo XI mission as stated in the previous chapter.

The translation of the original program written in BASIC to Lisp produces Code 8-1 as follows:

```
;code 8-1
(define (init-variables, fuel velocity height)
    (setq fuel 250.0) ;fuel units capacity
    (setq velocity 300.0) ;descent velocity in m/s
    ;generate a random height between 2500 and 6000 metres
    (while (<= height 2500)
        (setq height (rand 6001))
    )

    (println "Eagle landing on the Moon")
    (println "Fuel: " fuel " Velocity: " velocity " Height: "
            height)
    (list fuel velocity height)
)

;moon gravity, mg = 1.62519 m/s2
(define (update-variables x v h a)
    (setq v (sub v (mul 2.0 (sub a 1.625))))
    (setq h (add (sub h (mul 2.0 v)) (mul 2.0 (sub a 1.625))))
    (setq x (sub x (abs a)))
```

```
    (setq fuel x velocity v height h)
    (println "Fuel: " x " Velocity: " v " Height: " h)
    true
)

(define (simulate ,data fuel velocity height thrust)
    (setq data (init-variables))
    (setq fuel (first data) velocity (nth 1 data) height
            (nth 2 data))
    (while (and (> height 0.0) (> fuel 0.0))
        (print "\nRocket Thrust (0-25)?: ")
        (setq thrust (float (read-line)))
        (update-variables fuel velocity height thrust)
    ) ;while end

    (if (<= fuel 0.0)
        (println "Mission failure: Out of rocket fuel.
                Unavoidable LM crash")
    )
    (if (>= velocity 5.0)
        (println "Mission failure: Excessive velocity. LM
        crash")
    )
    (println "End simulation")
)
```

The listing of the program begins with the function *(init-variables)* that allows it to initialize the variables *fuel*, *velocity* and *height*. This last one is chosen randomly by means of the function *(rand)*, using a while loop that only ends when the initial height of the LM is placed between 2500 and 6000 m. Then the function prints these values and returns them as a list.

The function *(update-variables)* is the mathematical core in the simulation. It receives the variables representing the system status and then returns the updated values of these variables after using the expressions shown in Table 8.1. Finally, the function *(simulate)* puts it all together: Makes a call to *(init-variables)* and then enters the main event loop of the simulation where a thrust value is requested to the user and the system status is updated. The main event loop only ends when the LM reaches the Moon's surface or when the LM runs out of fuel. Finally, some *if* statements inform the user about the simulation outcome. Let us see an abbreviated run of the program:

> (simulate)
: Eagle landing on the Moon
Fuel: 250 Velocity: 300 Height: 5041
Rocket Thrust (0-25)?: 10

Fuel: 240 Velocity: 283.25 Height: 4491.25
Rocket Thrust (0-25)?:

...

...

Rocket Thrust (0-25)?: 1
Fuel: 45 Velocity: 3.75 Height: -0.12
End simulation

The reader can download Code 8-1 directly from the Book's website in order to become familiar with the program. Chances are that several crashes will be the initial outcomes of the simulation, including also some situations where all the fuel has been exhausted, especially when the LM is initially located near 6000 m high. It is not easy to land a spacecraft on the Moon!

Paraphrasing the words of John F. Kennedy at Rice University in 1962, "*We choose to go to the moon in this decade and do the other things, not because they are easy, but because they are hard*", we have decided for this chapter to develop a fuzzy system able to get a smooth touchdown on the Moon surface. The next sections of this chapter will show how to build a FRBS that substitutes the human user in the previous LM simulation. Entering the realm of fuzzy control.

8.2.1   Fuzzy Control

In general, a control system is an arrangement of physical components designed to alter, to regulate, or to command, through a control action, another physical system so that it exhibits certain desired characteristics or behavior (Ross 2010). A good example of control system is the brake system in a car: the set of disks, brake pads, brake liquid, etc. is the "arrangement of physical components", while the car is the "another physical system" that must behave adequately (in this case to diminish its speed or to stop it completely). To perform the adequate "control action" we need yet a model, a complete "understanding" of the system that allows those physical components to perform the task. In the case of the car, the model for braking it is installed well inside our brain when we learn to drive. Usually in standard control theory these models are built by means of differential equations. Interestingly, traditional control engineers tended to treat their mathematical models of physical systems as exact and precise though they knew that the models were neither (Seising 2007). Take a falling body in terrestrial gravity as an example. Kinematics equations as the ones described in the previous section are exact if we forget the atmosphere and the mass of the falling body. If we take into account these factors, then differential equations are needed. If we consider also the shape of the solid, then aerodynamics enters into play and differential equations are no more an exact tool for describing the movement of the solid. Then we have no other option than using simulations at the wind tunnel. And this is a very expensive resource.

The fuzzy control model exposed in the next section is not only able to get a smooth touchdown on the Moon surface, but it would work also in the presence of an atmosphere, taking also into account the variable mass of the LM due to fuel consumption.

Fuzzy control systems satisfy completely Ross' definition, but the control action is obtained not by means of a set of mathematical equations but through the use of a Fuzzy Rule Based System (FRBS). In fact, a fuzzy control system is basically a FRBS where two especially important features are always present:

- The output of every fuzzy control system is an action.
- Every fuzzy control system uses time as a fundamental magnitude.

Control systems use sensors for measuring the desired information that describe the status of the system to control, such as velocity, temperature, density, humidity, etc. The obtained data from these sensors are the crisp inputs to a FRBS that, as shown in Fig. 7.6, ultimately produces a crisp output. In control systems this output represents an action that must be applied to the system to be controlled in order to modify o maintain its desired characteristics or behavior. Both classic control systems and fuzzy control systems result into an action, but while classic ones use mathematical equations, fuzzy based ones do use expert knowledge fuzzy rules.

After the action is applied on the system to control, the effects of the action cause changes in the behavior of the system. These changes, reflected in the system state variables, are again measured by the sensors and then a new input to the controller is created. This circular flow of information between sensor reading, controller and output action gives name to closed-loop control systems, the ones we are interested in this section. A pivotal design matter in closed-loop control systems is the concept of sampling time, the difference of time between two consecutive readings obtained from the sensors. While sampling time is usually of no importance in general FRBS, no fuzzy control system (FCS) could be developed without a clock whose ticks manage the readings of the sensors. In some cases the interval of time can extend to minutes or even hours while in other cases sampling time will be in the rank of milliseconds. The importance of sampling time will be evident after reflecting on the following example.

Imagine we are designing a fuzzy control system for an intelligent insulin pump for people suffering diabetes. Let us suppose that after a glucose reading obtained by a glucose meter the model suggests an action based in administering a subcutaneous quantity x of insulin to the patient. With a sampling time equal to 5 min another glucose reading will show about the same levels of glucose because the previously administrated insulin, including the more rapid acting types of this hormone, has not still started to show its physiological action. Then, after the second reading, the model recommends another subcutaneous quantity x of insulin. The net result is a patient receiving $2x$ units of insulin in 5 min. This probably will cause hypoglycemia on the patient and in extreme cases it could mean his/her exitus. On the other hand, a sampling time of, let us say, 1 h, probably would translate into weak control actions, producing a failure in lowering blood glucose levels towards normal values. This example suggests that determining an adequate sampling time in a control system is one of the most important parameters of its design.

Fuzzy Control has produced such a quantity of successful results, patents and commercial products that by itself has already proven the advantages of the theory of fuzzy sets and fuzzy logic. Examples run from exact focus determination and correct exposure in cameras (Chen and Pham 2000) to automatic gearboxes in cars (Yamaguchi et al. 1993) to modern docking systems in spacecrafts (Dabney 1991), to name only a few. In fact, real practical applications of Fuzzy Logic are virtually endless today and Fuzzy Control is taught in every major college of engineering throughout the world.

In the previous chapter we already introduced an example of fuzzy control when designing an air-conditioner controller, but we did it while explaining the inner workings of a FRBS. An in depth discussion of Fuzzy Control is out the scope of this book, but the interested reader can easily find specific texts about this branch of engineering and mathematics, e.g. Passino and Yurkovich (1998). Even so, developing a control model for landing a simplified version of the LM on the Moon will show the reader more than a basic insight into fuzzy control.

8.2.2 Controlling the Eagle

As we stated at the beginning of Sect. 7.7, Buzz Aldrin was providing a continuous readout of LM altitude and velocity to Neil Armstrong in order to get a smooth touchdown of the Eagle on the Moon. It seems then natural to use these parameters as the input data for developing the LM fuzzy control model. Let us see the corresponding linguistic variables for each of them.

The first one, named *height*, is composed by seven fuzzy sets with the following linguistic labels: near-zero, very-small, medium, medium-high, high, very-high and extra-high. Just note that the range of this linguistic variable extends from 0 to 8000 m. It's a huge range as can be seen in Fig. 8.3, so in order to include it completely in this page the two rightmost fuzzy sets have not been drawn at scale, as suggested by the obliquely dashed lines. You also have probably realized that the supports of the corresponding fuzzy sets are shorter and shorter as we approximate

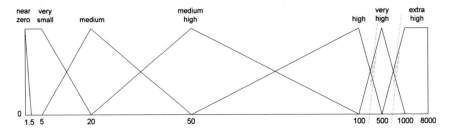

Fig. 8.3 Linguistic variable representing height of LM over the Moon's surface

to the left of the linguistic variable until arriving to a very short support fuzzy set named "near-zero". This is so because the control becomes critical when we approach the Moon surface and we shall need to have an excellent linguistic description of the model when developing its rules.

In Code 8-2a we can see this linguistic variable translated into Lisp expressions:

```
;code 8-2a
(setq h1 '(near-zero 0 0 0 1.5))
(setq h2 '(very-small 0 0 5 20))
(setq h3 '(medium 5 20 20 50))
(setq h4 '(medium-high 20 50 50 100))
(setq h5 '(high 50 100 100 500))
(setq h6 '(very-high 100 500 500 1000))
(setq h7 '(extra-high 500 1000 8000 8000))
(setq height '(h1 h2 h3 h4 h5 h6 h7))
```

The following input linguistic variable, named *velocity*, is composed again by seven fuzzy sets named very-very-small, very-small, small, medium, quick, very-quick and top (meaning top velocity). The range extends from −1.0 to 300 m/s.

As can be seen in Fig. 8.4, the fuzzy sets belonging to *velocity* are arranged in a similar way as in *height*, showing a skew towards the left of the universe of discourse where they are defined. At first it can seem strange that the leftmost fuzzy set has a support defined between −1 and 1 m/s (for clarity, the label of this fuzzy set, "very-very-small" has been omitted in the figure). This is so to cover those possible cases where the LM has suffered from excessive thrust in the final touchdown phase and is not descending but in fact is elevating over the lunar soil. The corresponding Lisp code for this linguistic variable is shown in Code 8-2b:

```
;code 8-2b
(setq v0 '(very-very-small -1.0 0.0 0.0 1.0))
(setq v1 '(very-small 0.0 0.0 0.0 5.0))
(setq v2 '(small 0.0 5.0 5.0 10.0))
```

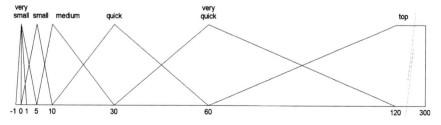

Fig. 8.4 Linguistic variable representing descent velocity of LM

```
(setq v3 '(medium 5.0 10.0 10.0 30.0))
(setq v4 '(quick 10.0 30.0 30.0 60.0))
(setq v5 '(very-quick 30.0 60.0 60.0 120.0))
(setq v6 '(top 60.0 120.0 300.0 300.0))
(setq velocity '(v0 v1 v2 v3 v4 v5 v6))
```

The output of the fuzzy control model is thrust, that is, how much gas must be ultimately expelled by the nozzle of the LM in order to modify its descent velocity. As previously assumed, a thrust p produces a positive acceleration a on the LM. The set of singletons for describing a suitable range of positive accelerations is named *thrust* and is composed by the following singletons: no-thrust (0), thrust-gently (0.6), maintain (1.625), small-minus (2.5), small (5), medium (10), strong (15) and very-strong (25), where the figures enclosed into parenthesis indicate the resulting acceleration associated to every single action.

The singleton "maintain" has a special meaning in this arrangement of single-tons: The Moon gravity causes a downwards acceleration exactly equal to 1.625 m/s^2 on a falling body, so a thrust p producing an upwards acceleration $a = 1.625$ m/s^2 will result into a body moving with no acceleration, that is, a body moving at a constant velocity in the Moon's gravitational field. This singleton is extremely convenient for linguistically describing the expert rules of the fuzzy control model for the LM.

Additionally, it should be noted that the chosen range of accelerations would produce a relatively comfortable descending experience to our simulated team of astronauts. Having into account that 1 g (terrestrial) = 9.8 m/s^2, this means that the maximum thrust of our model is approximately equivalent to 2.55 g, less than the usual 3.0 g acceleration produced by the past Shuttle missions and even less than the stark trust of the already historical Saturn V rocket, at 4.0 g (Fortescue et al. 2011). Also important in the last phase of descent are the singletons "no-thrust" and "thrust-gently". Figure 8.5 shows the intervening singletons.

The corresponding Lisp code for this set of singletons is given by the Lisp expressions in Code 8-2c:

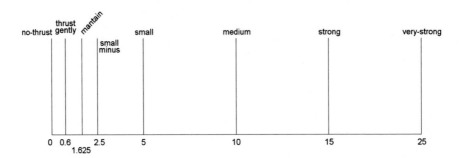

Fig. 8.5 Output singletons representing lunar module thrust

```
;code 8-2c
(setq thrust '(
    (no-thrust 0.0)
    (thrust-gently 0.6)
    (maintain 1.625) ;equivalent to moon gravity
    (small-minus 2.5)
    (small 5.0)
    (medium 10.0)
    (strong 15.0)
    (very-strong 25.0))
)
```

For building the rules of the LM fuzzy control model we combine every fuzzy set from the linguistic variable *height* with every fuzzy set contained in the linguistic variable *velocity*, associating an adequate singleton representing thrust to every rule. This results into a set of 49 expert rules that completely describes the system. Code 8-2d shows the corresponding Lisp code:

```
;code 8-2d
(setq rules-LEM '((height velocity thrust)
    (near-zero very-very-small thrust-gently AND-product)
    (near-zero very-small thrust-gently AND-product)
    (near-zero small small AND-product)
    (near-zero medium maintain AND-product)
    (near-zero quick medium AND-product)
    (near-zero very-quick very-strong AND-product)
    (near-zero top very-strong AND-product)

    (very-small very-very-small thrust-gently AND-product)
    (very-small very-small maintain AND-product)
    (very-small small small-minus AND-product)
    (very-small medium small AND-product)
    (very-small quick small AND-product)
    (very-small very-quick strong AND-product)
    (very-small top very-strong AND-product)

    (medium very-very-small thrust-gently AND-product)
    (medium very-small thrust-gently AND-product)
    (medium small maintain AND-product)
    (medium medium medium AND-product)
    (medium quick medium AND-product)
    (medium very-quick strong AND-product)
    (medium top very-strong AND-product)
```

```
(medium-high very-very-small thrust-gently
AND-product)
(medium-high very-small maintain AND-product)
(medium-high small maintain AND-product)
(medium-high medium medium AND-product)
(medium-high quick medium AND-product)
(medium-high very-quick strong AND-product)
(medium-high top very-strong AND-product)

(high very-very-small thrust-gently AND-product)
(high very-small thrust-gently AND-product)
(high small maintain AND-product)
(high medium medium AND-product)
(high quick medium AND-product)
(high very-quick medium AND-product)
(high top very-strong AND-product)

(very-high very-very-small no-thrust AND-product)
(very-high very-small no-thrust AND-product)
(very-high small no-thrust AND-product)
(very-high medium maintain AND-product)
(very-high quick maintain AND-product)
(very-high very-quick medium AND-product)
(very-high top very-strong AND-product)

(extra-high very-very-small no-thrust AND-product)
(extra-high very-small no-thrust AND-product)
(extra-high small no-thrust AND-product)
(extra-high medium maintain AND-product)
(extra-high quick medium AND-product)
(extra-high very-quick strong AND-product)
(extra-high top strong AND-product))
)
```

When the LM is located at high altitude the rules simply try to moderate its descent velocity. If altitude becomes smaller and velocity is still near its maximum the model will react applying very strong thrust. Soon the LM falls at approximately uniform speed under the influence of the singleton named "maintain". The most interesting phase of flight happens when the LM approaches the Moon surface. Here the first 28 expert rules of the model show their power, especially those that contain the fuzzy sets "very-very-small", "very-small" and "small" from the linguistic variable *velocity*. In fact, under normal descent conditions, only twelve rules get the LM on the Moon.

This is interesting because the number of rules in the model could be simplified. As an example, when the LM is located at or near top height, the following rules could be eliminated:

```
(very-high very-very-small no-thrust AND-product)
(very-high very-small no-thrust AND-product)
(very-high small no-thrust AND-product)
(extra-high very-very-small no-thrust AND-product)
(extra-high very-small no-thrust AND-product)
(extra-high small no-thrust AND-product)
```

The reader can play with the model, trying to eliminate those rules that "never work", thus obtaining a more efficient set of rules, but make no mistake: the real world is full of instances where things that would never happen ultimately do happen. Depending on how critical a control application is robustness is far better than efficiency.

After building the knowledge database of the model is easy to code the simulation and fuzzy control program for the LM because we only need to take the fundamentals of code 8-1 and connect it with the fuzzy model description. Code 8-3 shows the resulting Lisp program for completing our Eagle landing simulation model:

```
;code 8-3
(define (update-variables x v h a)
    (setq v (sub v (mul 2.0 (sub a 1.625))))
    (setq h (add (sub h (mul 2.0 v)) (mul 2.0 (sub a 1.625))))
    (setq x (sub x (abs a)))
    (list h v x);returns current height, velocity and fuel
)

(define (simulate, fuel v h force list-data n)
    ;initialize variables:
    (setq n 0)
    (setq force 0.0)
    (setq fuel 500.0);fuel capacity
    (setq v 300.0);descent velocity in m/s

    (setq h 0.0)
    ;generate a random height between 2500 and 6000 metres:
    (while (<= h 2500)
        (setq h (rand 6001))
    )

    (println "Eagle landing on the Moon")
    (while (and (> h 0.0) (> fuel 0.0))
```

```
    (println "Fuel: " fuel " Height: " h " Velocity: " v)
    (print "\nPress return to continue... ")
    (read-line)

    (if (<= v -1.0)
      (setq force 0.0) ; if LEM elevates at v>1 m/s, apply no
      thrust
        (setq force (fl-inference rules-LEM h v))
    ) ; end if
      (print "\nThrust by fuzzy model: " force)
    (print "\nPress return to continue... ")
    (read-line)

    (setq list-data (update-variables fuel v h force))
    (setq h (first list-data))
    (setq v (nth 1 list-data))
    (setq fuel (last list-data))
    (++ n)
  ) ; while end

  (if (<= fuel 0.0)
      (println "Mission failure: Out of rocket fuel.
                Unavoidable LEM crash")
  )
  (if (>= v 5.0)
      (println "Mission failure: Excessive velocity. LEM
      crash")
  )
  (println "End simulation. Final velocity: " v " n: " n)
)
```

Only these few lines of code put together all the parts:

```
(if (<= v -1.0)
  (setq force 0.0) ; if LEM elevates at v>1 m/s, apply no
  thrust
  (setq force (fl-inference rules-LEM h v))
) ; end if
```

This conditional construction manages the internals of the simulation. If the LM velocity is less than −1.0 m/s, that is, if the LM is elevating, then no thrust is applied and the model will wait until the Moon gravitational field produces again a positive descending speed. Otherwise, a call to the function *(fl-inference)* is made and the crisp output of the fuzzy control module is transferred to the variable force. Trying the program at the Lisp prompt is easy, pressing the return key for every iteration:

> **(simulate)**
: Eagle landing on the Moon
Fuel: 500 Height: 3000 Velocity: 300
Press return to continue...

Thrust by fuzzy model: 15
Press return to continue...
Fuel: 485 Height: 2480.25 Velocity: 273.25

...

Press return to continue...
Thrust by fuzzy model: 2.432348398
End simulation. Final velocity: 0.8998958331 n: 40

When the initial height is $h_0 = 3000$ m the final landing velocity on the Moon obtained by the fuzzy control mode is less than 1.0 m/s, thus meeting the NASA real standards for the LM (Klump 1971). Other initial height values can result in slightly higher final landing speeds. The reader is encouraged to improve the model using two strategies: modifying the knowledge database (fuzzy sets and/or set of fuzzy rules) or improving the resolution of sampling-time from 2 to 1 s. Changing the sampling time will need a redesign of the knowledge database. Moreover, the sampling time used on the Apollo missions for the LM was bounded between 1 and 2 s and some lags in the systems had to be had into account. I think the use of fuzzy control theory to the complete LM landing procedures on the Moon is fascinating and a project on this matter would be a technical delight to the interested reader. For this undertaking it is nice to know that NASA provides enough documentation on the Internet (NASA 2015).

8.2.3 Interpreting Results

The first thing to note, aside the final velocity at the end of the simulation, is the number of iterations produced in the main event loop. When falling from 3000 m high, the number of iterations is n = 40, so, having into account every loop pass equals to 2 s this results into a descent flight time of almost one minute and a half.

When running the simulation the user can numerically observe the actual height, descent velocity and remaining fuel. After some simulations are made it is difficult to have a clear picture of the system's behavior, so obtaining a chart where these magnitudes are plotted is a good way to gain more knowledge of the system. NewLisp incorporates a complete set of graphical functions but since this would translate into more complex code in the book we shall use another strategy: to slightly transform Code 8-3 in order to redirect the program's output to disk instead to the console and then use a standard spreadsheet program to plot the file contents. Code 8-4 shows the transformation to be done inside the function *(simulate)*:

```
;code 8-4
(device (open "LEM-OUT.csv" "write"))
(println "Fuel;Height;Velocity;Thrust")
(println fuel ";" h ";" v ";" force)

(while (and (> h 0.0) (> fuel 0.0))
    (if (<= v -1.0)
      (setq force 0.0)
      (setq force (fl-inference rules-LEM h v))
    )
    (setq list-data (update-variables fuel v h force))
    (setq h (first list-data))
    (setq v (nth 1 list-data))
    (setq fuel (last list-data))
    (println fuel ";" h ";" v ";" force)
    (++ n)
);while end
(close (device))
```

As the reader can see, it is only a question of conveniently using the functions *(open)*, *(print)* and *(close)*. Thus, the values of fuel, height, velocity and thrust are written to a file named LEM-OUT.csv that can easily be imported in many spreadsheet applications. From this point, graphically representing the output data is straightforward. Figure 8.6 shows thrust along descent time from the fuzzy control model:

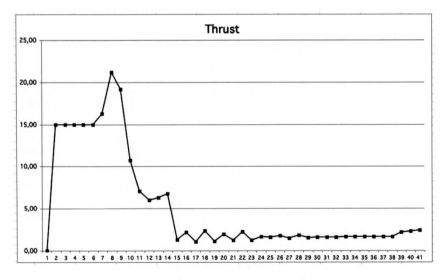

Fig. 8.6 Thrust along time as calculated by the LM fuzzy control descent model

As can be seen the model starts the descent flight applying no thrust and then, just in the second main loop pass of the simulation it produces an interval of continuous thrust at p = 15, raising it until reaching a peak about p = 22 and then initiating a smooth thrust descent that ends at values of p around 6-7. Later it still falls until stabilizing at p values between 1 and 3. The effects of this trust on the rest of LM system variables can be appreciated on Figs. 8.7, 8.8 and 8.9.

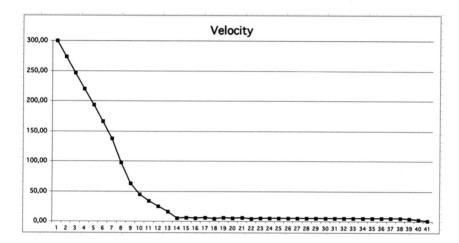

Fig. 8.7 LM obtained descent velocity along time

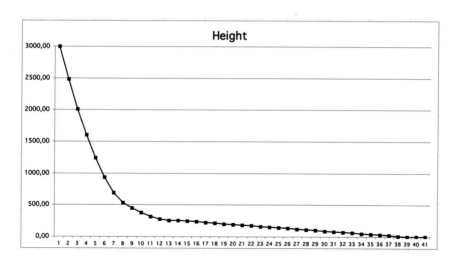

Fig. 8.8 Height over the Moon surface along descent time

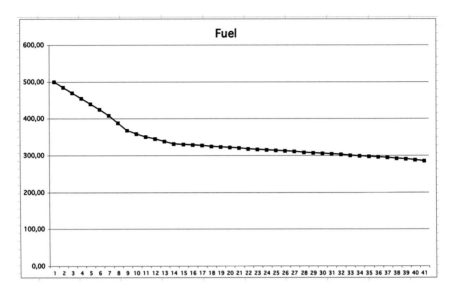

Fig. 8.9 Fuel consumption in fuel units along descent time

It is interesting to note how the control model gets initially a quick decrease of descending velocity and then it limits itself to maintain a virtually constant and moderate descent speed. A final slight increase in thrust produces a smooth, soft lunar landing. Fuel consumption is relatively low because the initial height. Consumption increases, obviously, for higher initial heights.

8.3   Double Stars in Astronomy: Speech Synthesis

When we observe the night sky from a site free from light pollution we can enjoy a wonderful sight formed by thousands of stars. It is not a generally known fact, but about half of them are double stars, that is, a set of two or more stars that orbit around a common center of masses. At the naked eye these systems appear as a single point of light, but using a telescope many of them can be split through the eyepiece, thus revealing their true double nature. These celestial objects result from the gravitational collapse of interstellar molecular clouds composed mainly by hydrogen that, due to initial inhomogeneities in density in the cloud, suffer a process of fragmentation. During this process, sections of the cloud collapse locally, forming protostars that eventually evolve into a set of two or more stars that orbit their common centre of masses with periods ranging from a few tens to a few million of years. Since aside metallicity, the main parameter for the life evolution of a star is its initial mass, such processes determine not only the separation and rotational period of the system, but also the spectral type of the components (Ostlie and Carrol 2006).

In strong contrast to optical pairs, i.e., apparent double stars that are simply formed due to line-of-sight coincidence, visual binaries are true gravitationally bounded pairs located in our galaxy. When observed through a number of years, they show different values of relative angle and angular separation between the stellar components, resulting into a curve when these parameters are plotted (Argyle 2004). In this section we are especially interested in visual binaries whose angular separation is equal or more to 1 arcsecond. With the exception of extremely good observing sites in the world such as the Canary Islands or Hawaii, this is usually the practical resolution limit of an optical telescope placed on Earth due to the atmosphere steadiness (well, lack of it), usually known in observational astronomy as "seeing".

Speech synthesis is, simply put, a computer-generated simulation of human speech that requires two phases: the first one consists in the grammatical generation of text on a certain subject matter and the second one is the audible generation of the obtained text. We are interested in the grammatical generation of double star reports using as input several intrinsic numerical parameters of this type of celestial objects. These parameters are: Separation (d), Visual magnitude of components (Mv_1, Mv_2) and Spectral class of every star ($s1$, $s2$). The aperture D in inches of the telescope is also taken into account to produce a human-like observing report. Expressing it formally, we are going to design a fuzzy logic based model F such as:

$$F(d, Mv_1, Mv_2, s1\ s2, D) \rightarrow Text\ report$$

As an example, when using Albireo, Beta Cygni, as the target double star ($d = 34.3$ arcseconds, $Mv_1 = 3$, $Mv_2 = 5.5$, $s_1 = $ K3, $s_2 = $ B9) observed through a four inches aperture telescope, the report generation obtained from our model F will be:

"Albireo is a really open double star that can be very easily split using any observing instrument. Difference of coloration is rather easy to observe and from an aesthetic point of view is a jewel in the sky. There is a medium difference of magnitude between components. The primary is a bright star, while the secondary is medium bright"

In order to obtain this type of speech synthesis we shall use a concatenation approach, that is, we shall use a fixed pattern of text containing several empty "pockets" where variable text will be instantiated depending on the numerical parameters from a given double star. The text structure will be the following one:

"**str1** is a **str2** double that can be **str3** split using **str4**. Difference of coloration is **str5** and from an aesthetic point of view is **str6**. There is a **str7** difference of magnitude between components. The primary is a **str8** star, while the secondary is **str9**"

The first variable string of text, **str1**, is the name of the double star. The rest of variable strings, **str2**, **str3**, **str4**, **str5**, **str6**, **str7**, **str8** and **str9** will be obtained from numerical values. Some of the numerical parameters of a double star produce a direct translation into text, such as angular distance between components or their visual magnitude. However some other information as difficulty to observe the celestial object in its true double nature or the perceived beauty of the stellar system at the eyepiece are subjective. Fuzzy logic based models are, as we already know, very well suited to this task (Argüelles and Trivino 2013).

In Chap. 7 we have used the Takagi-Sugeno (TK) model for explaining Fuzzy Rule Based Systems (Takagi and Sugeno 1985). Although the TK model was initially developed for Fuzzy Control Systems, it is of general application in FRBS when the required output is not an action but a numerical description about a certain system. Until now we have built models with two crisp inputs and a crisp output. It is time now to see how we can aggregate multiple inputs in FRBS and how we can use numerical output as new input to other FRBS. For building our intended double star speech synthesis system we shall use the architecture shown in Fig. 8.10.

The meaning of the complete set of numerical parameters (enclosed in boxes in the figure) is as follows:

- Separation: When astronomers speak about separation between stars in a double system they are not using kilometers or miles, but arcseconds, that is, an angular separation expressed in seconds of arc. Every degree has 60 min of arc, each of them divided again into 60 s of arc. Thus an arcsecond equals to 1/3600th of a degree. This angular distance is certainly small and is approximately equivalent to the view obtained from a one Euro coin located 5 km away from us. Measuring double stars was a frenetic astronomical activity back in centuries XVIII and XIX, and still today many amateur astronomers continue to systematically measure doubles stars. There is an important reason behind this task: Traditionally, the only way to discover the mass of stars is by means of observing double stars (Argyle 2004).
- Primary magnitude: In astronomy, the term "magnitude" refers to bright. The smaller the number, the brighter a star is perceived by an observer's eye. As a reference, our Sun has a −27 magnitude, being obviously the brighter celestial

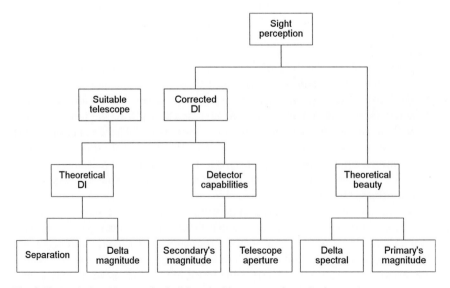

Fig. 8.10 Logical architecture for building double star speech synthesis reports

object as seen from Earth. Sirius, the brighter star from the North Hemisphere aside the Sun has a magnitude of −1.4. The scale is logarithmic and every unit in magnitude scale means about 2.512 times difference in bright from two stars. Under a really dark observing place the naked eye limit is established around magnitude 6.5, that is, fainter stars are not visible without using an observing instrument such a pair of binoculars or a telescope. To make things a bit more complex, astronomers distinguish between absolute and apparent magnitude. We are concerned, as catalogues of double stars are, only to apparent magnitude, that is, the bright of a star observed from Earth. In double stars terminology "primary magnitude" refers to the brightest star of the pair.

- Secondary magnitude: As the reader can infer from the above, this refers to the apparent magnitude of the fainter star of the pair.
- Delta magnitude: It is the numerical subtraction between primary and secondary magnitude values.
- Telescope aperture: The diameter of a telescope is the most important numerical parameter of these observing instruments because it determines both their maximum theoretical angular resolution and their limiting magnitude. As stated several paragraphs above, the usual practical observing resolution limit for a telescope placed on Earth is about one arcsecond. Observatory first class instruments such as the 10.4 m GRANTECAN Telescope in Canary Islands can observe celestial objects as faint as magnitude 25. The Hubble space telescope, entirely free from atmospheric limitations, is able to reach magnitude 31, although it must be said that observations made with these telescopes are made with electronic detectors.
- Delta spectral: We could extend pages about the concept of spectral class in stars but suffice is to know that when we speak about spectral class we are referring mainly to the temperature of the atmosphere of a star, that is, the temperature it has in its external surface. From the hottest to the colder stars astronomers have created a crisp classification based on capital letters: O, B, A, F, G, K, M (every astronomy student knows the mnemonics to learn it: "Oh, Be A Fine Girl, Kiss Me"). With the discovery of even colder, Brown Dwarfs stars the International Astronomical Union, IAI, has extended this classification to the letters L, T (the students have extended the mnemonics, too: "Late Tonight") but we shall not include them in our FRBS model. Since this is a crisp classification and Astronomy is a very traditional science, the observation of a continuum of spectral classes in stars had brought a division of ten subclasses for every spectral class, numerated from 0, the hottest star in the subclass, to 9, the colder one. As an example, the G spectral class is divided into G0, G1, G2, G3, G4, G5, G6, G7, G8 and G9 (our Sun is a G2 star, by the way). The hotter a star is, the bluer it seems to our eyes while on the other hand the colder a star is, the redder it appears rendered in the sky. In other words: O and B stars appears blue to us, while K and M stars seem orange-red and red to our eyes, depending also on the color sensitivity of the observer. With "Delta spectral" we mean the numerical distance in spectral classes between the components of a double star.

For example if the main component is B1 and the companion is B8, its Delta spectral, B8–B1 equals to 7.

- Theoretical DI (Difficulty Index), TDI: When using an optical telescope for observing double stars, some of them are more easy to split at the eyepiece into their components that others. Two numerical quantities describe mainly the degree of difficulty to visually get the split: angular separation and delta magnitude. Fuzzy logic has proven to be a valuable tool for assessing such a difficulty by means of a number in the closed interval [0–100] (Argüelles 2011). As we can see in Fig. 8.10, the theoretical DI builds on two previous parameters. In the Computational Theory of Perceptions (Zadeh 1999, 2001) such an element from the architecture shown is called a second order perception. All the previous parameters, which can be obtained directly by means of measurable data, are called first order perceptions. This and the next parameters are called second order perceptions.
- Detector capabilities: The Theoretical DI for a double star is intrinsic to the system and it does not dependent on the used observing instrument. Depending on the telescope's aperture and how dim is the secondary star in the pair, a factor K is obtained from a dedicated FRBS.
- Corrected DI, CDI: This parameter is obtained by the simple equation *Corrected DI = K x Theoretical DI*. This corrected DI has into account the aperture of the used telescope in the observation of a given double star.
- Suitable telescope aperture: This parameter is the output result of a dedicated FRBS that takes into account (inputs) both the Theoretical DI and the Detector capabilities. It will give the user a recommendation about a suitable size of telescope to observe a given double star.
- Theoretical beauty: Build upon the primary magnitude and delta spectral of the celestial objects under study, the theoretical, intrinsic beauty of a double star is obtained by another FRBS using the expert knowledge of seasoned double star observers.
- Sight perception (final sight): This is obtained by the last FRBS in the model, combining Theoretical beauty and Corrected Difficulty Index.

As we can appreciate, all the architecture shown in Fig. 8.10 can be described as a dance of numerical values that starts with first order perceptions and evolves into second order perceptions by means of several FRBS. After all the computations are made then it will be time for translating numbers into natural language descriptions.

8.3.1 Generating Suitable Linguistic Variables

Angular separation is the first parameter that comes to mind when speaking about double stars. In our model we deal with a numerical range from 1 to 100 arcseconds. When the angular separation between two stars in a double is big, it does not matter a difference of one or 2 arcseconds for getting an easy split at the eyepiece,

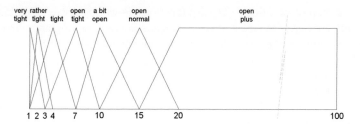

Fig. 8.11 Graphical representation of the linguistic variable "separation"

but when dealing with tight pairs, a difference of only 0.1 arcseconds can become critical. Wilhelm Struve, one of the greatest double stars observers from all times, experienced this phenomenon while making a classification for doubles in his *Catalogus Novus* (1827) depending on separation: Type I for doubles with less than 4″ separation, Type II for doubles between 4″ and 8″, Type III between 8″ and 16″ and Type IV for separations between 16″ and 32″. He realized the problems arising for such a sharp classification and again divided Type I into *Vicinae*, *Pervicinae* and *Vicinissimae*. This is, ultimately, the reason because the linguistic variable representing angular separation "compresses" its fuzzy sets towards left, as shown in Fig. 8.11.

Expressed into Lisp, this linguistic variable is expressed by Code 8-5a:

```
;code 8-5a: angular separation in double stars
(setq s1 '(very-tight 1.0 1.0 1.0 3.0))
(setq s2 '(rather-tight 1.0 2.0 2.0 4.0))
(setq s3 '(tight 1.0 4.0 4.0 7.0))
(setq s4 '(open-tight 3.0 7.0 7.0 10.0))
(setq s5 '(a-bit-open 7.0 10.0 10.0 15.0))
(setq s6 '(open-normal 10.0 15.0 15.0 20.0))
(setq s7 '(open-plus 15.0 20.0 100.0 100.0))
(setq separation '(s1 s2 s3 s4 s5 s6 s7))
```

An important remark on the used Lisp structure for representing fuzzy sets must be made now. Maybe it has been an itch for the reader from chapter five of this book. Let us have a look at the first fuzzy set in Code 8-5a:

```
(setq s1 '(very-tight 1.0 1.0 1.0 3.0))
```

The question is: Why not to use the following, simpler structure:

```
(setq s1 '(1.0 1.0 1.0 3.0))
```

After all, it would have allowed a more compact and efficient set of Lisp expressions not only for representing fuzzy sets and linguistic variables, but also an

easier work on programming FuzzyLisp functions. Powerful answer: Flexibility and
adaptability are one of the most important factors not only in programming but also
in Artificial Intelligence. At first sight, the symbol `very-tight` seems super-
fluous, but if we reflect a bit, we shall soon realize that as a symbol, it can point to
practically anything in Lisp. It can point to an atom, a list, a list of lists, a file, an
image…, you name it. Describing fuzzy sets in this form we always have a handle
available for extending their functionality. Since we are now interested in linguistic
descriptions of double star maybe it will not be a bad idea that these symbols point
to string of text, as shown in Code 8-5b:

```
;code 8-5b
;associated linguistic descriptions to fuzzy sets in
separation:
(setq very-tight "very tight")
(setq rather-tight "rather-tight")
(setq tight "tight")
(setq open-tight "open-tight")
(setq a-bit-open "bit open")
(setq open-normal "open")
(setq open-plus "really open")
```

In this way, every fuzzy set from the linguistic variable "separation" not only
stores a membership function, but also it has associated a textual description. This
will be the key for speech synthesis later in this chapter.

The linguistic variable "delta-magnitude" is formed by four fuzzy sets, as shown
in Fig. 8.12.

Translated into Lisp, its fuzzy sets and the associated linguistic descriptions are
shown in Code 8-5c:

```
;code 8-5c: delta magnitude in double stars:
(setq d1 '(very-small 0.0 0.0 0.0 1.0))
(setq d2 '(medium 0.0 1.25 1.25 2.5))
(setq d3 '(rather-big 1.0 2.5 2.5 4.0))
(setq d4 '(very-big 2.5 5.0 9.0 9.0))
```

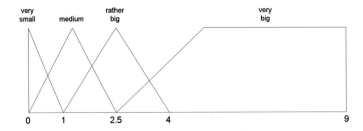

Fig. 8.12 Graphical representation of the linguistic variable "delta-magnitude"

```
(setq delta-m '(d1 d2 d3 d3 d4))
;associated linguistic descriptions:
(setq very-small "very small")
(setq medium "small")
(setq rather-big "medium")
(setq very-big "rather big")
```

Now, for obtaining the Theoretical Difficulty Index, we need first a suitable set of singletons, shown in Code 8-5d:

```
;code 8-5d: output singletons for calculating Theoretical
DI
(setq actions-di '(
      (very-easy 0.0)
      (rather-easy-plus 10.0)
      (rather-easy 20.0)
      (something-easy 40.0)
      (something-difficult 60.0)
      (rather-difficult 75.0)
      (very-difficult 100.0))
)
```

And finally, Code 8-5e shows the expert rules to complete the knowledge base for the Theoretical Difficulty Index:

```
;code 8-5e: Expert rules
(setq rules-separation '((separation delta-m actions-di)
      (very-tight very-small rather-difficult AND-product)
      (very-tight medium very-difficult AND-product)
      (very-tight rather-big very-difficult AND-product)
      (very-tight very-big very-difficult AND-product)

      (rather-tight very-small something-difficult AND-
      product)
      (rather-tight medium rather-difficult AND-product)
      (rather-tight rather-big very-difficult AND-product)
      (rather-tight very-big very-difficult AND-product)

      (tight very-small something-easy AND-product)
      (tight medium something-difficult AND-product)
      (tight rather-big rather-difficult AND-product)
      (tight very-big very-difficult AND-product)

      (open-tight very-small rather-easy AND-product)
      (open-tight medium something-easy AND-product)
```

```
(open-tight   rather-big   something-difficult   AND-
product)
(open-tight very-big rather-difficult AND-product)

(a-bit-open very-small very-easy AND-product)
(a-bit-open medium rather-easy AND-product)
(a-bit-open rather-big something-easy AND-product)
(a-bit-open very-big something-difficult AND-product)

(open-normal very-small very-easy AND-product)
(open-normal medium very-easy AND-product)
(open-normal rather-big rather-easy AND-product)
(open-normal very-big rather-easy AND-product)

(open-plus very-small very-easy AND-product)
(open-plus medium very-easy AND-product)
(open-plus rather-big very-easy AND-product)
(open-plus very-big rather-easy-plus AND-product))
)
```

Trying this FRBS at the Lisp prompt is easy. For the double star Castor, *Alpha Geminorum*, with separation $d = 2''$, primary magnitude $Mv_1 = 2.5$ and secondary magnitude $Mv_2 = 3.5$ we shall have:

> (fl-inference rules-separation 2 1)
: 79.09

This resulting value indicates us that a double star with these parameters for angular separation and difference of apparent magnitudes is a bit hard to split at the eyepiece.

Before describing the rest of linguistic variables of the complete model for speech synthesis we should comment that when complexity in programming increases, "divide and conquer" is an excellent strategy. It is thus suggested to split the knowledge database and the final FuzzyLisp application into different files, as suggested in Fig. 8.13.

Fig. 8.13 Software engineering structure for dealing with complex models

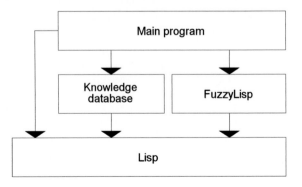

As an example, we can name as "dstar-knowledge-db.lsp" a file where all the definitions of fuzzy sets, associated linguistic descriptions, linguistic variables and expert rules are stored and then have another file named, for example, "speech-dstar-main.lsp" where all the inferences and user interface will be implemented. Aside the comments, the first two lines of such a file should be the following ones:

```
;speech-dstar-main.lsp: A program for speech synthesis on
;double stars
(load "fuzzylisp.lsp")
(load "dstar-knowledge-db.lsp")
```

In this way, the main program can make direct calls to Lisp, to the knowledge database or to FuzzyLisp functions. Naturally, Lisp is always at the foundation base of the application, but using this structure the code is simpler to write and understand, resulting into less programming errors and thus becoming a better programming approach. After having said this, we shall continue describing the knowledge database for double stars. Now it is the turn for the FRBS that describes the Detector Capabilities of the optical system.

When we speak about "detector capabilities" we are referring to the sensibility of the human retina to light. If the secondary component of a double star is faint and the telescope's aperture is small then a given double star will be always hard to split at the eyepiece, especially if the separation between the components of a double star is small. Thus, we design another FRBS to obtain a real number K in the range [1,3]. This K value will be later multiplied to the already obtained Theoretical Difficulty Index, TDI, resulting into the so named "Corrected Difficulty Index", CDI. Let us go with the linguistic variables for this FRBS. Figure 8.14 shows the linguistic variable "secondary magnitude" describing the bright of the fainter component in a double star.

Code 8-6a shows the translation to Lisp code of this linguistic variable and the linguistic descriptions associated to its fuzz sets:

```
;code 8-6a: secondary-magnitude LV
(setq m21 '(very-bright -1.5 -1.5 -1.5 1.375))
(setq m22 '(bright -1.5 1.375 1.375 4.25))
```

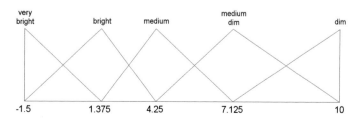

Fig. 8.14 Graphical representation of the linguistic variable "secondary-magnitude"

```
(setq m23 '(medium 1.375 4.25 4.25 7.125))
(setq m24 '(medium-dim 4.25 7.125 7.125 10.0))
(setq m25 '(dim 7.125 10.0 10.0 10.0))
(setq secondary-magnitude '(m21 m22 m23 m24 m25))
;associated linguistic descriptions to fuzzy sets in
;secondary-magnitude:
(setq very-bright "very bright")
(setq bright "bright or rather bright")
(setq medium "medium bright")
(setq medium-dim "dim")
(setq dim "rather or very dim")
```

The second linguistic variable in this FRBS is the aperture of the used telescope, expressed by its optics diameter (lens or mirror) in inches, the usual unit for expressing aperture in observational astronomy. Figure 8.15 shows it.

Code 8-6b expresses this in Lisp language. Do note that for the fuzzy sets of this linguistic variable there are no associated linguistic descriptions. Instead we shall need descriptions for recommended telescope aperture later.

```
;code 8-6b: telescope aperture LV
(setq ap1 '(a-small 2.0 2.0 2.0 4.0))
(setq ap2 '(a-medium 2.0 4.0 4.0 9.0))
(setq ap3 '(a-large 4.0 9.0 12.0 12.0))
(setq aperture '(ap1 ap2 ap3))
```

Since from this FRBS we desire to obtain a real value K in the closed interval [0–3] we shall design the output singletons accordingly, as shown in Code 8-6c:

```
;code    8-6c:    Singletons    for    calculating    detector
capability
;(expressed by a real value K)
(setq singletons-di-plus '(
     (s-low 1.0)
     (s-almost-low 1.15)
     (s-medium-minus 1.25)
```

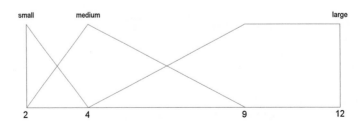

Fig. 8.15 Graphical representation of the linguistic variable "aperture"

```
                (s-medium 1.35)
                (s-medium-high 2.0)
                (s-high 3.0))
    )
```

the resulting expert rules are shown in Code 8-6d:

```
;code 8-6d: Expert rules
(setq rules-di-plus
            '((secondary-magnitude aperture singletons-
               di-plus)
    (very-bright a-small s-low AND-product)
    (very-bright a-medium s-low AND-product)
    (very-bright a-large s-low AND-product)

    (bright a-small s-almost-low AND-product)
    (bright a-medium s-low AND-product)
    (bright a-large s-low AND-product)

    (medium a-small s-medium AND-product)
    (medium a-medium s-low AND-product)
    (medium a-large s-low AND-product)

    (medium-dim a-small s-medium-high AND-product)
    (medium-dim a-medium s-medium-minus AND-product)
    (medium-dim a-large s-low AND-product)

    (dim a-small s-high AND-product)
    (dim a-medium s-medium AND-product)
    (dim a-large s-almost-low AND-product))
    )
```

Now, let us try this FRBS at the Lisp prompt. If we use a four inches aperture telescope for observing Castor ($Mv_2 = 3.5$), then we shall have:

> (fl-inference rules-di-plus 3.5 4)
: 1.0

When testing the previous FRBS, we obtained a TDI = 79.09 for the double star Castor. Taking into account the K = 1.0 value now obtained, the Corrected Difficulty Index will obviously be CDI = 79.09 × 1.0 = 79.09. The reason for maintaining the difficulty Index in this case is because Castor is a rather bright double star, so the detection capability of our retina is more than enough to observe Castor with a four inches aperture telescope. However, if we observe Castor through a 2.5 in. aperture telescope, things would change:

> **(fl-inference rules-di-plus 3.5 2.5)**
: 1.22

And then, with K = 1.22, the obtained CDI will be CDI = 79.09 × 1.22 = 96.76, showing that under such circumstances Castor becomes a lot harder to split because a 2.5 in. telescope gathers less stellar light (dimmer resulting image) and has less resolution than a four inches telescope.

For those cases where the resulting CDI exceeds 100.0 we shall use the following algorithm expressed in pseudo-code:

```
(calculate CDI = TDI x K)
(if obtained CDI > 100 then CDI = 100)
```

This algorithm is, as we know, certainly easy to implement, and we shall do it in the main program of this project.

For obtaining the Theoretical Beauty we shall use as inputs the Primary's magnitude, that is, the apparent visual magnitude of the brightest star in the pair, and its Delta Spectral, that is, the numerical distance between spectral lasses, as discussed in the previous section. It happens that the more contrast in coloration, that is, the bigger its delta spectral between components a double star has, the more beauty is perceived for an observer (Adler 2006). It also happens that when delta spectral is very small when the components are both bright, the resulting beauty at the telescope's eyepiece is easy to appreciate.

Translating into Lisp the concept of Primary's magnitude can't be easier. We only need a line of code for it:

```
;code 8-7a: primary magnitude LV
(setq primary-magnitude secondary-magnitude)
```

That is, we shall use the same fuzzy sets and associated linguistic descriptions as expressed by Code 8-6a (Fig. 8.14), assigning them to the symbol `primary-magnitude`. On the other hand, the linguistic variable "delta-spectral" contains five fuzzy sets defined on an universe of discourse ranging from 0 to 70, being equivalent to the numerical distance between the spectral class O0 and M9. Figure 8.16 shows graphically this linguistic variable.

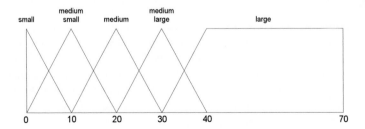

Fig. 8.16 Graphical representation of the linguistic variable "delta-spectral"

The Lisp representation of this linguistic variable is straightforward, as shown in Code 8-7b:

```
;code 8-7b: delta-spectral LV
(setq ds1 '(ds-small 0.0 0.0 0.0 10.0))
(setq ds2 '(ds-medium-small 0.0 10.0 10.0 20.0))
(setq ds3 '(ds-medium 10.0 20.0 20.0 30.0))
(setq ds4 '(ds-medium-large 20.0 30.0 30.0 40.0))
(setq ds5 '(ds-large 30.0 40.0 70.0 70.0))
(setq delta-spectral '(ds1 ds2 ds3 ds4 ds5))
;associated linguistic descriptions to fuzzy sets
;in delta-spectral:
(setq ds-small "virtually not existent")
(setq ds-medium-small "not easy at all to observe")
(setq ds-medium "neither easy not hard to observe")
(setq ds-medium-large "rather easy to observe")
(setq ds-large "easily appreciated")
```

We shall create seven output singletons for describing the Theoretical Beauty of the pair, expressed in the closed interval [0–100] as can be seen in Code 8-7c:

```
;code 8-7c: Singletons for calculating Theoretical Beauty
(setq singletons-Tbeauty '(
     (stb-top-class 100.0)
     (stb-high-plus 80.0)
      (stb-high 75.0)
     (stb-normal 50.0)
     (stb-low 25.0)
     (stb-uninteresting 0.0))
)
```

and now, the adequate expert rules for this FRBS are shown in Code 8-7d:

```
;code 8-7d: Expert rules
(setq rules-Tbeauty '((primary-magnitude delta-spectral
singletons-Tbeauty)
     (very-bright ds-small stb-top-class AND-product)
     (very-bright ds-medium-small stb-top-class AND-
     product)
     (very-bright ds-medium stb-high AND-product)
     (very-bright ds-medium-large stb-top-class AND-
     product)
     (very-bright ds-large stb-top-class AND-product)
```

```
(bright ds-small stb-top-class AND-product)
(bright ds-medium-small stb-high AND-product)
(bright ds-medium stb-normal AND-product)
(bright ds-medium-large stb-high-plus AND-product)
(bright ds-large stb-top-class AND-product)

(medium ds-small stb-high AND-product)
(medium ds-medium-small stb-normal AND-product)
(medium ds-medium stb-normal AND-product)
(medium ds-medium-large stb-top-class AND-product)
(medium ds-large stb-high-plus AND-product)

(medium-dim ds-small stb-normal AND-product)
(medium-dim ds-medium-small stb-low AND-product)
(medium-dim ds-medium stb-low AND-product)
(medium-dim ds-medium-large stb-uninteresting AND-
product)
(medium-dim ds-large stb-uninteresting AND-product)

(dim ds-small stb-low AND-product)
(dim ds-medium-small stb-low AND-product)
(dim ds-medium stb-uninteresting AND-product)
(dim ds-medium-large stb-uninteresting AND-product)
(dim ds-large stb-uninteresting AND-product))
)
```

Following our example with Castor, its Theoretical Beauty will be obtained with the following expression at the Lisp prompt:

> **(fl-inference rules-Tbeauty 2.5 1)**
: *87.72*

Observing again Fig. 8.10 we shall realize that there is only one FRBS to go. This one takes as inputs the obtained value of theoretical-beauty and the CDI value. The design of this FRBS is based on the idea that the theoretical-beauty value cannot be improved. That is, depending on the Difficulty Index it only can be worsened. If a double star is easy to observe the resulting Sight Perception will be the same as its theoretical-beauty value. On the other hand, if the CDI is high then the final Sight Perception will have a lower value than its corresponding theoretical-beauty. With these ideas in mind, we can create a linguistic variable for representing the theoretical-beauty, as shown in Fig. 8.17.

As usually, translating this arrangement of fuzzy sets into Lisp is easy, as shown in Code 8-8a:

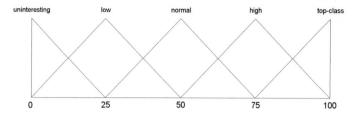

Fig. 8.17 Graphical representation of the linguistic variable "theoretical-beauty"

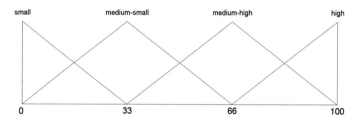

Fig. 8.18 Graphical representation of the linguistic variable "corrected DI"

```
;code 8-8a: theoretical-beauty LV
(setq tb1 '(tb-uninteresting 0.0 0.0 0.0 25.0))
(setq tb2 '(tb-low 0.0 25.0 25.0 50.0))
(setq tb3 '(tb-normal 25.0 50.0 50.0 75.0))
(setq tb4 '(tb-high 50.0 75.0 75.0 100.0))
(setq tb5 '(tb-top-class 75.0 100.0 100.0 100.0))
(setq theoretical-beauty '(tb1 tb2 tb3 tb4 tb5))
```

The other input linguistic variable, Corrected DI (CDI) can be modeled in an identical fashion, as expressed by Fig. 8.18.

Again, translating this LV into Lisp is immediate, as shown in Code 8-8b:

```
;code 8-8b: corrected-DI LV
(setq di1 '(di-small 0.0 0.0 0.0 33.0))
(setq di2 '(di-medium-small 0.0 33.0 33.0 66.0))
(setq di3 '(di-medium-high 33.0 66.0 66.0 100.0))
(setq di4 '(di-high 66.0 100.0 100.0 100.0))
(setq corrected-DI '(di1 di2 di3 di4))
```

Now we only need to create a suitable set of singletons for representing the intended output, that is, the sight-perception value, expressed also in the closed interval [0–100]. Code 8-8c shows it:

```
;code 8-8c: Singletons for calculating Sight Perception
(setq singletons-Sight '(
```

```
        (sgh-top-class 100.0)
        (sgh-high 75.0)
        (sgh-normal 50.0)
        (sgh-low 25.0)
        (sgh-uninteresting 0.0))
)
```

And finally, the rules for relating theoretical-beauty, CDI and sight perception are described in Code 8-8d:

```
;code 8-8d: Expert rules
(setq  rules-Sight  '((theoretical-beauty  corrected-DI
singletons-Sight)
        (tb-uninteresting di-small sgh-uninteresting AND-
        product)
        (tb-uninteresting di-medium-small sgh-uninteresting
        AND-product)
        (tb-uninteresting di-medium-high sgh-uninteresting
        AND-product)
        (tb-uninteresting  di-high  sgh-uninteresting  AND-
        product)

        (tb-low di-small sgh-low AND-product)
        (tb-low di-medium-small sgh-low AND-product)
        (tb-low di-medium-high sgh-uninteresting AND-product)
        (tb-low di-high sgh-uninteresting AND-product)

        (tb-normal di-small sgh-normal AND-product)
        (tb-normal di-medium-small sgh-normal AND-product)
        (tb-normal di-medium-high sgh-low AND-product)
        (tb-normal di-high sgh-low AND-product)

        (tb-high di-small sgh-high AND-product)
        (tb-high di-medium-small sgh-high AND-product)
        (tb-high di-medium-high sgh-normal AND-product)
        (tb-high di-high sgh-normal AND-product)

        (tb-top-class di-small sgh-top-class AND-product)
        (tb-top-class  di-medium-small  sgh-top-class  AND-
        product)
        (tb-top-class di-medium-high sgh-high AND-product)
        (tb-top-class di-high sgh-high AND-product))
)
```

Table 8.2 Summarized results of theoretical difficult index, corrected difficulty index, theoretical beauty and sight perception for castor (Alpha Geminorum)

Theoretical DI	Corrected DI	Theoretical beauty	Sight perception
79.09	79.09	87.72	62.72

For testing this last FRBS, and for obtaining the final sight perception we shall use the obtained values of theoretical beauty = 87.72 and CDI = 79.09 at the Lisp prompt:

> **(fl-inference rules-Sight 87.72 79.09)**
: 62.72

Table 8.2 summarizes the obtained numerical values for the second order perceptions in our model.

8.3.2   Managing Linguistic Expressions

After obtaining the numerical values for TDI, CDI, Theoretical beauty and Sight perception it is time to develop the code for speech synthesis, taking numerical data as input and obtaining text string data as output. The key function to cross the bridge between numbers and text strings is named *(extract-description)* and is shown in Code 8-9:

```
;code 8-9
  (define (extract-description lv x,
            a-list n i location mu-value sub-list)
    (setq a-list (fl-lv-membership2? lv x))
    (setq n (length a-list) i 0 location 0 mu-value 0.0)
    (list n i)

    (while (< i n)
            (setq sub-list (nth i a-list));extracts sublist
            (if (> (last sub-list) mu-value)
            (begin
                    (setq mu-value (last sub-list))
                    (setq location i))
            );end if
            (++ i)
    );while end
    (eval (first (nth location a-list)))
  )
```

This function takes a linguistic variable and a crisp value belonging to its universe of discourse as inputs and after the internal processing is done it returns the linguistic description (text string) associated to the fuzzy set with maximum membership degree for such x value. Let us try an example at the Lisp prompt in order to observe how it works in practice:

> **(extract-description separation 2.7)**
: "rather-tight"

While not a especially complex function by itself, it is one of the key components in the strategy for this fuzzy based model of speech synthesis. Now in Fig. 8.19 we can observe, selected with a thick vertical bar located at right in the selected boxes, what linguistic variables we are interested in for generating the final text report.

Almost all the required LVs have been already coded into Lisp with three exceptions: The difficulty to split a double star (expressed from its Corrected DI), another variable for commenting a suitable telescope (when using the linguistic variable Telescope aperture" we were referring to the telescope we actually use, the model will tell us a telescope recommendation for observing a given pair), and finally, another linguistic variable for describing the perceived final sight at the eyepiece. Code 8-10 shows the required Lisp code:

```
;code 8-10: Additional linguistic variables:
;LV-exp-diff is for expressing how hard is to split a double
star:
```

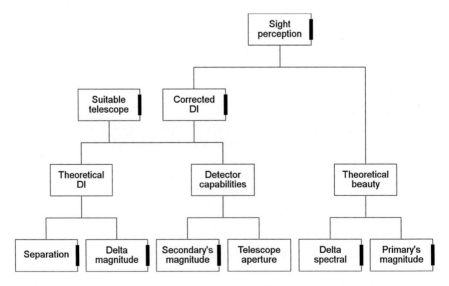

Fig. 8.19 Linguistic variables involved in the speech synthesis (those with a *vertical bar* at their right)

```
(setq exp-diff1 '(expd-very-easy 0.0 0.0 0.0 25.0))
(setq exp-diff2 '(expd-easy 0.0 25.0 25.0 50.0))
(setq exp-diff3 '(expd-not-hard 25.0 50.0 50.0 75.0))
(setq exp-diff4 '(expd-moderate 50.0 75.0 75.0 100.0))
(setq exp-diff5 '(expd-hard 75.0 100.0 100.0 100.0))
(setq LV-exp-diff
      '(exp-diff1 exp-diff2 exp-diff3 exp-diff4 exp-diff5))
;associated linguistic descriptions to fuzzy sets in
LV-exp-diff:
(setq expd-very-easy "very easily")
(setq expd-easy "easily")
(setq expd-not-hard "not especially hard to")
(setq expd-moderate "hard to")
(setq expd-hard "hardly")


;LV-exp-telescope is for expressing a suitable telescope:
(setq exp-tel1 '(expt-very-easy 0.0 0.0 0.0 25.0))
(setq exp-tel2 '(expt-easy 0.0 25.0 25.0 50.0))
(setq exp-tel3 '(expt-not-hard 25.0 50.0 50.0 75.0))
(setq exp-tel4 '(expt-moderate 50.0 75.0 75.0 100.0))
(setq exp-tel5 '(expt-hard 75.0 100.0 100.0 100.0))
(setq LV-exp-telescope
      '(exp-tel1 exp-tel2 exp-tel3 exp-tel4 exp-tel5))
;associated linguistic descriptions to fuzzy sets in
LV-exp-diff:
(setq expt-very-easy "any observing instrument")
(setq expt-easy "small telescopes")
(setq expt-not-hard "small and medium-size telescopes")
(setq expt-moderate "good telescopes under good seeing
skies")
(setq expt-hard "only first-class instruments under ex-
cepcional seeing")


;LV-final-sight is for expressing the final observed sight
at the eyepiece:
(setq fsight1 '(boring 0.0 0.0 0.0 20.0))
(setq fsight2 '(boring-plus 0.0 20.0 20.0 40.0))
(setq fsight3 '(snormal 20.0 40.0 40.0 60.0))
(setq fsight4 '(snormal-plus 40.0 60.0 60.0 80.0))
(setq fsight5 '(good 60.0 80.0 80.0 100.0))
(setq fsight6 '(very-good 80.0 100.0 100.0 100.0))
(setq LV-final-sight
      '(fsight1 fsight2 fsight3 fsight4 fsight5 fsight6))
;associated linguistic descriptions to fuzzy sets in
LV-final-sight:
```

```
(setq boring "a frankly uninteresting double")
(setq boring-plus "a not especially interesting pair")
(setq snormal "a rather normal double")
(setq snormal-plus "a beautiful pair")
(setq good "a top-class double")
(setq very-good "a jewel in the sky")
```

Now, for example:

> **(extract-description LV-final-sight 87)**
: *"a top-class double"*

We are certainly in the right track for the intended speech synthesis model, but there is still an important detail that we must solve: Spectral classes are given in a combination of letters and numbers, such as B4 or G2 and we require a numerical value for the "Delta spectral" component in the model. For solving this question, we shall first enumerate all the possible spectral classes, as shown in Code 8-11a:

```
;code 8-11a: Lisp symbol for representing spectral classes:
(setq s-classes '(
"O0" "O1" "O2" "O3" "O4" "O5" "O6" "O7" "O8" "O9" "B0" "B1" "B2" "B3" "B4"
"B5" "B6" "B7" "B8" "B9" "A0" "A1" "A2" "A3" "A4" "A5" "A6" "A7" "A8" "A9"
"F0" "F1" "F2" "F3" "F4" "F5" "F6" "F7" "F8" "F9" "G0" "G1" "G2" "G3" "G4"
"G5" "G6" "G7" "G8" "G9" "K0" "K1" "K2" "K3" "K4" "K5" "K6" "K7" "K8" "K9"
"M0" "M1" "M2" "M3" "M4" "M5" "M6" "M7" "M8" "M9") )
```

and now, Code 8-11b shows a function named *(find-class-diff)* that takes two spectral classes expressed as strings and then returns its numerical distance:

```
;code 8-11b: returns numerical distance between two spec-
tral classes
   (define (find-class-diff class1 class2, p q)
     (setq p (find class1 s-classes))
     (setq q (find class2 s-classes))
     (abs (- p q))
)
```

As can be seen, this is just another simple, yet powerful function. Let us try it:

> **(find-class-diff "A2" "B3")**
: *9*

Almost all the ingredients for the final model are now ready. We still need an user input function for asking the required numerical dataset. Function *(data-in)* in Code 8-12 takes charge of it:

```
;code 8-12: Entering input data from the keyboard
(define (data-in, star-name d m1 m2 sp1 sp2 a)
(print "Name of the star: ")
            (setq star-name (read-line))
(print "Separation between components: ")
            (setq d (float (read-line)))
(print "Magnitude first component: ")
            (setq m1 (float (read-line)))
(print "Magnitude second component: ")
            (setq m2 (float (read-line)))
(print "Spectral class first component: ")
            (setq sp1 (read-line))
(print "Spectral class second component: ")
            (setq sp2 (read-line))
(print "Telescope aperture in inches: ")
            (setq a (float (read-line)))

    (list star-name d m1 m2 sp1 sp2 a)
)
```

This function is so simple that it does not need further explanation. Only to comment that it returns a list formed by the data gathered in the input process.

Now it is time to write another function for running all the FRBS in the model at a time. This function, named *(run-the-FRBS)* is shown in Code 8-13:

```
;code 8-13: running all the FRBS in the model
(define (run-the-FRBS, data-list sp-diff delta-mag DI K
DI-OK TB)
    (setq data-list (data-in))
    (setq sp-diff (find-class-diff (nth 4 data-list)
            (nth 5 data-list)))
    (setq delta-mag (sub (nth 3 data-list) (nth 2 data-
            list)))

    ;calculate Difficulty Index:
    (setq DI (fl-inference rules-separation (nth 1 data-
    list)
      delta-mag))

    ;calculate K factor:
    (setq K (fl-inference rules-di-plus (nth 3 data-list)
            (nth 6 data-list)))

    ;calculate corrected DI:
    (setq DI-OK (mul DI K))
```

```
   (if (> DI-OK 100.0)
        (setq DI-OK 100.0)
   )
   ;calculate theoretical Beauty:
   (setq TB (fl-inference rules-Tbeauty (nth 2 data-list)
   sp-diff))

   ;calculate final Beauty (Sight):
   (setq FB (fl-inference rules-Sight TB DI-OK))

   ;return numerical results as a list:
   (append data-list (list DI DI-OK TB FB))
)
```

The function starts calling *(data-in)* and then returns a list formed by all the required information for start the process of pure speech synthesis. The output is a list whose elements are structured in the following way: Name of the double star, separation, primary's magnitude, secondary's magnitude, spectral class of primary star, spectral class of secondary, used telescope aperture, TDI and CDI. Let us try it!

> (run-the-FRBS)
: Name of the star: Castor
Separation between components: 2
Magnitude first component: 2.5
Magnitude second component: 3.5
Spectral class first component: A1
Spectral class second component: A2
Telescope aperture in inches: 4
("Castor" 2 2.5 3.5 "A1" "A2" 4 79.09090909 79.09090909 87.7173913 62.7173913)

Now we are only one step away from finishing our speech synthesis model. The function (print-report) puts it all together, calling *(run-the-FRBS)* and then making successive calls to *(extract-description)* in order to finally compose the output linguistic report. Code 8-14 shows this main function:

```
;code 8-14: main function
(define (print-report,
            numerical-data sep mv1 mv2 spc1 spc2 apert
            TDI CDI TB FS str1 str2 str3 str4 str5 str6
            str7 str8 str9)
   (setq numerical-data (run-the-FRBS))
   (setq str1 (nth 0 numerical-data));get star's name
   (setq sep (nth 1 numerical-data))
   (setq mv1 (nth 2 numerical-data))
   (setq mv2 (nth 3 numerical-data))
```

```
(setq spc1 (nth 4 numerical-data))
(setq spc2 (nth 5 numerical-data))
(setq apert (nth 6 numerical-data))
(setq TDI (nth 7 numerical-data))
(setq CDI (nth 8 numerical-data))
(setq TB (nth 9 numerical-data))
(setq FS (nth 10 numerical-data))

;obtain the strings:
(setq str2 (extract-description separation sep))
(setq str3 (extract-description LV-exp-diff CDI))
(setq str4 (extract-description LV-exp-telescope CDI))
(setq str5 (extract-description delta-spectral
   (find-class-diff spc1 spc2)))
(setq str6 (extract-description LV-final-sight FS))
(setq str7 (extract-description delta-m (abs (sub mv1
mv2))))
(setq str8 (extract-description primary-magnitude
mv1))
(setq str9 (extract-description secondary-magnitude
mv2))

   ;finally, concatenate the output report:
   (println str1 " is a " str2 " double star that can be " str3
          " split using " str4 ". Difference of coloration
          is " str5 " and from an aesthetic point of view
          is "
          str6 ". There is a " str7 " difference of magnitude
          between components. The primary is a " str8 "
          star, while the secondary is " str9 "."
      );end println
)
```

Now let us enjoy its working at the Lisp prompt:

> (print-report)
:Name of the star: Castor
Separation between components: 2
Magnitude first component: 2.5
Magnitude second component: 3.5
Spectral class first component: A1
Spectral class second component: A2
Telescope aperture in inches: 4
Castor is a rather-tight double star that can be hard to split using good tele-scopes under good seeing skies. Difference of coloration is virtually not existent

Table 8.3 List of selected double stars for testing the model

Star name	Constellation	Sep	Mv_1	Mv_2	Sp_1	Sp_2
Sigma Orionis	Orion	12.9	4	7.5	O9	B2
Albireo	Cygnus	34.3	3	5.5	K3	B9
O. Struve 111	Orion	2.6	6	10	A5	B5
Rigel	Orion	9.4	0	7	B8	B5
Struve 747	Orion	36	5.5	6.5	B0	B1
Mintaka	Orion	52.8	2	6.5	O9	B2
Zeta Ori	Orion	2.5	1.9	4	O9	B0
Iota Ori	Orion	11	2.8	6.9	O5	B9
Eta Ori	Orion	1.5	3.1	3.4	B1	B2
Beta Cep	Cepheus	13	3.2	7.9	B2	A2
Delta Cyg	Cygnus	2.5	2.9	6.3	B9	F1
Lambda Ori	Orion	4	3.6	5.5	O8	B0
STF 422	Taurus	7	5.9	6.8	G9	K6
Epsilon Per	Perseus	9	2.9	7.6	B0	A2
Theta 1 Ori	Orion	9	6.4	7.9	B0	B0
Algieba	Leo	4.4	2.5	3.5	K0	G7
Theta Per	Perseus	20	4.1	9.9	F7	M1
Eta Lyrae	Lyra	28	4.4	9.1	B3	B3
Castor	Gemini	2	2.5	3.5	A1	A2
Wasat	Gemini	6.3	3.5	8	F0	F0

and from an aesthetic point of view is a beautiful pair. There is a small difference of magnitude between components. The primary is a bright or rather bright star, while the secondary is medium bright.

Depending on your computer, you can copy the report into a new text document and select the "Speech" option directly from the Operating System. You will hear the report directly from your computer.

The reader can play the model with double star data from catalogues easily downloadable from the Internet or with some example data as shown in Table 8.3.

8.4 Spirometry Analysis in Medicine: Floating Singletons

Chronic Obstruction Pulmonary Disease (COPD) is a severe condition caused by long-term exposure to lung irritants such as dust or micro granular material. Since many activities in industrial processes in quarries and other extractive industries generate micro particles (silicates, coal dust, etc.), the resulting atmospheres at many industrial facilities are aggressive to the human lungs. As a consequence, many people working in such environments end up developing COPD

(Santo-Tomas 2011). It is usual that the medical services at such companies carry out periodic revisions concerning the health status of their respective working forces. Spirometric analysis is a fundamental test in such medical revisions.

An spirometry is a clinical test where a device (spirometer) records the volume of air exhaled by a patient and plots it as a function of time, producing a curve that shows the lung function, specifically the volume and flow of air that can be exhaled. The patient inhales maximally and then exhales as rapidly and completely as possible. This technique is the most common one of the pulmonary function tests, being a suitable clinical test for detecting COPD. In fact, spirometry is the only medical test able to detect COPD years before shortness of breath develops (Hyatt et al. 2009). A typical spirometric curve can be observed in Fig. 8.20.

The most important numerical parameters obtained by means of a spirometry are the following ones:

- FVC: Forced Vital Capacity: It is the whole volume of air that can be exhaled by a patient after full inspiration in the test.
- FEV1: Forced Expiratory Volume in one second: It is the volume of air that can be exhaled in the first second of a spirometric test. Both FVC and FEV1 are measured in litres.
- TI: Tiffenau Index: It is a relationship between FEV1 and FVC values given by the following expression: $TI = 100 \times (FEV1/FVC)$

From the height, x, in metres and age, y, in years of an individual, theoretical values of FEV1 and FVC for men, in litres, can be obtained using the following expressions (Morris et al. 1971):

$$Theoretical\ FVC = 5.76x - 0.026y - 4.34 \qquad (8\text{-}1)$$

$$Theoretical\ FEV1 = 4.3x - 0.026y - 2.49 \qquad (8\text{-}2)$$

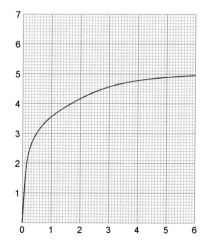

Fig. 8.20 A typical spirometry. The *horizontal axis* represents time in seconds, while the *vertical axis* represents litres of exhaled air. This spirometry show moderate COPD, with FEV1 = 3.6 l and FVC = 4.95 l (see text)

Comparing the experimental *FEV1* and *FVC* values obtained from a spirometry with the theoretical values, lung obstruction is detected applying the following, traditional, algorithm:

If ((*Observed FVC* > *80 % theoretical FVC*) and (*Observed FEV1* > *80 % theoretical FEV1*) and (*TI* > *80 %*))
        then spirometry is normal
        else the individual suffers an obstructive pulmonary disease.

Similar algorithms are given, for example, by Qaseem et al. (2011). However, expressions (8-1) and (8-2) are in fact equations representing two regression lines from a general population, thus being a "sharp" approximation for representing observed cases from the real world. Physicians know that the requirements for detecting COPD in workers exposed to ambient dust must be more strict than the simple application of the aforementioned algorithm, and test results must be evaluated in light of the patient's clinic history, physical examination and pertinent laboratory findings. Moreover, medical practice shows that smoking is a strong factor that worsens COPD. In this section of the chapter we are going to explore a sophisticated FRBS architecture that gathers the experience of physicians and also has into account the smoking factor. This architecture uses floating singletons (Argüelles 2013).

8.4.1   Introducing Floating Singletons

Basically speaking, floating singletons are a special type of singletons belonging to a first FRBS where their precise location on the real axis is only known at run-time. At design time some intervals are established in the first FRBS where these special singletons are allowed to float and later, the defuzzification of a complementary, second FRBS fixes their location. That is, a floating singleton, *sf*, has the ability to move inside an universe of discourse defined in a closed interval $[a,b]$. A set of traditional singletons s_i defines a set of $I_i = i - 1$ internal intervals in $[a,b]$ as can be observed in Fig. 8.21a, b:

We say that sf_i is a floating singleton by-left if:

$$s_{i-1} \leq sf_i \leq s_i \rightarrow sf_i \in I_{i+1} \tag{8-3}$$

with $sf_1 = s_1$. Conversely, sf_i is a floating singleton by-right if:

$$s_i \leq sf_i \leq s_{i+1} \rightarrow sf_i \in I_i \tag{8-4}$$

with $sf_n = s_n$ (see Fig. 8.21a, b). Now we can expose a more restricted definition: a floating singleton, sf_i, is a special type of singleton that has the ability to move inside its assigned interval I_i until run-time, that is, when the model is run at the

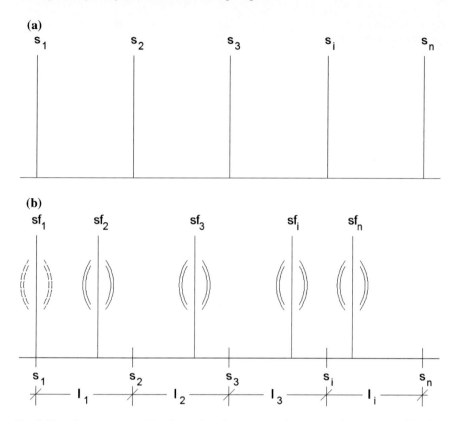

Fig. 8.21 a, b At *top*, a traditional set of singletons, s_i can be seen. At *bottom* a set of floating singletons sf_i "by-left" is shown. Note the position of traditional singletons s_i on the horizontal axis at *bottom*

computer. The floating nature of floating singletons is described by means of the existence of a second FRBS composed by fuzzy rules of the type:

$$\text{if p } is\ M_g\ and\ q\ is\ N_h\ then\ z_i = t_{i,j} \tag{8-5}$$

where $t_{i,j} \in [s_i, s_{i+1}]$ are a set of singletons $t_{i,j}$, expressed as consequents. Here we must remark that every singleton $t_{i,j}$ is defined on the interval I_i. This can be expressed graphically by means of the Fig. 8.22.

Now, the final position of every floating singleton sf_i comes from the defuzzification of the singletons t_i, expressed, as we already know, by Eq. 7-22. As can be seen, in this arrangement of models, inferences are made first for the second FRBS whose rules are given by (8-5), and then, the defuzzification of t_i is made for "fixing" the position of the floating singletons. Finally, after making the inferences for the first FRBS, the final defuzzification process is made.

Fig. 8.22 Graphical
interpretation of floating
singletons sf$_i$, and their
relationship with standard
singletons t$_{i,j}$

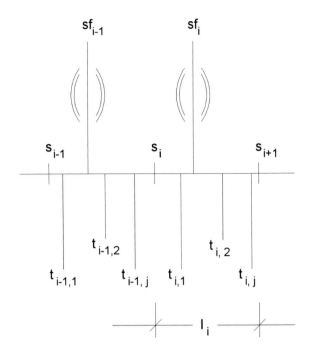

8.4.2   Modeling Pulmonary Obstruction with FuzzyLISP

For developing a model representing pulmonary obstruction after obtaining the
FVC and FEV1 values from a real spirometry, we shall first create two magnitudes,
f1 and f2, expressed by:

$$f1 = 100(FVC\,from\,spirometry)/(theoretical\,FVC) \tag{8-6}$$

$$f2 = 100(FEV1\,from\,spirometry)/(theoretical\,FEV1) \tag{8-7}$$

where the theoretical FVC and theoretical FEV1 values are obtained from
expressions (8-1) and (8-2), respectively. The following step is to create two linguistic variables, Pct-FVC and Pct-FEV1, for expressing all the possible values of
f1 and f2. These linguistic variables will allow us to create a first FRBS composed
by expert rules of the type:

$$if\,f1_i\,is\,Pct\text{-}FVC\,and\,f2_i\,is\,Pct\text{-}FEV\,then\,Obstruction\,is\,sf_i \tag{8-8}$$

The key point now is the definition of the floating singletons sf_i for describing
pulmonary Obstruction. Since at design time they "float" we do not give fixed real
numbers for them, but some intervals. Table 8.4a shows the floating singletons
involved in our model:

As can be observed in the table, these floating singletons are expressed by a set of intervals I_i. Please do note also that the last floating singleton, "Extreme" is fixed at [100], that is, we are creating a model with floating singletons by-right. In this way, an instance from rules given by (8-8) could be:

$$if\ f1_i\ is\ Severe\ and\ f2_i\ is\ Normal\ then\ Obstruction\ is\ Important[65,85] \qquad (8\text{-}9)$$

So "Important" is a floating singleton that can adopt numerical values between 65 and 85. Now, how do we get a fixed value for "Important" in such a way that we can make fuzzy logic inferences from the rules given by (8-8)? The answer comes from a second FRBS that takes into account smoking. The model will ask for the number of cigarettes/day and the number of years of smoking associated to the person passing the spirometric test. If he or she is not a smoker, both values will be null, and then Table 8.4a would convert to Table 8.4b.

Figure 8.23 shows the whole architecture of the intended model for describing pulmonary obstruction.

As can be seen, the second FRBS takes the number of smoking years and the number of cigarettes per day as crisp numerical inputs, ultimately generating an

Table 8.4a List of floating singletons for describing pulmonary obstruction

Floating singleton name	Floating interval I_i
Null	[0,15]
Appreciable	[15,65]
Important	[65,85]
Very-important	[85,100]
Extreme	[100,100]

Table 8.4b List of singletons for describing pulmonary obstruction in non smokers

Singleton name	Singleton defined at
Null	0
Appreciable	15
Important	65
Very-important	85
Extreme	100

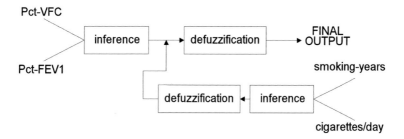

Fig. 8.23 FRBS architecture for spirometric and COPD analysis

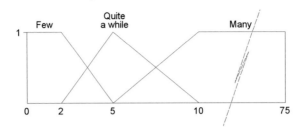

Fig. 8.24a Graphical representation of the linguistic variable "smoking-years".

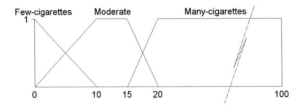

Fig. 8.24b Graphical representation of the linguistic variable "cigarettes-day"

output that converts into standard singletons, ready for being used in the defuzz-ification for the first FRBS. For representing the number of cigarettes/day and the number of years of smoking we shall create two linguistic variables named, respectively, "smoking-years" and "cigarettes-day". The former, composed by three fuzzy sets, is shown in Fig. 8.24a.

Figure 8.24b shows the graphical representation of the Linguistic Variable "cigarettes/day".

Translating these variables to Lisp is immediate, as can be appreciated in Code 8-15:

```
;code 8-15: Linguistic variables for floating
;singletons management
;smoking-years:
(setq sy1 '(Few 0.0 0.0 2.0 5.0))
(setq sy2 '(Quite-a-while 2.0 5.0 5.0 10.0))
(setq sy3 '(Many 5.0 10.0 75.0 75.0))
(setq smoking-years '(sy1 sy2 sy3))
;cigarettes-day:
(setq cd1 '(Few-cigarettes 0.0 0.0 0.0 10.0))
(setq cd2 '(Moderate 0.0 10.0 15.0 20.0))
(setq cd3 '(Many-cigarettes 15.0 20.0 100.0 100.0))
(setq cigarettes-day '(cd1 cd2 cd3))
```

Now we need to define the floating singletons as described in Table 8.4a. This is shown in code 8-16:

```
;code 8-16: Definition of all the floating singletons
 intervals
;first floating singletons interval:
(setq Sf1 '(
     (sf1-left 0.0)
     (sf1-medium 8.0)
     (sf1-right 15.0))
)

;second floating singletons interval:
(setq Sf2 '(
     (sf2-left 15.0)
     (sf2-medium 40.0)
     (sf2-right 65.0))
)

;third floating singletons interval:
(setq Sf3 '(
     (sf3-left 65.0)
     (sf3-medium 75.0)
     (sf3-right 85.0))
)

;fourth floating singletons interval:
(setq Sf4 '(
     (sf4-left 85.0)
     (sf4-medium 92.5)
     (sf4-right 100.0))
)

;fifth floating singletons interval:
(setq Sf5 '(
     (sf5-left 100.0)
     (sf5-medium 100.0)
     (sf5-right 100.0))
)
```

For each of these intervals we shall need a dedicated set of expert rules joining
the linguistic variables "smoking-years" and "cigarettes-day" in the antecedents as
follows:

$$\text{if } x_i \text{ is smoking-years } and \text{ } y_i \text{ is cigarettes-day } then \text{ singleton is } t_i \qquad (8\text{-}10)$$

In this way we shall be able to determine the floating singletons into standard
singletons. Code 8-17 shows all the required expert rules:

```
;code 8-17: Expert rules for determining the floating
singletons:
(setq rules-s1 '((smoking-years cigarettes-day Sf1)
     (Few Few-cigarettes sf1-left AND-product)
     (Few Moderate sf1-left AND-product)
     (Few Many-cigarettes sf1-medium AND-product)

     (Quite-a-while Few-cigarettes sf1-left AND-product)
     (Quite-a-while Moderate sf1-medium AND-product)
     (Quite-a-while Many-cigarettes sf1-right AND-product)

     (Many Few-cigarettes sf1-medium AND-product)
     (Many Moderate sf1-right AND-product)
     (Many Many-cigarettes sf1-right AND-product))
)

(setq rules-s2 '((smoking-years cigarettes-day Sf2)
     (Few Few-cigarettes sf2-left AND-product)
     (Few Moderate sf2-left AND-product)
     (Few Many-cigarettes sf2-medium AND-product)

     (Quite-a-while Few-cigarettes sf2-left AND-product)
     (Quite-a-while Moderate sf2-medium AND-product)
     (Quite-a-while Many-cigarettes sf2-right AND-product)

     (Many Few-cigarettes sf2-medium AND-product)
     (Many Moderate sf2-right AND-product)
     (Many Many-cigarettes sf2-right AND-product))
)

(setq rules-s3 '((smoking-years cigarettes-day Sf3)
     (Few Few-cigarettes sf3-left AND-product)
     (Few Moderate sf3-left AND-product)
     (Few Many-cigarettes sf3-medium AND-product)

     (Quite-a-while Few-cigarettes sf3-left AND-product)
     (Quite-a-while Moderate sf3-medium AND-product)
     (Quite-a-while Many-cigarettes sf3-right AND-product)

     (Many Few-cigarettes sf3-medium AND-product)
     (Many Moderate sf3-right AND-product)
     (Many Many-cigarettes sf3-right AND-product))
)
```

```
(setq rules-s4 '((smoking-years cigarettes-day Sf4)
      (Few Few-cigarettes sf4-left AND-product)
      (Few Moderate sf4-left AND-product)
      (Few Many-cigarettes sf4-medium AND-product)

      (Quite-a-while Few-cigarettes sf4-left AND-product)
      (Quite-a-while Moderate sf4-medium AND-product)
      (Quite-a-while Many-cigarettes sf4-right AND-product)

      (Many Few-cigarettes sf4-medium AND-product)
      (Many Moderate sf4-right AND-product)
      (Many Many-cigarettes sf4-right AND-product))
)

(setq rules-s5 '((smoking-years cigarettes-day Sf5)
      (Few Few-cigarettes sf5-left AND-product)
      (Few Moderate sf5-left AND-product)
      (Few Many-cigarettes sf5-medium AND-product)

      (Quite-a-while Few-cigarettes sf5-left AND-product)
      (Quite-a-while Moderate sf5-medium AND-product)
      (Quite-a-while Many-cigarettes sf5-right AND-product)

      (Many Few-cigarettes sf5-medium AND-product)
      (Many Moderate sf5-right AND-product)
      (Many Many-cigarettes sf5-right AND-product))
)
```

Let us test at the Lisp prompt all these sets of rules for a person that started to smoke ten years ago and smokes seven cigarettes every day:

> **(fl-inference rules-s1 10 7)**
: 12.9

> **(fl-inference rules-s2 10 7)**
: 57.5

> **(fl-inference rules-s3 10 7)**
: 82

> **(fl-inference rules-s4 10 7)**
: 97.75

> **(fl-inference rules-s5 10 7)**
: 100

Table 8.4c List of singletons for describing pulmonary obstruction for x = 10 smoking years and y = 7 cigarettes/day	Singleton name	Singleton defined at
	Null	12.9
	Appreciable	57.5
	Important	82
	Very-important	97.75
	Extreme	100

In this way, for x = 10 years and y = 7 cigarettes/day, the resulting singletons (we could say they have just finished their floating nature after inferences are made in the second FRBS) are shown in Table 8.4c.

Now we need to establish the linguistic variables Pct-FVC and Pct-FEV1, shown in Fig. 8.25. The fuzzy sets distribution in both of them is identical and only the linguistic labels do differ, as can be seen next, in the code, where the "fvc" and "fev" prefixes are used:

As usually, Code 8-18 put these definitions into Lisp:

```
;code 8-18, linguistic variables Pct-FVC and Pct-FEV1:
(setq fvc1 '(fvc-Severe 20.0 20.0 30.0 50.0))
(setq fvc2 '(fvc-Moderate 30.0 50.0 50.0 70.0))
(setq fvc3 '(fvc-Slight 50.0 70.0 70.0 85.0))
(setq fvc4 '(fvc-Normal 70.0 85.0 100.0 120.0))
(setq fvc5 '(fvc-Excellent 100.0 120.0 150.0 150.0))
(setq Pct-FVC '(fvc1 fvc2 fvc3 fvc4 fvc5))

(setq fev11 '(fev-Severe 20.0 20.0 30.0 50.0))
(setq fev12 '(fev-Moderate 30.0 50.0 50.0 70.0))
(setq fev13 '(fev-Slight 50.0 70.0 70.0 85.0))
(setq fev14 '(fev-Normal 70.0 85.0 100.0 120.0))
(setq fev15 '(fev-Excellent 100.0 120.0 150.0 150.0))
(setq Pct-FEV1 '(fev11 fev12 fev13 fev14 fev15))
```

In the first section of this book we learnt that Lisp is a great computer language where there is not a sharp frontier between code and data. A language that is even

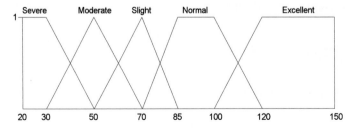

Fig. 8.25 Graphical representation of the linguistic variables "Pct-FVC" and "Pct-FEV1"

able to generate its own code. Now this is a pretty opportunity to demonstrate it. For obtaining the required inferences from the expert rules enunciated in (8-8) we now need that Lisp generate the code for describing the final singletons. It is time to introduce the most important function for our spirometric analysis model as shown in Code 8-19:

```
;code 8-19: generating code for final singletons
(define (create-floating-singletons years cgr-day,
                    s1 s2 s3 s4 s5 l1 l2 l3 l4 l5)
   (setq s5 (fl-inference rules-s5 years cgr-day))
   (setq l5 (list 'Extreme s5))
   (setq s4 (fl-inference rules-s4 years cgr-day))
   (setq l4 (list 'Very-Important s4))
   (setq s3 (fl-inference rules-s3 years cgr-day))
   (setq l3 (list 'Important s3))
   (setq s2 (fl-inference rules-s2 years cgr-day))
   (setq l2 (list 'Appreciable s2))
   (setq s1 (fl-inference rules-s1 years cgr-day))
   (setq l1 (list 'Null s1))
   (list l5 l4 l3 l2 l1)
)
```

Let us try the function *(create-floating-singletons)* using again the values x = 10 smoking years and y = 7 cigarettes/day from the previous example:

> **(setq Obstruction (create-floating-singletons 10 7))**
  : *((Extreme 100) (Very-Important 97.75) (Important 82) (Appreciable 57.5) (Null 12.9))*

Please, do appreciate how the list now stored in the symbol "Obstruction" is completely equivalent to the evaluation of the following Lisp code:

```
(setq Obstruction '(
      (Extreme 100)
      (Very-Important 97.75)
      (Important 82)
      (Appreciable 57.5)
      (Null 12.9))
)
```

Now we can finally describe in Code 8-20 the expert rules corresponding to the first FRBS, as expressed by (8-8):

```
;code 8-20
(setq rules-obstruction '((Pct-FVC Pct-FEV1 Obstruction)
      (fvc-Severe fev-Severe Extreme AND-product)
```

```
(fvc-Severe fev-Moderate Extreme AND-product)
(fvc-Severe fev-Slight Extreme AND-product)
(fvc-Severe fev-Normal Important AND-product)
(fvc-Severe fev-Excellent Important AND-product)

(fvc-Moderate fev-Severe Extreme AND-product)
(fvc-Moderate fev-Moderate Extreme AND-product)
(fvc-Moderate fev-Slight Very-Important AND-product)
(fvc-Moderate fev-Normal Important AND-product)
(fvc-Moderate fev-Excellent Important AND-product)

(fvc-Slight fev-Severe Extreme AND-product)
(fvc-Slight fev-Moderate Very-Important AND-product)
(fvc-Slight fev-Slight Appreciable AND-product)
(fvc-Slight fev-Normal Appreciable AND-product)
(fvc-Slight fev-Excellent Null AND-product)

(fvc-Normal fev-Severe Extreme AND-product)
(fvc-Normal fev-Moderate Important AND-product)
(fvc-Normal fev-Slight Appreciable AND-product)
(fvc-Normal fev-Normal Null AND-product)
(fvc-Normal fev-Excellent Null AND-product)

(fvc-Excellent fev-Severe Extreme AND-product)
(fvc-Excellent fev-Moderate Appreciable AND-product)
(fvc-Excellent fev-Slight Appreciable AND-product)
(fvc-Excellent fev-Normal Null AND-product)
(fvc-Excellent fev-Excellent Null AND-product))
)
```

All the ingredients for the design of a fuzzy-logic based model of spirometric analysis are already prepared. We now only need a final function that puts it all together. Such a function, named *(main)* incorporates a tiny user interface and is shown in Code 8-21:

```
;code 8-21. Main function for the spirometric analysis
model
(define (main, height age fvc fev1 tfvc tfev1 pctfvc pctfev1 TI
          years cigarettes Obstruction)
   (print "Height in metres: ") (setq height (float (read-
   line)))
   (print "Age in years: ") (setq age (float (read-line)))
   (print "Spirometry FVC in litres: ")
   (setq fvc (float (read-line)))
   (print "Spirometry FEV1 in litres: ")
```

```
              (setq fev1 (float (read-line)))
   (print "Years of smoking: ") (setq years (float (read-
   line)))
   (print "Number of cigarettes/day: ")
              (setq cigarettes (float (read-line)))

   ;calculate theoretical tfvc, tfev1 and then pctfvc and
   pctfev1:
   (setq tfvc (sub (sub (mul 5.76 height) (mul 0.026 age))
   4.34))
   (setq tfev1 (sub (sub (mul 4.3 height) (mul 0.026 age))
   2.49))
   (setq pctfvc (mul (div fvc tfvc) 100.0))
   (setq pctfev1 (mul (div fev1 tfev1) 100.0))
   (setq TI (mul (div fev1 fvc) 100.0))

   ;finally run the fuzzy model:
   (setq  Obstruction  (create-floating-singletons  years
   cigarettes))
   (println "Pulmonary obstruction index: "

       (fl-inference rules-obstruction pctfvc pctfev1))

       (list tfvc tfev1 pctfvc pctfev1 TI);returns partial
       results
)
```

Let us try the model at the Lisp prompt with some real data from a spirometric test for a 32 years old man, 1.80 m height and with FVC = 4.82 l, FEV1 = 4.15 l spirometric results. We assume this individual is a ten years old smoker and smokes, as an average, 7 cigarettes/day:

> (main)
: Height in metres: 1.8
Age in years: 32
Spirometry FVC in litres: 4.82
Spirometry FEV1 in litres: 4.15
Years of smoking: 10
Number of cigarettes/day: 7
Pulmonary obstruction index: 12.9
(5.196 4.418 92.76366436 93.93390675 86.09958506)

Aside obtaining a Pulmonary Obstruction Index, *(main)* returns also intermediate results from the input data as a list. In this example these are: Theoretical FVC: 5.196, Theoretical FEV1: 4.418, %FVC: 92.76, %FEV1: 93.93 and Tiffenau

Index = 86.10. Let us quote the traditional algorithm mentioned some pages before in this chapter for evaluating spirometries:

"If ((*Observed FVC > 80 % theoretical FVC*) and (*Observed FEV1 > 80 % theoretical FEV1*) and (*TI > 80 %*))
        then spirometry is normal
        else the individual suffers an obstructive pulmonary disease"

Under such algorithm the spirometry in the example is normal, but the inclusion of the smoking parameters in the fuzzy model reveals what physicians know from experience: Even obtaining good spirometric results, smokers usually show a certain degree of Chronic Obstruction Pulmonary Disease. For long-time smokers a spirometry always reveals an obstructive pulmonary disease because smoking is the first cause of COPD in the world. The fuzzy model developed in this section helps physicians to reveal COPD before it shows in a spirometric test, resulting into a Pulmonary Obstruction Index bigger than zero (12.9 in this case).

8.5 As a Summary

This is a different summary with respect to summaries in the previous chapters. Aside the introduction to floating singletons only some paragraphs above (floating singletons can be understood as an advanced topic in this work), there have not been theoretical concepts in this chapter, so this section will be short. We have learnt how to design some medium-complexity projects, including examples of simulation/control, speech synthesis and a case of expert diagnosis system in medicine.

Now I would like to seize the reader's attention in order to remember some little material from the first chapter in Sect. 1.3 when we spoke about neurons in our brains and the incredible work developed by chemical reactions in the synapses when we think. We realized that thinking can be seen as an extremely quick exchange of electrons between atoms of chemical elements and molecules of neurotransmitters in a huge number of individual chemical reactions. In a similar way, artificial intelligence maybe will be only a sophisticated and relatively simple handling of arithmetic processes happening at really high speeds in a parallel architecture of processes. In this second decade of century XXI we are still learning the basics of the game, but even now, just after having read this book, we are ready to examine the number of simple algebraic operations required to operate the models developed in this book.

With this goal in mind we only require a line of Lisp code inserted in the FuzzyLisp function (*fl-set-membership?*), the most "low-level" function in

Table 8.5 Number of calls to the function (*fl-set-membership?*) and the resulting number of simple arithmetic calls for running the fuzzy-logic based projects described in this chapter

	Moon landing	Castor	Spirometry
Function calls	27,440	900	520
Arithmetic calls	109,760	3600	2080

FuzzyLisp that, as the reader well knows, returns the membership degree of an element x in a fuzzy set X. That line of Lisp code reads as follows:

```
(setq *number-calls* (add *number-calls* 1))
```

where *number-calls* is a global variable that can be initialized at the Lisp prompt before running any of the projects developed in this chapter:

> **> (setq *number-calls* 0)**
> *: 0*

After running any of the projects, we could type *number-calls* at the Lisp prompt and the language will tell us how many times the function has been required in the computations. When the resulting membership degree is located in the open interval (0,1), the function (*fl-set-membership?*) requires two subtractions, one multiplication and one division, that is, as a superior bounding limit we can estimate that every call to this function requires four arithmetic operations. Armed with this trap we can calculate the whole number of arithmetic calculations for, (a) a Moon landing sequence from 3000 m high, (b) a linguistic report for the double star Castor, and c) the spirometry example described some paragraphs above. Results are given in Table 8.5.

The Moon landing project requires about 30 times more arithmetic calculations than the other ones because it is an iterative process at a 2 s sampling interval, but the important thing is that all the expert knowledge embedded in the fuzzy rules of the models and all the inferences made for running them are in fact reduced to a finite set of arithmetic operations.

Is this set of arithmetic operations "Artificial Intelligence"? Since Fuzzy Logic is a branch of AI we must concede that certainly it is. Now let us imagine for a moment the development of future complex artificial intelligence models based on fuzzy logic, neural networks, genetic algorithms and other AI strategies not even formulated yet. Now let us add computing power in the order of tens or hundred of millions of arithmetic operations per second concentrated in small hardware computing-units. Without doubt it will be "Artificial Intelligence" too. Let us try to see 50 years, or still better, two centuries into the future. The question evolves and now it is not a discussion about if it is or it is not Artificial Intelligence. The question, probably, will be "is Artificial Intelligence distinguishable from natural intelligence"?

Time will tell.

References

Adler, A.: Pretty double stars for everyone. In: Sky and Telescope. http://www.skyandtelescope.com/observing/objects/doublestars/3304626.html (2006). Acceded 2015

Argüelles, L.: Systemes de classification des etoiles doubles au moyen des techniques de logique floue. Obs. Trav. **78**, 2–6 (2011)

Argüelles, L.: Introducing floating singletons with an example application in clinic spirometry. In: 8th Conference of the European Society for Fuzzy Logic and Technology (EUSFLAT 2013) (2013)

Argüelles, L., Trivino, G.: I-Struve: Automatic linguistic descriptions of visual double stars. Eng. Appl. Artif. Intel. **26**(9), 2083–2092 (2013)

Argyle, R.: More than one sun. In: Observing and Measuring Double Stars. Springer, Berlin (2004)

Chen, C., Pham, T.: Introduction to Fuzzy Sets, Fuzzy Logic, and Fuzzy Control Systems. CRC Press, Boca Raton (2000)

Dabney, R.: Autonomous Rendezvous and Capture System Design. NASA (1991)

Fortescue, P., Swinered, G., Stark, J.: Spacecraft Systems Engineering. Wiley, New York (2011)

Hyatt, R., Scanlon, P., Nakamura, M.: Interpretation of Pulmonary Function Tests: A Practical Guide. Lippincott Williams & Wilkins (2009)

Klump, A.: Apollo Guidance, Navigation and Control. MIT (1971)

Morris, J.F. Koski, A., Johnson, L.C.: Spirometric standards for healthy, non-smoking adults. Am. Rev. Respir. Dis. **103**, 57–67 (1971)

NASA: Apollo Lunar Module Documentation. https://www.hq.nasa.gov/alsj/alsj-LMdocs.html. Acceded Feb 2015

Ostlie, D., Carroll, B.: An Introduction to Modern Astrophysics. Benjamin Cummings, San Francisco (2006)

Passino, K., Yurkovich, S.: Fuzzy Control. Addison Wesley, Reading (1998)

Qaseem, A., et al.: Diagnosis and management of stable chronic obstructive pulmonary disease: a clinical practice guideline update from the American College of Physicians, American College of Chest Physicians, American Thoracic Society, and European Respiratory Society. Ann. Intern. Med. **155**(3), 179–191 (2011)

Ross, T.: Fuzzy Logic with Engineering Applications. Wiley, New York (2010)

Santo-Tomas, L.H.: Emphysema and chronic obstructive pulmonary disease in coal miners. Curr. Opin. Pulm. Med. **17**(2), 123–125 (2011)

Seising, R.: The Fuzzification of Systems. Springer, Berlin (2007)

Yamaguchi, H., Narita, Y., Takahashi, H., Katou, Y.: Automatic transmission shift schedule control using fuzzy logic. SAE technical paper 930674 (1993)

Takagi, T., Sugeno, M.: Fuzzy identification of systems and its applications to modeling and control. IEEE Trans. Syst. Man Cybernet. **15**(1), 116–132 (1985)

Zadeh, L.A.: From computing with numbers to computing with words—from manipulation of measurements to manipulation of perceptions. IEEE Trans. Circuits Syst. **45** (1999)

Zadeh, L.A.: A new direction in AI: towards a computational theory of perceptions of measurements to manipulation of perceptions. AI Mag. **22**(1), 73–84 (2001)

Appendix A
NewLisp Versus Common Lisp

NewLisp has been used as the computing vehicle for developing this book. Being a Lisp dialect, it contains the overall Lisp programming paradigm and shares the standard Lisp features. However, as it happens with languages and dialects, it differs in details. The most important ones are discussed in this appendix when compared with Common Lisp. For further details, you can consult the online NewLisp documentation and/or the documentation of your Common Lisp compiler.

A.1   Arithmetic

As we saw in Chap. 1, NewLisp has different functions for distinguishing between integer and real arithmetic operations. When handling integers, the functions are +, -, * and /, so for example: *(+2 7) → 9, (-4 3) → 1, (*3 4) → 12, (/12 4) → 3* and so on. On the other hand, if we wish to operate with real numbers, we need to use the functions *(add), (sub), (mul) and (div)*, and then, for example: *(add 3.1 6.9) → 10, (sub 7.2 3) → 4, (mul 2.5 4.3) → 10.75* and *(div 7 4) → 1.75*. Common Lisp uses only the +, -, * and / functions, but does it in a different way. It reads the supplies data first, and then decide if the arithmetic operation involves integer numbers, real numbers, or a combination of both. For example, *(+2 7) → 9, (-4 3) → 1, (*3 4) → 12, (/12 4) → 3*, as in NewLisp, but *(+2 7.5) → 9.5, (-5 1.5) → 3.5, (*2 2.2) → 4.4* and *(/2.2 2.2) → 1.0*, but, for example:

CL> (/1 3)
: 1/3

When using integers in a division resulting into a rational number, Common Lisp returns fractions, thus, it never loses precision. If we wish that Common Lisp returns the decimal expansion of a rational number we shall need to tell it explicitly, writing at least one of the numbers in decimal format:

CL> (/ 1 3.0)
: 0.33333334

© Springer International Publishing Switzerland 2016
L. Argüelles Méndez, *A Practical Introduction to Fuzzy Logic using LISP*,
Studies in Fuzziness and Soft Computing 327,
DOI 10.1007/978-3-319-23186-0

A practical and useful rule of thumb is to always use decimal points in Common Lisp when handling real numbers, as it happens in Fuzzy Logic. For example, for describing a set of young people, you can write in NewLisp:

$$(setq\ age1\ `(young\ 15\ 25\ 25\ 35))$$

but in Common Lisp is advisable to always write:

$$(setq\ age1\ `(young\ 15.0\ 25.0\ 25.0\ 35.0))$$

A.2   Fundamental Lisp Functions

As we saw in Chap. 2, there are many Lisp functions, but the number of functions that build up the core of the language are relatively small. Table A.1 shows some fundamental Lisp functions and their equivalence in both versions of Lisp.

From these functions, some of them show differences in their behavior. After writing, for example, *(setq a '(x y z))*, then in NewLisp, *(last a)* → *z, but in Common Lisp, (last a)* → *(z)*, that is, instead of returning the last element as an atom, Common Lisp returns the last element as a list. However, for both Lisp

Table A.1 Fundamental functions of NewLisp and Common Lisp

NewLisp function	Common Lisp equivalent
(quote)	*(quote)*
(eval)	*(eval)*
(atom?)	*(atom)*
(symbol?)	*(symbolp)*
(number?)	*(numberp)*
(list?)	*(lisp)*
(first)	*(car)*
(rest)	*(cdr)*
(nth)	*(nth)*
(last)	*(last)**
(length)	*(length)*
(cons)	*(cons)**
(list)	*(list)*
(assoc)	*(assoc)*
(append)	*(append)*
(reverse)	*(reverse)*
(pop)	*(pop)*
(rotate)	*n/a*
(random)	*(random)**

Those with different behavior are marked with an asterisk

dialects: *(nth 2 a)* → *z*. This suggest the following code in Common Lisp for obtaining the same behavior:

```
(defun last1 (list)
       (nth (- (length list) 1) list))
```

or, alternatively:

```
(defun last1 (list)
   (car (last list)))
```

In both cases, now from Common Lisp, *(last1 a)* → *z*.

The case of *(cons)* shows also a difference between dialects. As an example, in NewLisp *(cons 'a b')* → *(a b)*, while in Common Lisp *(cons 'a 'b)* → *(a · b)* that is, in Common Lisp we get a dotted pair while in NewLisp we get a standard list. This must be handled with care. For example, after writing the following in Common Lisp: *(setq x (cons 'a 'b))* → *(a · b)*, and then: *(car x)* → *a*, *(cdr x)* → *b*, but *(last x)* → *(a · b)*. On the other hand, the Lisp expression *(cons '(this is) '(not hard))* → *((this is) not hard)* produces the same result in both dialects.

The NewLisp function *(rotate)* is not implemented in standard Common Lisp, but here is an example for moving the first element of a list to its end:

```
(defun left-rotate (list)
       (append (rest list) (list (first list)))))
```

and now, remembering the previous assignment *(setq a '(x y z))*, we have *(left-rotate a)* → *(y z x)*.

Finally, the function *(random)* is differently implemented in both dialects, but the goal is identical, that is, to produce a random number. In NewLisp we have, for example: *(random)* → *0.7830992238*, that is, it produces a random real number between 0 and 1, and no parameter is needed to call the function. On the other hand, in Common Lisp we need to pass an integer parameter to the function and then the language will return an integer between 0 and the integer passed as argument. For example, for generating a number between 0 and 100 we only need to write: *(random 100)* → *67*.

A.3 Defining Functions

From Chap. 3 we remember that the Basic anatomy of a NewLisp function has the following structure:

(define (function-name arguments)
   (lisp-expression$_i$)
)

In Common Lisp, it takes the following form:

```
(defun function-name (arguments)
    (lisp-expressionᵢ)
)
```

As an example, in Code 3-3 we introduced a function for calculating the Body Mass Index. In Common Lisp it would be as follows:

```
(defun BMI (mass height)
      (/ mass (* height height))
)
```

Interesting things happen when we need to use local variables in functions. The following code is Code 3-9 expressed in Common Lisp.

```
(defun bmi2 (mass height)
    (let
      ((result) (advice))
    )

    (setq result (/ mass (* height height)))

    (if (< result 18.5)
      (setq advice "You are excessively thin")
    )
    (if (and (>= result 18.5) (< result 24.9))
      (setq advice "Congrats, your weight is ok")
    )
    (if (and (>= result 24.9) (< result 29.9))
      (setq advice "You should try some diet and exercise
                    because you have some overweight")
    )
    (if (>= result 29.9)
       (setq advice "Warning, you suffer obesity.
                    Speak with your doctor")
    )
      advice
)
```

As you can observe, the internal variables result and advice are declared by means of the Common Lisp reserved word let, telling the language that the symbols located inside its matching parenthesis are local variables. Testing the function, we have, for example:

CL > (BMI2 75 1.80)
: *"Congrats, your weight is ok"*

Speaking about variables, we must add that global variables must be declared explicitly in Common Lisp by means of the keyword `defparameter`. As an example, if we wish to create a global linguistic variable *glucose* for describing glucose levels in blood in mg/dl, we could write:

```
(defparameter gl_1 '(Very-low 60 (60 1) (60 1) 82))
(defparameter gl_2 '(Rathar-low 60 (87.5 1) (87.5 1) 115))
(defparameter gl_3 '(Low 60 (115 1) (115 1) 165))
(defparameter gl_4 '(Medium-low 115 (165 1) (165 1) 219.5))
(defparameter gl_5 '(Medium-high 165 (219.5 1) (219.5 1) 274))
(defparameter gl_6 '(High 219.5 (300 1) (300 1) 300))
(defparameter glucose '(gl_1 gl_2 gl_3 gl_4 gl_5 gl_6))
```

A.4   Iteration

Along this book we have extensively used the `while` NewLisp keyword in order to help people with previous experience in other programming languages where "while loops" are always used. However, in Common Lisp the `do` macro is the fundamental iteration operator and the `while` keyword is usually not included in the language. The following code implements `while` loops in Common Lisp:

```
;a definition of while in Common Lisp
(defmacro while (test &rest body)
  '(do ()
       ((not,test))
     ,@body)
)
```

Now, let us rewrite, for example Code 3-17 in Common Lisp:

```
(defun my-find (atm lst)
  (let
    ((n) (i) (aux) (rest))
  )

  (setq n (length lst))
  (setq i 0)
  (setq result nil)
```

```
    (while (< i n)
      (progn
              (setq aux (nth i lst))
              (if (eq aux atm)
                  (setq result (cons i atm))
              )
              (setq i (+ 1 i))
      ) ;progn end
    ) ;while end
    (list (car result) (cdr result))
  )
```

CL> (my-find 'RED '(YELLOW RED GREEN BLUE WHITE))
: (1 RED)

Several additional things must be noted in this Common Lisp function:

- The *(eq)* function. In NewLisp we use the comparator operator *"="* for comparing any type of values. In Common Lisp the comparator operator *"="* is only used for comparing numerical values. Everything else must be compared with the *(eq)* function.
- The use of *(progn)*: In NewLisp we use *(begin)*. In Common Lisp, *(progn)* is used instead.
- The last line differs between both languages: In NewLisp we used simply `result`. However in CommonLisp we need to use the expression (`list (car result) (cdr result)`) in order to avoid a dotted pair as a function result since it would return *(1 . RED)* otherwise.

A.5   nil and true

In NewLisp *nil* and *true* are Boolean constants. In Common Lisp *nil* has an additional rol as a list terminator. For example, in NewLisp we have: *(cons 'x nil)* → *(x nil)*, while in Common Lisp: *(cons 'x nil)* → *(x)*.

As usually, "practice makes perfect", so the reader used to NewLisp will need some days in order to get used to Common Lisp. On the other hand, experienced Common Lisp programmers will find the NewLisp code in this book easy to follow and I suspect they will soon rewrite the most important functions for adapting them to their programming styles.

As additional resources, the interested reader can move to Common Lisp using the excellent introductory book "Common Lisp: A Gentle Introduction to Symbolic Computation", by David Touretzky, at about the same level of the introduction to NewLisp in this book. At a higher level, the de-facto standard in Common Lisp is the great book "ANSI Common Lisp", by Paul Graham.

Appendix B
Glossary of FuzzyLisp Functions

This Appendix alphabetically shows all the functions that build up FuzzyLisp. For every function the information is structured as follows:

- Name of the function: Gives the name of the function.
- Explanation: Explains how the function works and what the function returns.
- Syntax: Offers the syntax of the function, that is, its name and all its required arguments.
- FuzzyLisp representation: either FLSSR (FuzzyLisp Standard Set Representation) or FLDSR (FuzzyLisp Discrete Set Representation)
- Example: Shows a practical example that helps to put the function in context.
- Source code number: Gives the source code number in X-Y format, where X is the number of the chapter and Y is the number of the code inside the chapter for quickly locating the source code in this book.

fl-3dmesh

Explanation: This function creates an ASCII output file in comma-separated values format (CSV) where every line adopts the following structure: x_i, y_i, z_i. Both x_i, y_i are input crisp values from the universes of discourse of their respective linguistic variables from a Fuzzy Rule Based System (FRBS). On the other hand, z_i is the inferred value from every possible pair (x_i, y_i). The output file is in fact a discretized geometrical 3D mesh.

Syntax: *(fl-3d-dmesh namefile set-of-rules nx ny)*

- *namefile*: file name for storing the output data on the computer's hard disk.
- *set-of-rules*: The complete set of expert rules of a FRBS expressed in list format.

© Springer International Publishing Switzerland 2016
L. Argüelles Méndez, *A Practical Introduction to Fuzzy Logic using LISP*,
Studies in Fuzziness and Soft Computing 327,
DOI 10.1007/978-3-319-23186-0

- *nx*: Resolution of the 3D mesh over the x-axis.
- ny: Resolution of the 3D mesh over the y-axis.

FuzzyLisp representation: FLSSR
Example(s): *(fl-3d-mesh "air-conditioner-controller.csv" rules-controller 20 20) → Writing 3Dmesh ... 3Dmesh written to file*
Source code number: 7-13.
Note: The FuzzyLisp function *(fl-3d-dmesh)* is suited to deal with FRBS where input linguistic variables are composed by fuzzy sets with discrete membership functions (FLDSR).

fl-alpha-cut
Explanation: *(fl-alpha-cut)* scans a trapezoidal or triangular membership function from left to right and returns the obtained alpha-cut *alpha* as a list, including the name of the original fuzzy set.
Syntax: *(fl-alpha-cut fset alpha)*

- *fset*: a list representing a fuzzy set with a continuous membership function, either a triangle or a trapezium.
- *alpha*: a real number representing the horizontal line $y = alpha$ for obtaining an alpha-cut. It is required that $\alpha \in [0,1]$.

FuzzyLisp representation: FLSSR
Example(s): *(fl-alpha-cut '(B1 7 10 12 15) 0.7) → (B1 9.1 12.9)*
Source code number: 6-4.

fl-belongs?
Explanation: *(fl-belongs?)* returns *true* if a crisp value *x* defined on the real axis belongs to the fuzzy set *fset*, else returns *nil*.
Syntax: (fl-belongs? fset x)

- *fset*: a list representing a fuzzy set with a continuous membership function, either a triangle or a trapezium.
- *x:* a real number.

FuzzyLisp representation: FLSSR
Example(s): *(fl-belongs '(medium 10.0 20.0 30.0 40.0) 23.0) → true; (fl-belongs '(medium 10.0 20.0 30.0 40.0) 3.0) → nil*
Source code number: 6-1.

fl-belongs2?
Explanation: *(fl-belongs2?)* is a sort of mix of the functions *(fl-belongs?)* and *(fl-set-membership?)*. If the crisp value *x* is contained in the support of *fset* it returns the membership degree of *x* to *fset*, otherwise, it returns *nil*.

Syntax: *(fl-belongs2? fset x)*

- *fset*: a list representing a fuzzy set with a continuous membership function, either a triangle or a trapezium.
- *x:* a real number.

FuzzyLisp representation: FLSSR
Example(s): After assigning, e.g., *(setq S '(medium 10 20 30 40)),* then *(fl-belongs2? S 22)* → *(medium 1)*, *(fl-belongs2? S 100)* → *nil*
Source code number: 6-3.

fl-db-new-field
Explanation: This function creates a new field in a CSV format database. The new field contains the fuzzified values from an already existing numerical field. The fuzzification is obtained by means of a given fuzzy set.
Syntax: *(fl-db-new-field lst sharp-field fz-field fset mode)*

- *lst*: list containing an entire database.
- *sharp-field*: a string representing the name of a numerical field in the database.
- *fz-field*: a string for naming the new field to create.
- *fset*: a list representing a fuzzy set, either in a FuzzyLisp standard set representation or in a discrete set representation.
- *mode*: an integer. A value of 1 means that *fset* has a FLSSR. A value of 2 means a FLDSR.

FuzzyLisp representation: FLSSR/FLDSR
Example(s): assuming the fuzzy set *BM* defined by *(setq BM '(bright-magnitude -1 -1 3 5))* and that all the rest of function parameters have been correctly initialized, the function call *(setq messier (fl-db-new-field messier "Magnitude" "fz-magnitude" BM 1))* creates a new field named *"fz-magnitude"* where all the numerical values from the field *"Magnitude"* have been fuzzified by the fuzzy set *BM*.
Source code number: 6-26.

fl-def-set
Explanation: *(fl-def-set)* defines and creates a fuzzy set by means of two alpha-cuts *a-cut1, a-cut2*. The returned fuzzy set has either a triangular or trapezoidal membership function.
Syntax: *(fl-def-set name a-cut1 a-cut2)*

- *name*: symbol for associating a name to the resulting fuzzy set.
- *a-cut1*: first alpha-cut expressed by a list in the following format *(extreme-left extreme-right alpha-cut-value)*.
- *a-cut2*: second alpha-cut with the same format as *a-cut1*. It is required that *a-cut1 < a-cut2*.

FuzzyLisp representation: FLSSR

Example(s): *(fl-def-set 'young '(15.0 35.0 0) '(25.0 25.0 1.0))* → *(young 15 25 25 35)*

Source code number: 6-5a.

fl-defuzzify-rules

Explanation: This function takes as input the list obtained from either *(fl-dtranslate-all-rules)* or *(fl-dtranslate-all-rules)* and then converts that fuzzy information into a crisp numerical value.

Syntax: *(fl-defuzzify-rules translated-rules)*

- *translated-rules*: list representing the output of either *(fl-dtranslate-all-rules)* or *(fl-dtranslate-all-rules)*

FuzzyLisp representation: FLSSR/FLDSR

Example(s): *(fl-defuzzify-rules (fl-translate-all-rules rules-controller 22 0.25))* → *-60*

Source code number: 7-11.

fl-discretize

Explanation: *(fl-discretize)* takes a fuzzy set with triangular or trapezoidal characteristic function and discretizes it with a resolution given by *steps*. In other words, it transforms a FuzzyLisp Standard Set Representation into a FuzzyLisp Discrete Set Representation.

Syntax: *(fl-discretize fset steps)*

- *fset*: a list representing a fuzzy set with a continuous membership function, either a triangle or a trapezium.
- steps: an integer representing the resolution of the discretization process.

FuzzyLisp representation: FLSSR → FLDSR

Example(s): *(fl-discretize '(B1 7 10 12 15) 4)* → *(B1 (7 0) (7.75 0.25) (8.5 0.5) (9.25 0.75) (10 1) (10.5 1) (11 1) (11.5 1) (12 1) (12.75 0.75) (13.5 0.5) (14.25 0.25) (15 0))*

Source code number: 6-6.

fl-discretize-fx

Explanation: This function discretizes any continuous function $y = f(x)$ in n steps between $x = a$ *and* $x = b$.

Syntax: *(fl-discretize-fx name fx steps a b)*

- *name*: Symbol for associating a name to the function's resulting fuzzy set.
- *fx*: mathematical continuous function to discretize, expressed in Lisp format as a list.
- *steps*: Integer value for expressing the required resolution in the discretization process.

- *a*: starting point for discretization. Real value.
- *b*: ending point for discretization. Real value.

FuzzyLisp representation: FLDSR

Example(s): After defining a bell-shaped continuous function to the symbol *f* by means of the expression *(setq f '(div (add 1.0 (cos (mul 2.0 pi (sub x 2.0)))) 2.0))*, then, e.g.: *(setq dBell (fl-discretize-fx 'Bell f 10 1.5 2.5))* → *(Bell (1.5 0) (1.6 0.09549150283) (1.7 0.3454915028) (1.8 0.6545084972) (1.9 0.9045084972) (2 1) (2.1 0.9045084972) (2.2 0.6545084972) (2.3 0.345491502) (2.4 0.09549150283) (2.5 0))*

Source code number: 6-8.

fl-dlv-membership2?

Explanation: This function returns as a list all the membership degrees of a crisp value *x* to every fuzzy set contained in a linguistic variable. All the fuzzy sets from the linguistic variable have a discrete characteristic function.

Syntax: *(fl-dlv-membership2? lv x)*

- *lv*: a list representing a linguistic variable composed by discrete fuzzy sets.
- *x*: a real number.

FuzzyLisp representation: FLDSR

Example(s): assuming the linguistic variable *lv-age-bells* has been adequately initialized, then the function call *(fl-dlv-membership2? lv-age-bells 23)* produces the following output: *((Young 0.1302642245) (Young+ 0.5478879113) (Mature 0) (Mature+ 0) (Old 0))*

Source code number: 6-25.

fl-dset-hedge

Explanation: This function applies a fuzzy hedge (linguistic modifier) to a fuzzy set.

Syntax: *(fl-dset-hedge dset hedge)*

- *dset*: a list representing a discrete fuzzy set.
- *hedge*: a Lisp symbol, either VERY or FAIRLY.

FuzzyLisp representation: FLDSR

Example(s): *(fl-dset-hedge (fl-discretize '(A1 7 10 12 15) 4) 'VERY)* → *(A1 (7 0) (7.75 0.0625) (8.5 0.25) (9.25 0.5625) (10 1) (10.5 1) (11 1) (11.5 1) (12 1) (12.75 0.5625) (13.5 0.25) (14.25 0.0625) (15 0))*

Source code number: 7-7.

fl-dset-membership?

Explanation: *(fl-dset-membership?)* returns the interpolated membership degree of a crisp value *x* defined on the real axis to the discrete fuzzy set *fset*. In practical terms the difference with the function *(fl-set-membership?)* is based on the type of

used representation for fuzzy sets. *(fl-dset-membership?)* is used for *FLDSR*, while
(fl-set-membership?) is used for FLSSR.
   Syntax: *(fl-dset-membership? dfset x)*

- *dfset:* a list representing a fuzzy set with a discrete membership function.
- *x*: a real number.

FuzzyLisp representation: FLDSR
Example(s): After assigning, e.g., *(setq dA (fl-discretize '(B1 7 10 12 15) 4)),*
then *(fl-dset-membership? dA 8.2)* → *(B1 0.4)*
   Source code number: 6-7a.

fl-dtruth-value-fuzzy-implication-p-q?

Explanation: This function returns the truth-value of a compound fuzzy
implication
$p \rightarrow q$.
   Syntax: *(fl-dtruth-value-fuzzy-implication-p-q? P Q x y)*

- *P*: a list representing a discrete fuzzy set associated to the predicate of a fuzzy
  proposition *p*.
- *Q*: a list representing a discrete fuzzy set associated to the predicate of a fuzzy
  proposition *q*.
- *x*: a real number for expressing the subject of a fuzzy proposition *p*.
- *y*: a real number for expressing the subject of a fuzzy proposition *q*.

FuzzyLisp representation: FLDSR
Example(s):): assuming the fuzzy set *P* defined by *(setq P (fl-discretize '(old
50 90 90 90) 4))* and another fuzzy set *Q* defined by *(setq Q (fl-discretize '(young
0 0 15 30) 4)),* the fuzzy implication "if John is old then Eva is young" when
John is 55 years old and Eva is 18 can be represented by the function call
(fl-dtruth-value-implication-p-q? P Q 55 18) → *1.*
   Source code number: 7-6b.

fl-dtruth-value-negation-p?

Explanation: This function returns the truth-value of the negation of a fuzzy
proposition.
   Syntax: *(fl-dtruth-value-negation-p?P x)*

- *P*: a list representing a discrete fuzzy set associated to the predicate of a fuzzy
  proposition *p*.
- *x*: a real number for expressing the subject of a fuzzy proposition *p*.

FuzzyLisp representation: FLDSR
Example(s): assuming the fuzzy set *P* defined by *(setq Q (fl-discretize '(young 0
0 15 30) 4)),* the fuzzy proposition "Eva is not young" when Eva is 18 years old can
be represented by the function call *(fl-dtruth-value-negation-p? Q 18)* → *0.2*
   Source code number: 7-5b.

fl-dtruth-value-p-and-q?

Explanation: This function returns the truth-value of a compound fuzzy proposition containing the logical connective "and".

Syntax: *(fl-dtruth-value-p-and-q?P Q x y)*

- *P*: a list representing a discrete fuzzy set associated to the predicate of a fuzzy proposition *p*.
- *Q*: a list representing a discrete fuzzy set associated to the predicate of a fuzzy proposition *q*.
- *x*: a real number for expressing the subject of a fuzzy proposition *p*.
- *y*: a real number for expressing the subject of a fuzzy proposition *q*.

FuzzyLisp representation: FLDSR

Example(s): assuming the fuzzy set *P* defined by *(setq P (fl-discretize '(old 50 90 90 90) 4))* and another fuzzy set *Q* defined by *(setq Q (fl-discretize '(young 0 0 15 30) 4))*, the fuzzy compound proposition "John is old and Eva is young" when John is 55 years old and Eva is 18 can be represented by the function call *(fl-dtruth-value-p-and-q? P Q 55 18)* → *0.125*

Source code number: 7-3b.

fl-dtruth-value-p-or-q?

Explanation: This function returns the truth-value of a compound fuzzy proposition containing the logical connective "or".

Syntax: *(fl-dtruth-value-p-or-q?P Q x y)*

- *P*: a list representing a discrete fuzzy set associated to the predicate of a fuzzy proposition *p*.
- *Q*: a list representing a discrete fuzzy set associated to the predicate of a fuzzy proposition *q*.
- *x*: a real number for expressing the subject of a fuzzy proposition *p*.
- *y*: a real number for expressing the subject of a fuzzy proposition *q*.

FuzzyLisp representation: FLDSR

Example(s): assuming the fuzzy set *P* defined by *(setq P (fl-discretize '(old 50 90 90 90) 4))* and another fuzzy set *Q* defined by *(setq Q (fl-discretize '(young 0 0 15 30) 4))*, the fuzzy compound proposition "John is old or Eva is young" when John is 55 years old and Eva is 18 can be represented by the function call *(fl-dtruth-value-p-or-q? P Q 55 18)* → *0.8*

Source code number: 7-4b.

fl-expand-contract-set

Explanation: This function expands or contracts a fuzzy set. The returned fuzzy set is still placed over its original position, but its support and nucleus are expanded or contracted accordingly.

Syntax: *(fl-expand-contract-set fset k)*

- *fset*: a list representing a fuzzy set with a continuous membership function, either a triangle or a trapezium.
- *k*: a real number.

FuzzyLisp representation: FLSSR
Example(s): *(fl-expand-contract-set '(a 0 1 1 2) 2.0)* → *(a -1 1 1 3)*, *(fl-expand-contract-set '(a -1 1 1 3) 0.5)* → *(a 0 1 1 2)*
Source code number: 6-18.

fl-fuzzy-add

Explanation: Returns a fuzzy number as the result of adding two fuzzy numbers *A*, *B*.
Syntax: *(fl-fuzzy-add name A B)*

- *name:* a symbol for associating a name to the function's resulting fuzzy number.
- *A*: first fuzzy number to add.
- *B*: second fuzzy number to add.

FuzzyLisp representation: FLSSR
Example(s): After defining two fuzzy numbers, e.g., *(setq A '(around-2 1.75 2 2.25))* and *(setq B '(around-5 4.8 5 5 5.2))*, then *(fl-fuzzy-add 'A+B A B)* → *(A+B 6.55 7 7 7.45)*
Source code number: 6-13a.

fl-fuzzy-add-sets

Explanation: This function returns a fuzzy number as the result of adding all the fuzzy numbers contained in a set of fuzzy numbers.
Syntax: *(fl-fuzzy-add-sets fsets name)*

- *fsets*: A list containing all the fuzzy numbers to add.
- *name*: a symbol for associating a name to the function's resulting fuzzy number.

FuzzyLisp representation: FLSSR
Example(s): After creating several fuzzy numbers, e.g., *(setq F1 '(set1 -2 0 0 2))*, *(setq F2 '(set2 3 5 5 7))*, *(setq F3 '(set3 6 7 7 8))*, *(setq F4 '(set4 7 9 11 12))*, *(setq F5 '(set5 8 10 10 12))*, then *(setq Fsets '(F1 F2 F3 F4 F5))*, and finally: *(setq SFs (fl-fuzzy-add-sets Fsets 'Sum-of-Fs))* → *(Sum-of-Fs 22 31 33 41)*
Source code number: 6-19.

fl-fuzzy-average

Explanation: This function returns a fuzzy number as the average of n fuzzy numbers contained in a set of fuzzy numbers.

Syntax: *(fl-fuzzy-average fsets name)*

- *fsets*: A list containing all the fuzzy numbers to average.
- *name*: a symbol for associating a name to the function's resulting fuzzy number.

FuzzyLisp representation: FLSSR

Example(s): *(fl-fuzzy-average Fsets 'Average)* → *(Average 4.4 6.2 6.6 8.2).* See the assignments for building *Fsets* in the entry for the function *(fl-fuzzy-add-sets).* Source code number: 6-20.

fl-fuzzy-div
Explanation: This function returns a fuzzy number as the result of dividing two fuzzy numbers *A*, *B*. *A* and *B* are represented by triangular or trapezoidal shaped membership functions. The resulting fuzzy number *A/B* is represented by means of a discrete characteristic function.

Syntax: *(fl-fuzzy-div name A B n)*

- *name*: a symbol for associating a name to the function's resulting fuzzy number.
- *A*: first fuzzy number involved in the *A/B* division process.
- *B*: second fuzzy number involved in the *A/B* division process.
- *n*: integer for expressing the resolution of the process.

FuzzyLisp representation: FLSSR → FLDSR

Example(s): After defining two fuzzy numbers, e.g., *(setq A '(around-2 1.75 2 2 2.25))* and *(setq B '(around-5 4.8 5 5 5.2)),* then *(fl-fuzzy-div 'B/A B A 5)* → *(B/A (2.133333333 0) (2.2 0.2) (2.269767442 0.4) (2.342857143 0.6) (2.419512195 0.8) (2.5 1) (2.584615385 0.8) (2.673684211 0.6) (2.767567568 0.4) (2.866666667 0.2) (2.971428571 0))*
Source code number: 6-15.

fl-fuzzy-factor
Explanation: This function takes a fuzzy number *A* and then multiplies it by a crisp number *k*, returning the fuzzy number *k.A*. In practical terms when *k >1* it performs a multiplication and when *k <1* it performs a division by *k*.

Syntax: *(fl-fuzzy-factor fset k)*

- *fset*: a list representing a fuzzy number with a continuous membership function, either a triangle or a trapezium (fuzzy interval).
- *K*: a real number.

FuzzyLisp representation: FLSSR

Example(s): After defining a fuzzy number, e.g., *(setq A '(A1 -2 3 3 8)),* then *(fl-fuzzy-factor A 3)* → *(A1 -6 9 9 24),* and *(fl-fuzzy-factor A 0.25)* → *(A1 -0.5 0.75 0.75 2)*
Source code number: 6-16.

fl-fuzzy-mult

Explanation: Returns a fuzzy number as the result of multiplying two fuzzy numbers A, B. A and B are represented by triangular or trapezoidal shaped membership functions. The resulting fuzzy number A.B is represented by means of a discrete characteristic function.

Syntax: *(fl-fuzzy-mult name A B n)*

- *name*: a symbol for associating a name to the function's resulting fuzzy number.
- *A*: first fuzzy number to multiply.
- *B*: second fuzzy number to multiply.
- *n*: integer for expressing the resolution of the process.

FuzzyLisp representation: FLSSR → FLDSR

Example(s): After defining two fuzzy numbers, e.g., *(setq A '(around-2 1.75 2 2 2.25))* and *(setq B '(around-5 4.8 5 5 5.2))*, then *(fl-fuzzy-mult 'AxB A B 5)* → *(AxB (8.4 0) (8.712 0.2) (9.028 0.4) (9.348 0.6) (9.672 0.8) (10 1) (10.332 0.8) (10.668 0.6) (11.008 0.4) (11.352 0.2) (11.7 0))*

Source code number: 6-14.

fl-fuzzy-shift

Explanation: This function shifts (moves horizontally) a fuzzy set towards left or right over the real axis X by an amount given by a real value *x*, returning the shifted fuzzy set.

Syntax: *(fl-fuzzy-shift fset x)*

- *fset*: a list representing a fuzzy set with a continuous membership function, either a triangle or a trapezium.
- *x*: a real number.

FuzzyLisp representation: FLSSR

Example(s): *(fl-fuzzy-shift '(young-plus 15.0 25.0 35.0 45.0) 5.0)* → *(young-plus 20 30 40 50)*

Source code number: 6-17.

fl-fuzzy-sub

Explanation: Returns a fuzzy number as the result of subtracting two fuzzy numbers A, B.

Syntax: *(fl-fuzzy-sub name A B)*

- *name:* a symbol for associating a name to the function's resulting fuzzy number.
- *A*: first fuzzy number involved in the A-B subtraction process.
- *B*: second fuzzy number involved in the A-B subtraction process.

FuzzyLisp representation: FLSSR
Example(s): After defining two fuzzy numbers, e.g., *(setq A '(around-2 1.75 2 2 2.25))* and *(setq B '(around-5 4.8 5 5 5.2))*, then *(fl-fuzzy-sub 'A-B A B)* → *(A-B - 3.45 -3 -3 -2.55)*
Source code number: 6-13b.

fl-inference

Explanation: This function is an automatic call to the functions *(fl-translate-all-rules)* and *(fl-defuzzify-rules)* in a sort of black box that directly transforms two input crisp values entering a Fuzzy Rule Based System (FRBS) into a resulting crisp value.
Syntax: *(fl-inference x y)*

- *x*: first crisp input numerical value to the FRBS.
- *y*: second crisp input numerical value to the FRBS

FuzzyLisp representation: FLSSR
Example(s): *(fl-inference rules-controller 22 0.25)* → *-60*
Source code number: 7-12.
Note: The FuzzyLisp function *(fl-dinference)* is suited to deal with FRBS where input linguistic variables are composed by fuzzy sets with discrete membership functions (FLDSR).

fl-int-div

Explanation: Returns a list representing the division of two intervals.
Syntax: *(fl-intv-div x1 x2 x3 x4)*

- *x1, x2*: real numbers expressing the left and right extremes of an interval *[x1, x2]*.
- *x3, x4*: real numbers expressing the left and right extremes of an interval *[x3, x4]*.

FuzzyLisp representation: n/a
Example(s): *(fl-intv-div 2 4 1 3)* → *(0.6666666667 4)*
Source code number: 6-12d.

fl-intv-add

Explanation: Returns a list representing the addition of two intervals.
Syntax: *(fl-intv-add x1 x2 x3 x4)*

- *x1, x2*: real numbers expressing the left and right extremes of an interval *[x1, x2]*.
- *x3, x4*: real numbers expressing the left and right extremes of an interval *[x3, x4]*.

FuzzyLisp representation: n/a
Example(s): *(fl-intv-add 2 4 1 3)* → *(3 7)*
Source code number: 6-12a.

fl-intv-mult
Explanation: Returns a list representing the multiplication of two intervals.
Syntax: *(fl-intv-mult x1 x2 x3 x4)*

- *x1, x2*: real numbers expressing the left and right extremes of an interval *[x1, x2]*.
- *x3, x4*: real numbers expressing the left and right extremes of an interval *[x3, x4]*.

FuzzyLisp representation: n/a
Example(s): *(fl-intv-mult 2 4 1 3)* → *(2 12)*
Source code number: 6-12c.

fl-intv-sub
Explanation: Returns a list representing the subtraction of two intervals.
Syntax: *(fl-int-sub x1 x2 x3 x4)*

- *x1, x2*: real numbers expressing the left and right extremes of an interval *[x1, x2]*.
- *x3, x4*: real numbers expressing the left and right extremes of an interval *[x3, x4]*.

FuzzyLisp representation: n/a
Example(s): *(fl-intv-sub 2 4 1 3)* → *(-1 3)*
Source code number: 6-12b.

fl-list-sets
Explanation: This function prints all the fuzzy sets belonging to a linguistic variable at the Lisp console.
Syntax: *(fl-list-sets lv)*

- *lv*: a list representing a linguistic variable.

FuzzyLisp representation: FLSSR
Example(s): assuming the linguistic variable *lv-age* has been adequately initialized, then the function call *(fl-list-sets lv-age)* produces the following output:
: *(young 0 0 15 30)*
: *(young-plus 15 30 30 45)*
: *(mature 30 45 45 60)*
: *(mature-plus 45 60 60 75)*
: *(old 60 75 90 90)*
Source code number: 6-22.

fl-lv-membership?
Explanation: This function prints all the membership degrees of a crisp value x to every fuzzy set contained in a linguistic variable at the Lisp console.
Syntax: *(fl-lv-membership? lv x)*

- *lv*: a list representing a linguistic variable.
- *x*: a real number.

FuzzyLisp representation: FLSSR
Example(s): *(fl-lv-membership? lv-age 32)* produces the following output:
: (young 0)
: (young-plus 0.8666666667)
: (mature 0.1333333333)
: (mature-plus 0)
: (old 0)
Source code number: 6-23.

fl-lv-membership2?
Explanation: This function returns as a list all the membership degrees of a crisp value x to every fuzzy set contained in a linguistic variable.
Syntax: *(fl-lv-membership2? lv x)*

- *lv*: a list representing a linguistic variable.
- *x*: a real number.

FuzzyLisp representation: FLSSR
Example(s): *(fl-lv-membership2? lv-age 32)* → *((young 0) (young-plus 0.8666666667) (mature 0.1333333333) (mature-plus 0) (old 0))*
Source code number: 6-24.

fl-set-complement-membership?
Explanation: This function returns the membership degree of a crisp value x to the complementary set of *fset*.
Syntax: *(fl-set-complement-membership? fset x)*

- *fset*: a list representing a fuzzy set with a continuous membership function, either a triangle or a trapezium.
- *x*: a real number.

FuzzyLisp representation: FLSSR
Example(s): *(fl-set-complement-membership? '(B1 7 10 12 15) 9)* → *(B1 0.3333333333)*
Source code number: 6-9.

fl-set-intersect-membership?

Explanation: *(fl-set-intersect-membership?)* returns the membership degree of the crisp value *x* to the intersection of fuzzy sets *fset1* and *fset2*.

Syntax: *(fl-set-intersect-membership?name fset1 fset2 x)*

- *name*: a symbol for associating a name to the function's resulting list.
- *fset1*: a list representing a fuzzy set with a continuous membership function, either a triangle or a trapezium.
- *fset2*: a list representing a fuzzy set with a continuous membership function, either a triangle or a trapezium.
- *x*: a real number.

FuzzyLisp representation: FLSSR

Example(s): After defining two fuzzy sets *A* and *B* e.g.: *(setq A '(Triangle 0 5 5 10))* and *(setq B '(Trapezium 5 10 15 20))*, then: *(fl-set-intersect-membership? 'AintB A B 8)* → *(AintB 0.4)*

Source code number: 6-11.

fl-set-membership?

Explanation: *(fl-set-membership?)* returns the membership degree of a crisp value *x* defined on the real axis to the fuzzy set *fset*.

Syntax: *(fl-set-membership? fset x)*

- *fset*: a list representing a fuzzy set with a continuous membership function, either a triangle or a trapezium.
- *x:* a real number.

FuzzyLisp representation: FLSSR

Example(s): *(fl-set-membership? '(young 12 20 28 36) 24)* → *1;* *(fl-set-membership? '(young 12 20 28 36) 54)* → *0*

Source code number: 6-2.

fl-set-union-membership?

Explanation: *(fl-set-union-membership?)* returns the membership degree of the crisp value *x* to the union of fuzzy sets *fset1* and *fset2*.

Syntax: *(fl-set-union-membership? name fset1 fset2 x)*

- *name*: a symbol for associating a name to the function's resulting list.
- *fset1*: a list representing a fuzzy set with a continuous membership function, either a triangle or a trapezium.
- *fset2*: a list representing a fuzzy set with a continuous membership function, either a triangle or a trapezium.
- *x*: a real number.

FuzzyLisp representation: FLSSR

Example(s): After defining two fuzzy sets *A* and *B* e.g.: *(setq A '(Triangle 0 5 5 10))* and *(setq B '(Trapezium 5 10 15 20))*, then: *(fl-set-union-membership? 'AuB A B 7.5)* → *(AuB 0.5)*

Source code number: 6-10.

fl-simple-defuzzification

Explanation: This function takes a fuzzy number and produces a crisp number for it with a simple algorithm.

Syntax: *(fl-simple-defuzzification fset mode)*

- *fset*: a list representing a fuzzy set with a continuous membership function, either a triangle or a trapezium.
- *mode*: an integer value from 1 to 4. This parameter gives increasing weight to the nucleus in the process of defuzzification.

FuzzyLisp representation: FLSSR

Example(s): *(fl-simple-defuzzification '(q 0 1 1 5) 1)* → *2*; *(fl-simple-defuzzification '(q 0 1 1 5) 4)* → *1.375*

Source code number: 6-21.

fl-translate-all-rules

Explanation: This function evaluates all the fuzzy rules contained in the knowledge database of a Fuzzy Rule Based System (FRBS), calling iteratively to the function *(fl-translate-rule)*.

Syntax: *(fl-translate-all-rules set-of-rules x y)*

- *set-of-rules*: The complete set of expert rules of a FRBS expressed in list format.
- *x*: first crisp input numerical value to the FRBS.
- *y*: second crisp input numerical value to the FRBS

FuzzyLisp representation: FLSSR

Example(s): *(fl-translate-all-rules rules-controller 22 0.25)* → *((0 0 0 0) (0 0 0 0) (0 1 0 0) (0 0 0 0) (0 0 0 0) (0 1 0 0) (0.8 0 0 0) (0.8 0 0 0) (0.8 1 0.8 -40) (0.2 0 0 0) (0.2 0 0 0) (0.2 1 0.2 -20) (0 0 0 -0) (0 0 0 0) (0 1 0 0))*

Source code number: 7-10.

Note: The FuzzyLisp function *(fl-dtranslate-all-rules)* is suited to deal with FRBS where input linguistic variables are composed by fuzzy sets with discrete membership functions (FLDSR).

fl-translate-rule

Explanation: This function takes an expert rule at a time from a Fuzzy Rule Based System (FRBS), performs the adequate inferences and translates the rule into membership degrees, that is, into numerical values.

Syntax: *(fl-translate-rule header rule x y)*

- *header*: first sublist from the body of rules in the FRBS where the enumeration of the used linguistic variables is expressed.
- *rule*: rule to translate from the FRBS in its adequate list format.
- *x*: first crisp input numerical value to the FRBS.
- *y*: second crisp input numerical value to the FRBS.

FuzzyLisp representation: FLSSR

Example(s): *(fl-translate-rule (first rules-controller) (nth 9 rules-controller) 22 0.25)* → *(0.8 1 0.8 -40)*

Source code number: 7-9.

Note: The FuzzyLisp function *(fl-dtranslate-rule)* is suited to deal with FRBS where input linguistic variables are composed by fuzzy sets with discrete membership functions (FLDSR).

fl-truth-value-fuzzy-implication-p-q?

Explanation: This function returns the truth-value of a compound fuzzy implication

$p \rightarrow q$.

Syntax: *(fl-truth-value-fuzzy-implication-p-q? P Q x y)*

- *P*: a list representing a fuzzy set associated to the predicate of a fuzzy proposition p.
- *Q*: a list representing a fuzzy set associated to the predicate of a fuzzy proposition q.
- *x*: a real number for expressing the subject of a fuzzy proposition p.
- *y*: a real number for expressing the subject of a fuzzy proposition q.

FuzzyLisp representation: FLSSR

Example(s): assuming the fuzzy set P defined by *(setq P '(old 50 90 90 90))* and another fuzzy set Q defined by *(setq Q '(young 0 0 15 30))*, the fuzzy compound proposition "John is old or Eva is young" when John is 55 years old and Eva is 18 can be represented by the function call *(fl-truth-value-implication-p-q? P Q 55 18)* → *1*. When John is 90 years old and Eva is 30, then *(fl-truth-value-implication-p-q? P Q 90 30)* → *0*

Source code number: 7-6.

fl-truth-value-negation-p?

Explanation: This function returns the truth-value of the negation of a fuzzy proposition.

Syntax: *(fl-truth-value-negation-p?P x)*

- *P*: a list representing a fuzzy set associated to the predicate of a fuzzy proposition *p*.
- *x*: a real number for expressing the subject of a fuzzy proposition *p*.

FuzzyLisp representation: FLSSR
Example(s): assuming the fuzzy set *P* defined by *(setq P '(old 50 90 90 90))*, the fuzzy proposition "John is not old" when John is 55 years can be represented by the function call *(fl-truth-value-negation-p? P 55)* → *0.875*
Source code number: 7-5.

fl-truth-value-p-and-q?
Explanation: This function returns the truth-value of a compound fuzzy proposition containing the logical connective "and".
Syntax: *(fl-truth-value-p-and-q?P Q x y)*

- *P*: a list representing a fuzzy set associated to the predicate of a fuzzy proposition *p*.
- *Q*: a list representing a fuzzy set associated to the predicate of a fuzzy proposition *q*.
- *x*: a real number for expressing the subject of a fuzzy proposition *p*.
- *y*: a real number for expressing the subject of a fuzzy proposition *q*.

FuzzyLisp representation: FLSSR
Example(s): assuming the fuzzy set *P* defined by *(setq P '(old 50 90 90 90))* and another fuzzy set *Q* defined by *(setq Q '(young 0 0 15 30))*, the fuzzy compound proposition "John is old and Eva is young" when John is 55 years old and Eva is 18 can be represented by the function call *(fl-truth-value-p-and-q? P Q 55 18)* → *0.125*
Source code number: 7-3.

fl-truth-value-p-or-q?
Explanation: This function returns the truth-value of a compound fuzzy proposition containing the logical connective "or".
Syntax: *(fl-truth-value-p-or-q?P Q x y)*

- *P*: a list representing a fuzzy set associated to the predicate of a fuzzy proposition *p*.
- *Q*: a list representing a fuzzy set associated to the predicate of a fuzzy proposition *q*.
- *x*: a real number for expressing the subject of a fuzzy proposition *p*.
- *y*: a real number for expressing the subject of a fuzzy proposition *q*.

FuzzyLisp representation: FLSSR

Example(s): assuming the fuzzy set *P* defined by *(setq P '(old 50 90 90 90))* and another fuzzy set *Q* defined by *(setq Q '(young 0 0 15 30))*, the fuzzy compound proposition "John is old or Eva is young" when John is 55 years old and Eva is 18 can be represented by the function call *(fl-truth-value-p-or-q? P Q 55 18)* → *0.8*

Source code number: 7-4.